Praise for Robert Zubrin

"Robert Zubrin is already one of the few individuals to have had a marked impact on U.S. space policy, with his compelling advocacy of a streamlined and affordable human mission to Mars. Having galvanized the Mars mission, he now turns his attention to the larger picture of humanity's place in space, writing with a very impressive clarity reminiscent of Asimov or McPhee. This book works among other ways as a brilliant bit of geography, or rather cosmography, in that after reading it you have a clearer idea than before of where you stand in the universe."

—Kim Stanley Robinson, author of *Red Mars, Green Mars,* and *Blue Mars*

"Bob Zubrin is the Tom Paine of space—and about time we had one, too. An engineer who knows how to do the feats he envisions, he unfurls an agenda for the expansion of world civilization, not just another plan for going further out. This lends his book the quality of a thoughtful manifesto, well worked out in scenarios for development that both make a profit and enlarge the human prospect."

—Gregory Benford, author of *Deep Time*

"Zubrin knows how to make things work, and he sees possibilities and alternatives everywhere."

—Dennis Overbye, *The New York Times Book Review*

"Zubrin systematically and convincingly destroys conventional wisdom about Mars travel."

—Michael D. Lemonick, *Newsday*

"His elegant problem-solving and skillful promotion have put Mars exploration back on the map."

—Mark Bowden, *Playboy*

"Zubrin shows how a flight to Mars has progressed from fantasy to . . . a reality that can be achieved by us. Zubrin is showing us the way."

—Buzz Aldrin

ALSO BY ROBERT ZUBRIN

The Case for Mars: The Plan to Settle the Red Planet and Why We Must (with Richard Wagner)

Islands in the Sky (with Stanley Schmidt)

ENTERING SPACE

Creating a Spacefaring Civilization

ROBERT ZUBRIN

Jeremy P. Tarcher/Putnam

a member of Penguin Putnam Inc.

New York

Most Tarcher/Putnam books are available at special quantity discounts for bulk purchases for sales promotions, premiums, fund-raising, and educational needs. Special books or book excerpts also can be created to fit specific needs. For details, write Putnam Special Markets, 375 Hudson Street, New York, NY 10014.

Jeremy P. Tarcher/Putnam
a member of
Penguin Putnam Inc.
375 Hudson Street
New York, NY 10014
www.penguinputnam.com

First trade paperback edition 2000

Published simultaneously in Canada

The Library of Congress has cataloged the hardcover edition as follows:

Zubrin, Robert.
Entering space : creating a spacefaring civilization / Robert Zubrin.
p. cm.
Includes bibliographical references.
ISBN 0-87477-975-8
1. Space colonies. 2. Outer space—Exploration. I. Title.
TL795.7.Z83 1999
919.9'04—dc21 99-24725 CIP

Printed in the United States of America
7 9 10 8 6

ISBN 1-58542-036-0 (trade paperback)

This book is printed on acid-free paper. ∞

BOOK DESIGN BY JUDITH STAGNITTO ABBATE
AND CLAIRE VACCARO

To my parents,

Charles Zubrin *and* Rosalyn Fallenberg Zubrin,

who gave me Life and Mind

Witness this new-made World, another Heav'n

From Heaven Gate not farr, founded in view

On the clear Hyaline, the Glassie Sea;

Of amplitude almost immense, with Starr's

Numerous, and every Starr perhaps a World

Of destined habitation.

–JOHN MILTON, *Paradise Lost*

CONTENTS

Acknowledgments ▪ *viii*
Introduction ▪ *ix*

TYPE 1: COMPLETING GLOBAL CIVILIZATION

Chapter One: On the Threshold of the Universe ▪ *3*
Chapter Two: The Age of Dinosaurs ▪ *21*
Chapter Three: The New Space Race ▪ *39*
Chapter Four: Doing Business on Orbit ▪ *58*

TYPE II: CREATING A SPACEFARING CIVILIZATION

Chapter Five: The View from the Moon ▪ *79*
Chapter Six: Mars: The New World ▪ *101*
Chapter Seven: Asteroids for Good and Evil ▪ *128*
Chapter Eight: Settling the Outer Solar System ▪ *157*

TYPE III: ENTERING GALACTIC CIVILIZATION

Chapter Nine: The Challenge of Interstellar Travel ▪ *187*
Chapter Ten: Extraordinary Engineering ▪ *224*
Chapter Eleven: Meeting ET ▪ *247*
Chapter Twelve: North to the Stars ▪ *274*

Appendix: Founding Declaration of the Mars Society ▪ *284*
Glossary ▪ *286*
References ▪ *291*
Index ▪ *295*

ACKNOWLEDGMENTS

I wish to acknowledge the extensive help I have received in developing this book from my good friend Richard Wagner, and from Mitch Horowitz, my very capable editor at Tarcher Putnam books. I am also thankful for useful ideas or information shared by many others, including Dr. Chris McKay of the NASA Ames Research Center, Dr. Everett Gibson of NASA Johnson Space Center, the late Professor Carl Sagan of Cornell University, Dr. Robert Bussard, Dr. Robert Forward, Mr. Alan Wasser, Dr. Martyn Fogg, Dr. Dana Andrews of Boeing Aerospace, Colonel Simon "Pete" Worden of the Ballistic Missile Defense Agency, Dr. Seth Shostak of the SETI Institute, and Professor Greg Benford of the University of California, Irvine. I also wish to express my thanks to my intrepid agent, Laurie Fox of the Linda Chester Literary Agency, who sold the book. Most of all I am thankful for the constant support of my wife, Maggie Zubrin, without which the completion of this work would have been impossible.

INTRODUCTION

Our time will be remembered, because this was
when we first set sail for other worlds.

—CARL SAGAN, 1987

THIS IS A BOOK about creating a spacefaring civilization—the next step in the development of human society. According to the best current archeological evidence, until around 50,000 years ago the human race, modern *Homo sapiens,* was confined to a small region surrounding the Rift Valley in eastern Africa. The climate was favorable, the game was fairly abundant, and nontechnological humans were more than adequate for the challenge of the environment. As a result, the Rift Valley dwellers were able to get by with a tool kit limited to little more than the same split-rock hand axes that had served their *Homo erectus* ancestors for the previous million years. For some unknown reason, however, a few bands of these people decided to leave this relative paradise and travel north to colonize Europe and Asia, eventually going on from there to cross the land bridge into the Americas.

They went *north,* into the teeth of the Ice Age, into direct competition with giant carnivores and stocky Neanderthals who had already adapted to life in the cold. They went north, into a world of challenge, where fruit, vegetables, and game were not available all year long and where efficient weapons, clothing, and housing were necessary. In abandoning Africa, they embraced a wider world that could be survived only through the development of technology. Thus was born *Homo technologicus,* man the inventor, amid ice and fire. Thus humanity transformed itself from an East African curiosity to the dominant species on this planet.

In a sense, the biblical tale of Genesis tells this story but has it backward. It was not eating of the Tree of Knowledge that forced humankind to leave Paradise. Rather, it was the abandonment of Paradise that forced humanity to seek the forbidden fruit.

Back in the *Sputnik* era, the Russian space visionary Nikolai S. Kardashev outlined a three-tier schema for classifying civilizations. Adopting Kardashev's scheme in slightly altered form, I define a Type I civilization as one that has achieved full mastery of all of its planet's resources. A Type II civilization is one that has mastered its solar system, while a Type III civilization would be one that has access to the full potential of its galaxy. The trek out of Africa was humanity's key step in setting itself on the path toward achieving the mature Type I status that the human race now approaches. The challenge today is to move on to Type II. Indeed, the establishment of a true spacefaring civilization represents a change in human status as fully profound—both as formidable and as pregnant with promise—as humanity's move from the Rift Valley to its current global society.

Space today seems as inhospitable and as worthless as the wintry wastes of the north might have appeared to an average resident of East Africa 50,000 years ago. But yet, like the north, it is the frontier whose possibilities and challenges will allow and drive human society to make its next great positive transformation.

FACING THE CHALLENGE

In the early 1400s, the Ming emperors of China initiated an ambitious program of global exploration. They constructed fleets of huge oceangoing vessels and sent them off on voyages of discovery. Sailing south and then west, the Chinese admirals explored Indonesia, then Java, and went on to discover India and then the Arabic civilizations of the Middle East. Then, turning south, they explored the east coast of Africa and discovered Madagascar. Had they been allowed to continue, in a very few years the Chinese fleets would have rounded the Cape of Good Hope and been in position to sail north and discover Europe. However, the Confucian bureaucrats who advised the emperor considered information about the outside world and other civilizations and philosophies to be intrinsically worthless and potentially destabilizing to the divine kingdom, and so they convinced the emperor to have his fleets recalled and the ships destroyed. As a result, Chinese civilization pulled inward, only to be discovered *itself* by European seafarers a century later.

By accepting the challenge of the outside world, Western civilization blossomed outward to dominate the globe. In contrast, the grand Chinese civilization grew demoralized in its stagnation and implicit acceptance of inferior global status and decayed, ultimately to be completely disrupted and remade by expansive Western influences.

Only twenty-five years ago, the United States, following in the footsteps of the Ming emperors, abandoned its own pioneering program of space exploration. Even as the Apollo astronauts were returning from the Moon, the Nixon administration issued orders to effectively burn the fleet, destroying the Saturn V rockets and the other technological wonders that NASA had just developed to open the universe to humanity. At that time, America's leaders could console themselves with the equivalent of the advice of the Ming court bureaucrats—exploration is too expensive, and nothing of value exists beyond what is already familiar.

Now we know better. The recent discoveries of numerous planets—potential homes for life—orbiting other stars, and of actual evidence for life on Mars, indicate forcefully that the universe is alive. Recent technological developments, ranging from demonstrations of prototype reusable launch vehicles to practical high-temperature superconducting wire, have made it clear that engineering solutions exist for all of the problems barring the way to establishing a spacefaring civilization. In other words, the universe is open—open for us, and open for others. Therefore, we must boldly face outward. To do otherwise is to knowingly stick our heads in the sand and accept the demoralizing notion of humanity as beings of a lesser order. To do otherwise is to abandon the tradition of pioneering—a tradition accepting of the challenge offered by new climates, new worlds, new technologies, and new ideas—that lifted humanity out of the Rift Valley and, more recently, gave birth to the dynamic career of Western civilization.

As John F. Kennedy said in 1961 when committing the United States to the race to the Moon: "A new ocean has opened and free men must sail it." The universe has presented us with its challenge. To remain who we are, we must accept. We must enter space.

ABOUT THIS BOOK

Entering Space concerns the epochal transformation of our global society into a spacefaring one. It is about the feasibility, the necessity, and the promise involved in transforming humanity from a Type I civilization to a mature Type II civilization, beginning to take on the challenge of the reach toward Type III status.

The vision and scope is vast, so we will take things one step at a time. Following the Kardashev scheme, the book is organized into three major sections corresponding to the three degrees of civilization.

In the first section, "Type I: Completing Global Civilization," chapter 1 reviews both our current situation and recent startling discoveries that have unmasked the universe, and by so doing put its challenge to us. In the next chapter, we will get into the nitty-gritty of current-day politics, finance, and engineering to discuss the causes of stagnation within the existing aerospace establishment that have thus far obstructed humanity from entering the real Space Age, and analyze why the status quo is vulnerable. Then, in chapter 3, we will look at the entrepreneurial underground that has sprung up to field a new array of launch systems that promise to break the current deadlock keeping humanity Earthbound. Finally, in chapter 4, we will discuss the vast set of commercial opportunities now opening up in near-Earth space that currently stoke the furnaces of the new entrepreneurial space-launch companies. These space-business opportunities include establishing interconnected worldwide communication satellite networks that will finally knit humanity into a global village, thereby completing the work of 50,000 years of human history in establishing a mature Type I society.

In the second section, "Type II: Creating a Spacefaring Civilization," chapter 5 discusses the possibilities inherent in human activity on the Moon, which is an ideal vantage point for peering out into the universe. Chapter 6 deals with the colonization of Mars, which, because it is the nearest planet that possesses all of the resources needed for establishing a new branch of human civilization, constitutes the central near-term goal for the breakout of humanity into the solar system and the establishment of humanity as a multi-planet species. Chapter 7 explores asteroids, those multitudes of small bodies lying beyond Mars that are rich in resources but also a threat to humankind's future existence. Chapter 8 explains the kinds of propulsion systems, well understood but not yet developed, that will enable humans to travel to the outer solar system and beyond. It then discusses the question of the colonization of the outer solar system—the vast realm of the giant planets, which are rich with the energy resources needed not only to power a Type II civilization nearly indefinitely, but provide it with the energy it needs to reach for Type III.

In the third and final section, "Type III: Entering Galactic Civilization," chapter 9 discusses the technological problems associated with interstellar travel. The central point here is that, while orders of magnitude more difficult than interplanetary travel, interstellar travel is fundamentally feasible using currently understood engineering approaches. Chapter 10 discusses the extraordinary types of engineering that humans will require as we move out among the stars. Chapter 11 then explores the issues associated with detecting and encountering other intelligent species in space. As fantastical as such discussion may seem, our current scientific understanding of the universe indicates forcefully that not merely one or two but multitudes of such extraterrestrial races must certainly exist, and that therefore, as humanity becomes starfaring, such contact is nearly inevitable. Finally, chapter 12 proceeds to speculate on the possible ultimate limits to humanity's progress.

I need to state here that I am an engineer, not a science fiction writer. I have written one previous book, *The Case for Mars: The Plan to Settle the Red Planet and Why We Must.* The human settlement of Mars is not so far away, in either space or time—I've worked on the design of both robotic and human Mars missions for Lockheed Martin—and all of the engineering required in *The Case for Mars* was pretty much brass-tacks stuff. Far from being speculative, the "Mars Direct" plan I outlined there has been adopted by NASA as the basis for its Design Reference Mission for near-term human Mars exploration. Except for some of the more futuristic material toward the end of the volume dealing with terraforming, I know how to make virtually everything talked about in that book work. Some of the material in this new book is even closer to reality—I am one of the founders of a company that has raised a considerable sum of money to build one of the rocketplanes discussed in chapter 3—but a great deal is clearly much further out. I have built working lab-test units to demonstrate the feasibility of machines that can make rocket propellant on Mars out of the Martian atmosphere, but no one has ever built anything that can be described as a lab-test version of a fusion rocket for interstellar flight. Nevertheless, the principles upon which fusion reactors and fusion rockets will work are clearly understood. They are not science fiction, just technologically immature.

Discussions of matters such as interstellar travel, the nature of extraterrestrials, and the ultimate limits of engineering must necessarily have a speculative character that goes well beyond anything involved in planning a Mars mission. However, willy-nilly the future is coming, and rigorously using the tools of reason, including those of the engineer, can allow us to see with our mind today things that are not yet visible to the eye.

Therefore, in this book, I have shunned the arsenal of science fiction and fantasy tricks, including wormhole travel, space or time warps, teleportation, null-space or hyperspace, psi-drive, and so on, and instead based my concepts for presently undeveloped technologies on the laws of physics and engineering as they are currently understood. The only exception to this comes in the ultimate chapter, which contains some speculation on what the limits or mutability of those laws might be.

After all, it is important not to be too conservative.

TYPE I

Completing Global Civilization

My guide and I came on that hidden road
to make our way back into the bright world;
and with no care for any rest, we climbed—
he first, I following—until I saw,
through a round opening, some of those things
of beauty Heaven bears. It was from there
that we emerged, to see—once more—the stars

—DANTE ALIGHIERI
Inferno
Canto XXXIV, Lines 133–139;
translated by Allen Mandelbaum

CHAPTER 1

On the Threshold of the Universe

The Earth is the cradle of mankind,
but one cannot stay in the cradle forever.

—KONSTANTIN TSIOLKOVSKY, 1895

And what would be the purpose of all this?
For those who have never known the relentless urge to explore
and discover, there is no answer. For those who have felt this urge, the
answer is self evident. For the latter there is no solution but to investigate
every possible means of gaining knowledge of the universe.
This then is the goal:
To make available for life every place where life is possible.
To make inhabitable all worlds as yet uninhabited,
and all life purposeful.

—HERMANN OBERTH, 1957

HUMANS ARE NOT native to the Earth. Our lack of proper biological adaptation to the prevailing terrestrial environment indicates that we originated elsewhere. We live on a planet with two permanent polar ice caps, a planet whose land masses in large majority are stricken with snow, ice, freezing nights, and killing frosts every year, and whose oceans' average temperature is far below that of our life's blood. The Earth is a cold place.

Our internal metabolism requires warmth. Yet we have no fur; we have no feathers; we have no blubber to insulate our bodies. Across most of this planet, unprotected human life for any length of time is as impossible as it is on the Moon. We survive here, and thrive here, solely by virtue of our technology.

All modern humans are the descendants of a very small band of people who lived in East Africa about 200,000 years ago. We find the earliest known remains of both *Homo sapiens* and its precursors in that region. In addition, detailed studies of the genetic material of current human populations show the greatest diversity in East Africa, with diversity decreasing in proportion to distance from that area. These statistics point unerringly to the central trunk of the human genetic tree.[1] Humans are not native to the frigid Earth, only to tropical Kenya. We colonized the rest.

The move outward from our birthplace did not occur quickly. For 150,000 years after the appearance of *Homo sapiens,* our ancestors remained in the tropics.[2] For the most part, this meant East Africa itself, although there is evidence for intermittent presence in southern Africa and the Middle East. In these regions, their hairless bodies and gracile limb structure provided the advantage of easy rejection of the waste heat generated by the active brains and bodies of the world's most intelligent animal. With the aid of a few simple crude stone implements inherited from their *Homo erectus* forebears, these early *Homo sapiens* were masters of their environment, and apparently saw little need to either move or change in any way. Indeed, the 150 millennia humanity spent in Africa was a period of almost total technological stagnation, with generation after generation living and dying doing things in exactly the same way as their parents, grandparents, and remote ancestors centuries, millennia, and tens of millennia before.

Such stagnation, next to which the pattern of culture in the most tradition-bound tribal society known today compares as an exponential explosion of revolutionary progress, appears even more incredible given the fact that all available paleontological evidence indicates that these people were biologically identical to modern humans, with the same brain and other physical capacities. Humans as we know them everywhere in the world, whether Yankee gadgeteers or Chinese peasants, are to one extent or another constantly experimenting, innovating, tinkering, trying new things. It seems impossible—it seems inhuman—but for 150,000 years early humanity's tool kit did not alter. We *change;* they didn't. In a very fundamental sense, those folks just weren't like us.

But then, for some reason—perhaps by choice, perhaps forced by population pressure resulting from humanity's own success in adapting to its native tropical environment—about 50,000 years ago some bands of these people left the African homeland to try their fortunes in the north. There they soon encountered the problems of life in the wintry wastes of Ice Age Europe and Asia. In this new and more challenging world, the old bag of tricks that had served static tropical man so well for so long no longer sufficed. Without the novel inventions of clothing, insulated shelter, and efficient control of fire, *Homo sapiens* could not survive a single winter in their new habitat. Inventing clothing meant inventing sewing. Shelters had to be either built de novo or won from powerfully built, stocky, cold-adapted Neanderthals or 1,500-kilogram cave bears. Moreover, these wanderers were no longer in a world where food could be reliably gathered all year long. Dealing with these challenges required fine-tooled weapons that could kill at a distance for combat and big game hunting and improved means of communication, planning, and coordination among *Homo sapiens* themselves. Thus, we were forced to develop language and other forms of symbolic communication. Within a few thousand years of their arrival in the north, we find our ancestors making all sorts of novel gear—a wide array of finely chipped and polished stone tools and weapons and bone tools, including sewing kits and fishing kits—and producing fine cave art and even musical instruments. The latter two innovations are especially significant. Many animals build shelters, and sea otters, chimpanzees, and crows have all been known to use simple tools. But creating symbolic art, that's something else. Of all the creatures of this Earth, only humans *paint.* The rendering and appreciation of visual images denotes a mental ability akin to that required to create and understand verbal images. In other words, it indicates the origin of language and with it, in all probability, stories, mythology, oral history, poetry, and songs. A qualitatively higher level of intellectual, and I would argue spiritual, development had been attained.

Moving into a more challenging environment to which it was not naturally adapted forced *Homo sapiens* to transcend itself. Instead of existing as a clever animal applying a fixed repertoire of abilities to deal with a fixed set of contingencies in a well-defined environment, we became a species whose fundamental means of dealing with the world is to constantly invent *new* abilities. *Homo sapiens* became *Homo technologicus,* man the inventor, and by so doing enabled itself to conquer all the environments of the world: deserts, forests, jungles, steppe, swamps, mountains, tundra, rivers, lakes, seas, oceans, and even the air.

By confronting the challenge of an alien environment we broke out of a

150,000-year rut to become something different, and, in my view, something better. Can we do it again?

It is a truism that necessity is the mother of invention. Thus, societies, like individuals, grow when challenged and stagnate when not. We see this pattern in human history again and again. Those societies that have achieved unchallenged dominance over their relevant domain have tended to crystallize into self-satisfied, static forms, with some classic examples being ancient Egypt and traditional China. "We are the world; we have everything there is to have, we know everything there is to know, we have done everything there is to do" is the proud slogan of such fundamentally dead cultures. In contrast, those societies that have been subject to stress have proved the most dynamic.

In the past, this progress-driving stress has taken primarily two forms: war, and what I call "frontier shock."

There can be no question that, in numerous times and places in the past, war or the threat of war has been a driving force for progress. The most obvious example of this is the arms race among various competing Western and semi-Western powers that has helped accelerate technological development within that international system for the past several centuries, most especially our own. An analogous arms race contributed to innovation within the conflicting societies of the classical Mediterranean world before their unification by the Romans. Military necessity has also sometimes given urgency to social reforms, such as mass education and public health measures.

Yet even ignoring its fundamentally horrific nature, as a driving force for progress war obviously has its limits. First, eventually one of the contending powers may win, thereby unifying the domain and removing the dynamic stress from the system. This happened to the classical Mediterranean world after its unification under the Roman Empire. We can see this today with the rapid degradation of the scientific and technological capabilities of the United States' national lab system after its victory in the Cold War. Second, a condition of perpetual conflict may frequently lead to the rise to power within a society of a warrior class, such as the Samurai, whose continuation in power requires maintaining the forms of warfare within a fixed mode, and thus technological stagnation. Third and most important, however, in the modern world the horrific nature of war cannot be ignored. Warfare is destructive of both the wealth and the human potential of a society, and as the level of technology advances, so does the level of destruction. With the advent of nuclear weapons and recombinant DNA–based bacteriology, warfare of a sufficiently serious nature to induce societal stress among leading states has become unthinkable, as it would lead to the collapse of civilization itself. Thus, in the modern age, the utility of warfare in driving human progress has more or less expired.

A much more interesting and dynamic force has been "frontier shock," the stress induced in a people when they are forced to confront new lands filled with new possibilities and new knowledge. Throughout human history the most progressive cultures have been those "Sea People," such as the Minoans, Phoenicians, Greeks, Diaspora Jews, Italian Renaissance city-states, the Hanseatic League, the Dutch, the British, and the Americans, whose leading elements have been primarily engaged in long-range (typically maritime) trade and/or exploration. Societies of "Land People" whose top elements have been drawn from a landed aristocracy ruling a fixed domain have had a much more limited view and thus generally been far more tradition bound and conservative.[3] The greatest stimulus occurs in those situations where not just a leading minority, but large fractions of a society's population are exposed to or immersed in the novel frontier environment where they are both forced and free to innovate. Thus, it is no coincidence that the blossoming of classical Greek culture occurred during and immediately following their age of Mediterranean colonization, or that the fantastic explosion of innovation in European culture that transformed unimpressive and relatively static Medieval Christendom into hyper-dynamic and globally dominant Western civilization occurred simultaneously with the West's age of discovery and colonization. The most extreme example of the stimulus of frontier shock is North American civilization, which was developed as a culture of innovation, anti-traditionalism, optimism, individualism, and freedom based on 400 years of formative interaction with the novel necessities and infinite possibilities posed by its vast and ever-changing frontier.[4]

But what of today? The world's physical environments have been mastered, the western frontier has been settled, the Cold War has been won. In 1990, President George Bush gave a speech in which he said that humanity had entered upon a "New World Order," and he was right. With the collapse of the Soviet bloc, the world has been more or less effectively unified under the committee sovereignty of the united West. Military stresses have thus largely been eliminated as a major driver of the world system. With the establishment and explosive growth of the Internet and other forms of global communication and rapid transportation, currently divergent human cultures will tend to fuse. This global unification and cultural fusion will probably result in a temporary flowering of the arts and some economic growth, but the problem is that the stresses in the system are being shorted out. The situation is comparable to that of a battery whose negative and positive terminals are connected by a wire. Energy is released in a flush as the charges unite, but after a while all the potentials are level and the battery is dead. Witness the stagnation and decay following the Pax Romana. Pax Mundana could be worse. Consider that modern medical science is currently closing in on an understanding of, and therefore the ability to defeat, the aging process—at the cellular level itself.[5] In the past, all human societies had the possibility of progress through the "changing of the guard," as one generation replaced another at the helm. In the future this might not be possible. With no reason for change, those in power in every social niche might stay there—forever.

Pax Terrestris yes. Pax Mundana no. Humanity does not need war, death, disease, decay, superstition, national or racial cults, archaic belief structures or despotisms, or any number of other residues of our primitive past against which many noble people have struggled through the ages. But humanity does need *challenge.* A humanity without challenge would be a humanity without change, without innovation, which fundamentally means a humanity without meaningful freedom. A humanity without challenge would be a humanity without humanity.

Furthermore, the "golden age" enjoyed by a static society is generally only a transitory phase on the path to hell. The resource base of any society is defined by its technology. If you fix the technology, you put finite limits on its economic foundations. The typical results are Malthusian forms of social control and eventual exhaustion and collapse. Thus, in his seminal work on world history, *The Evolution of Civilizations,* historian Carroll Quigley[6] identified seven major stages in the development of societies: (1) mixture, (2) gestation, (3) expansion, (4) conflict, (5) universal empire, (6) decay, and (7) invasion. Bush's timely announcement of the "New World Order" could appear to indicate that Western (essentially modern global) civilization has currently reached Stage 5. Should we choose to continue in the footsteps of such historical analogs, Stage 6 would soon follow.

In 1992, philosophy Professor Francis Fukuyama wrote a widely read book entitled *The End of History,* in which he posited that with the unification of the world resulting from the West's victory in the Cold War, human history had essentially "ended."[7] In 1997, *Scientific American* writer James Horgan published a more interesting best-seller entitled *The End of Science,* in which he held that all the really big discoveries to be made in science already have been made, and thus the enterprise of scientific discovery must soon grind to a halt.[8] (The day after I finished reading Horgan's book in February 1998, a group of astronomers announced that they had found a fifth fundamental force in nature.) In his book, Horgan interviewed Fukuyama and asked him what he thought of those who doubt we have reached the end of human history. "They must be space travel buffs," Fukuyama replied in derision. Indeed.

The Earth's challenges have largely been met, and the planet is currently in the process of effective unification. I believe this marks the end, not of human history, but of the *first phase* of human history, our development into a mature Type I civilization. It is not the end of history because, if we choose to embrace it, we have in space a new frontier offering endless challenge—an infinite frontier, filled with worlds waiting to be discovered and history waiting to be made by myriad new branches of human civilization waiting to be born.

The opening of the space frontier, the creation of a spacefaring civilization, is thus the critical task facing our age. Compared to it, all other human enterprises of the present day are of trivial significance. Our success in this endeavor will determine whether we stand at the beginning of human history or the end. It will determine whether humanity continues as a truly human species. Failure is unacceptable.

FAILURE IS unacceptable, yet we seem to be failing. The world's space programs, begun so proudly in the eras of *Sputnik* and the Apollo Moon launches, appear to be in a state of retreat verging on rout. The Russian program has collapsed, and the American effort, which has been going in circles for the past twenty years, has lost much support and is set for a fall the next time something goes wrong with the Shuttle or Space Station programs.

Consider the following: From 1961 to 1973, the United States launched a total of more than thirty robotic lunar and planetary missions and ten piloted Apollo lunar missions. From 1974 to 1986 we launched six robotic and no manned missions beyond Earth orbit, while from 1987 to the present an additional ten robotic and no piloted exploration missions were flown. Russian mission statistics follow a similar trend. While the demise of the Soviet program might be explained by the deterioration of that nation's economy (oversimplistically—since material conditions in the

Soviet Union were much worse in the 1950s when their program was launched), in the United States the opposite is the case. The U.S. economy today is more than double the size of the 1960s economy, per capita income is higher, and we face no major military threat that drains our resources. While politicians complain about the incapacity of the national budget to support space programs, neither we nor anyone else have ever been so rich or more able to afford to initiate a great new age of exploration. The flush nature of the U.S. economy is ironically illustrated by the fact that our current political leadership is apparently willing to accept a situation where we are spending about the same amount of dollars on space in real terms as we did in the 1961–1973 era, while accomplishing perhaps 1 percent as much. Surprising as it may seem, the average NASA budget in 1998 dollars during the heroic age of 1961–1973 was about $16 billion per year, only 20 percent more than it is today. During that period NASA not only launched the Mercury, Gemini, Apollo, Skylab, Ranger, Surveyor, and Mariner missions, but did all the development for the Pioneer, Viking, and Voyager missions as well. In addition, the space agency developed hydrogen/oxygen rocket engines, multi-staged heavy-lift launch vehicles, nuclear rocket engines, space nuclear reactors and radioisotope generators, spacesuits, in-space life support systems, orbital rendezvous techniques, interplanetary navigation technologies, deep-space data transmission techniques, reentry technologies, soft-landing rocket technologies, a space station, and more. In other words, virtually the entire bag of tricks that enables space exploration missions today was developed during that 1961–1973 period, and despite continued comparable expenditures, very little of importance has been developed since. In fact, in numerous important respects, such as our current lack of heavy-lift launch vehicles and space nuclear power and propulsion systems, our space capabilities today are inferior to what they were in 1973.

The U.S. space program of the 1960s was vastly more productive than that of today because it had *drive,* imparted to it by a focused goal that made its reach exceed its grasp—landing humans on the Moon. There can be no progress without a goal, and lacking one NASA has floundered for the past twenty-five years. If U.S. political leaders want to increase the taxpayer's return on their space dollar, by far the most important and effective thing they could do would be to give the space agency a challenge worthy of it—committing the nation to establish humans on Mars within a decade. As explained in my book *The Case for Mars,* such a goal is entirely feasible and could serve to galvanize today's space program into another period of grand accomplishment.

An army standing still costs almost as much to support as one in motion. By failing to mobilize the nation's space capabilities for a serious push out into the solar system, political opponents of human Mars exploration are not saving money; they are *wasting* it. Why are they willing to do so?

Well, as noted above, the United States today has plenty of money to waste. But as long as they are spending it, one would think that today's politicians would desire something in return. John F. Kennedy demanded *results* from the space program. The nation's current officialdom doesn't seem to care. Why not?

It is clear that an essential element giving urgency to the space programs of the 1960s was the Cold War competition between the United States and the Soviet Union. That is not to say, as frequently has been claimed, that the Cold War *caused* the Apollo program. There were many other ways that the young, action-oriented President Kennedy could have responded to the failure of his Bay of Pigs invasion of Cuba in the spring of 1961. For example, he could have repeated the invasion, using U.S. Marines backed by air cover instead of poorly equipped and trained Cuban exiles. He could have engaged in other geopolitical military moves. If he wanted to do something space related, he could have announced the initiation of an early "Star Wars" type anti-missile program, or accelerated the development of the then-current "Dyna-soar" military spaceplane. Any of these moves would have been a more traditional and focused response to a geopolitical threat than the initiation of a program to send humans to the Moon.

No, the Apollo program was not caused by the Cold War. The Apollo program was caused by an *idea,* originating in the minds of early-twentieth-century visionaries like Robert Goddard, Konstantin Tsiolkovsky, and Hermann Oberth, and widely promoted by a subsequent generation of visionaries including Wernher von Braun and Arthur C. Clarke. That idea, the imperative for human expansion into space, captured the minds of a subset of the public, including some of those in power, and through them mobilized the political energies made available by the Cold War during the early 1960s for its service. As made clear in an "Apollo 25 Years Later" article by Hugh Sidey appearing in *Time* magazine in July 1994, Kennedy believed in the necessity of humanity, and in particular America, taking on the challenge of the space frontier and used the tension with the Russians as a tool to acquire political support for such an initiative. (In fact, Sidey reports that Kennedy actually decided to send Americans to the lunar surface three days *before* the Bay of Pigs.)

The fact that Kennedy himself was moved by the *idea,* and his appreciation of the need for challenge, and not just by the *Russian threat,* is also made clear by his own speeches, such as his brilliant and enduring address delivered to Rice University in September 1962. Listen to how intimately he united two passions, American national pride and the call of space:

> We choose to go to the Moon! We choose to go to the Moon in this decade and do the other things, not because they are easy but because they are hard, because that goal will serve to organize and measure the best of our energies and skills, because that challenge is one that we are willing to accept, one we are unwilling to postpone, and one which we intend to win . . . This is in some measure an act of faith and vision, for we do not know what benefits await us . . . But space is there and we are going to climb it.

Yet the American political system was and is predominantly composed of minds considerably less profound than that of John F. Kennedy. For such people, the Cold War competition with the Russians provided the decisive rationale required to mobilize their support for the program.

It was to eradicate this motivating force that a group of State Department and National Security officials initiated, negotiated, and rammed through the ratification of the Outer Space Treaty of 1967. This treaty forbade any nation from claiming sovereignty over any extraterrestrial body, thereby eliminating international competition as a major supporting imperative for space exploration. While some have made excuses for this treaty, citing various points of alleged merit, the intent of its authors was to remove space from the highly charged domain of Cold War competition, thereby allowing the space program to be shut down in order to make its funding available for other projects. This is made clear in formerly classified documents obtained under the Freedom of Information Act in 1997 by Alan Wasser of the National Space Society, published here for the first time. In one of these documents, a December 9, 1966, letter from Assistant Secretary of State Henry Owen to National Security Advisor Walt Whitman Rostow, Owen states:

> *Walt:*
>
> *1. Here are two copies of the final draft of our space paper, as it is being distributed to members of the Space Council—McNamara, Webb, etc. The Vice President wishes it to be discussed at the Council.*

2. It will encounter strong opposition from NASA and Ed Welsh {secretary of the Space Council}. Nonetheless, I believe it is right, for two reasons:

(a) Moving toward a more cooperative relation with the USSR in this field will reinforce our over-all policy toward the Soviets.

(b) More importantly: It will save money [emphasis in original], *which can go to (i) foreign aid, (ii) domestic purposes—thus mitigating the political strain of the war in Vietnam.*

3. If the proposals in this memo are left to be fought out by the space marshals and their clients, we will lose. Therefore:

(a) I urge you to get into the fight personally—let the Vice President, Schultz (BOB), and others know how you feel.

(b) Send a copy to someone on the domestic side of the White House staff (feel free to use this covering memo, if you wish) to ensure that someone from that side, representing the constituency whose interests are most directly affected, gets into the fight.

Henry Owen

Owen's cover letter was appended to a secret memo entitled "Space Goals after the Lunar Landing" prepared by the State Department and released for high-level discussion by Secretary of State Dean Rusk. That memo motivating the proposed Outer Space Treaty read in part:

> . . . we see no compelling reasons for early, major commitments to such [space exploration] goals, or for pursuing them at the forced pace that has characterized the race to the moon. Moreover, if we can de-emphasize or stretch out additional costly programs aimed at the moon and beyond, resources may to some extent be released for other objectives . . .
>
> . . . whether our over-all space effort can prudently be conducted at a more deliberate pace in the future may depend in part on de-fusing the space race between the U.S. and the Soviets. . . .

It will be observed that the Outer Space Treaty of 1967 did little to lessen the U.S.-Soviet confrontation itself, which the respective national security establishments of the time continued to pursue avidly by the use of massive military force in Vietnam and Czechoslovakia. Frankly, I believe that if foreign policy officials wish to have an arena in which they can display their "toughness" and "resolve," it would be much better that they did it in outer space. The collateral damage would certainly be less. Apparently, however, the "Best and the Brightest" thought otherwise.

And their strategy worked. Within two years of the treaty's ratification in 1967, U.S. space funding dropped by 26 percent. Within four years it was down 45 percent. Within six years it was down 60 percent. While rising GNP since the early 1980s has allowed U.S. space absolute expenditures to gradually drift back up to Apollo levels, they remain a much smaller portion of the national budget, and, more important, the apparent urgency for accomplishment has been removed.

The Outer Space Treaty of 1967 was a tragedy because it drained away the energy the remaining twenty years of Cold War could have provided to space exploration. Had this not occurred, had the momentum of Apollo been allowed to continue, the United States would have moved to establish permanent bases on the Moon and Mars by the 1980s, and humanity might well be a multi-planet species today. However, the damage done, the 1967 Treaty is today mostly of academic significance, as even if it were repealed, whatever driving force for space exploration and development international competition might offer has since been obliterated by the collapse of the Soviet Union itself. Instead, those large space projects that remain, such as the International Space Station, are limping along on the comparatively tepid basis of international cooperation.

Of course, cooperation can be a wonderful thing, as it can enable a group to achieve together what none could achieve alone. Indeed, from a material point of view, the fact that all the world's major space programs can now be united in their efforts provides an unprecedented opportunity for humanity to accomplish great feats in space in the very near future. But in itself, cooperation can never serve as a spur to progress.

I recently had the opportunity to stand in the stadium at Rice University, exactly where Kennedy gave his 1962 speech. The decor is vintage 1960s—the place probably looks about the same as it did then. If you listen carefully, you can almost hear the Boston-accented cadences of the young president's oratory echoing through the stadium and the responding roars of the transported crowd. But the stands are empty, and the echoes are fading.

We stand at the top of a hill, having just succeeded in a laborious ascent. Before us stands the entrance to the Pax Mundana, whose gilded gateway arch is inscribed with its proud slogan: "Unity, Uniformity, Stability." The path through the downward slope on the entrance's other side begins pleasantly enough, but at its nether end we perceive another gate, one of iron.

Is there another path?

I believe that there is. Its traveling requires some effort. But it leads *up,* up from our little hill into spectacular mountains and beyond, into the greatest age of hope and light that humanity has ever known.

THE WOULD-BE Pax Mundana has a weakness. Like all other self-satisfied societies, it rests upon a conceit: that there is nothing of importance *outside.* Ancient Egypt, the Roman Empire, and the Chinese Middle Kingdom all enjoyed this belief, and suffered destruction in the process of its refutation. The less smug world of medieval Christendom was not so petrified, however, and in discovering the outside used the shock of encounter to break its internal and external chains and blossom. Our Western society, having just experienced a period of 500 years of expansion, is also in an uncrystallized form. While the primary motive forces of the previous period of progress are no longer with us, many of its powerful institutions and traditions still persist. The scientific renaissance of the past five centuries has endowed us with big searching eyes, and they have only just begun to close. We are still a society receptive to the stimulation that could be offered by new frontier shock, provided that it comes in time.

Therein lies both the crisis and the hope of the present day. With victory won, the regimental banners and heroic bugles of the previous epoch are on their way to the pawnshop. But the restless spirit that once followed them has not yet accepted oblivion, and lingers on, listening for a new call to action and to life. And though still muted, that call has come. Even as the Earth surrendered, the space frontier presented its challenge.

The challenge has come in the form of a series of remarkable discoveries and innovations over the past ten years that have changed the relationship between the human future and the rest of the universe. These began in 1987 with the discovery by Professor W. Chu of the University of Texas of high-temperature superconductivity. As explained in chapter 9, high-temperature superconductivity offers the potential of breakthrough technology in the area of magnetic sails and magnetic confinement fusion propulsion systems that may someday enable low-cost interplanetary and, ultimately, interstellar flight. Then, starting in the early 1990s, astronomers began to detect planets orbiting other stars. As a result, it is now clear that planetary systems—potential homes for life—are the rule in the universe rather than the exception. In fact, we now know of more planets *outside* of our solar system than within it, an imbalance that is increasing with each day. Also during the early 1990s, evidence piled up to the point where it is now conclusive that asteroid impacts on Earth have been responsible not just for the extinction of the dinosaurs but for other mass extinctions as well. The significance of this was driven home on March 9, 1998, when the *New York Times* and other major national newspapers reported that a mile-wide asteroid would pass within 50,000 kilometers of the Earth in 2028, presenting a finite possibility of an impact that would obliterate human civilization. Refined calculations by Jet Propulsion Lab scientists published the following day calmed things down a bit, as their findings indicated that the asteroid (dubbed 1997XF11) would likely miss Earth by as much as 800,000 kilometers. However, the experience was sobering: What if the answer turns out otherwise next time? What could we do about it? The message here is that life on Earth is part of a larger cosmic system, which we ignore at our peril.

In 1993, the U.S. Strategic Defense Initiative Organization (SDIO) successfully demonstrated a prototype of a *fully reusable* space-launch vehicle called the DC-X—showing a clear map toward a new space age in which travel to space could become almost as cheap and as commonplace as air travel is today. In 1994, the SDIO followed up by launching a low-cost probe to the Earth's Moon, finding evidence for the presence of water, the staff of life and the basis of chemical industry, a finding subsequently confirmed by NASA's *Lunar Prospector* probe in 1998. Then in 1996, NASA's Galileo probe uncovered evidence for what appears to be an *ocean* of liquid water under the ice-covered surface of Jupiter's moon Europa. Also in 1996, NASA scientists revealed direct evidence for the presence of relics of microbial life in ancient Martian rocks that were ejected from the Red Planet by meteoric impact millions of years ago. In 1997, this evidence was strongly supplemented by NASA's *Pathfinder* craft on the surface of Mars itself, which, by landing in an ancient flood plain, proved the existence of aqueous environments that could have supported the evolution and development of microbial life on Mars in the distant past. As if that weren't enough, in 1998 NASA's Mars Global Surveyor probe imaged extensive systems of dry riverbeds and produced topographic data strongly indicative of a former northern *ocean,* thus revealing potential past homes for bacterial life across large regions of the Red Planet. As humble as such Martian microbes might be, the implications drawn from their existence are spectacular. The processes that lead to the origin of life are not particular to the Earth. Combined with the fact that we now know that most stars have planets, and that virtually every star has a region surrounding it (near or far depending on the brightness of the star) that can support the type of liquid-water environments that gave birth to life on Earth and Mars, the implication is that a very large number of stars currently possess planets that have given rise to life.

Now the history of life on Earth is one of continual development from simple forms to more complex forms, with the more advanced forms manifesting ever-increasing degrees of activity, intelligence, and capability to evolve to even more advanced forms at an accelerated rate. If life is a general phenomenon in the cosmos, then so is intelligence. If the evidence of bacterial fossils presented in Martian meteorite ALH 84001 holds up—and it's holding up quite well—the implication is clear: We are not alone.

Collectively, these discoveries will soon make it apparent what sort of birthright humanity is abandoning, if we abandon space. We are being put to the test.

Like the Romans, who once mistakenly thought that their empire ruled "all the world that mattered," humans until recently could be content in their belief that they were already the lords of the only relevant piece of cosmic real estate. We now know that such self-satisfied belief was ignorance. We realize now that the universe "that matters" is far vaster than our one little world. A few years ago it was possible for a scientifically educated person to believe that our galaxy contained only one inhabited planet. The evidence is now before us that we live in a system containing billions of habitable and inhabited worlds. A few years ago, no one knew that incoming extraterrestrial objects, asteroids, have had a decisive influence on the survival and evolution of life on Earth. Now we know, and in knowing are faced with the fact that humanity's span on Earth can only be made secure if we gain control of the solar system's flight traffic. A few years ago, "practical" people with full access to all relevant facts could reasonably assert that the necessary costs involved in space travel were so large as to make the notion of a spacefaring civilization a chimera. But now we know that technologies can be brought into existence that can make this wider universe accessible to us, a universe that, therefore, in all probability, is already being accessed by others.

Under such circumstances, to be content with the Pax Mundana, humanity must not only blind itself, but lobotomize itself as well.

We stand on the threshold of the universe, considering whether we should step forward or step back. The question has been posed to us: Will humanity retreat and allow itself to be, and to see itself as, mere passengers adrift in a sea of stars? Or will we step forward and, in taking hold of our solar system, take charge of our destiny, a species fully capable of contending with the challenges to come? The choice is ours.

FOCUS ON: THE FAILURE OF THE MINGS

Today Western civilization dominates the globe, but it hardly had to turn out that way. Six hundred years ago, the West's progenitor, Christendom, was little more than a poor, semi-anarchic, embattled fringe in northwest and central Europe, wedged perilously between the vast Islamic and Tartar Khanate domains and the Atlantic Ocean. In A.D. 1400, by far the most powerful civilization in the world was not the West, or even its rather more impressive neighbors, but the Ming Empire of China.

The Mings were the richest, the most populous, the most knowledgeable, and the best organized nation on Earth. Having defeated the Mongol heirs of Genghis Khan in the mid 1300s, Ming China dominated Asia and set its course for expansion, both on land and on sea. For the latter venture they were superbly well equipped. In contrast to the diminutive vessels of contemporary Europeans and Arabs, which rarely exceeded 15 meters in length, the 400-ship Ming navy included craft as long as 135 meters with 45 meters of beam divided into numerous watertight compartments, which made them virtually unsinkable. Carrying as many as nine masts equipped with batten-shaped cotton sails, and fitted with huge sternpost rudders, the giant Ming junks had excellent sailing qualities both for beating up into the wind and for sailing down. Furthermore, the Ming mariners possessed the magnetic compass and excellent skills making maps with compass bearings, allowing them to reliably revisit all ports of call they encountered.

In 1405, the Ming Emperor Yung Lo ordered his magnificent navy to embark on a series of exploratory ventures to show the flag and extract submission from all they encountered. In command of this incredible expeditionary fleet of 250 ships (the *smallest* of which was 55 meters long and carried five masts) and 28,000 men was Admiral Cheng Ho, a Chinese Muslim of humble origins.

Cheng Ho's first expedition reached Java, Sumatra, Ceylon, and Calicut. On the four following expeditions he reached Siam, Malacca, the East Indies, Bengal, the Maldive Islands, and the Persian Gulf sultanate of Ormuz. Detachments of his fleet visited Ryukyu and Brunei, Aden on the Red Sea, Mogadishu in Somaliland, Malindi, and the Zanzibar coast. On his sixth expedition, which lasted from 1421 to 1422, Cheng Ho visited thirty-six countries starting in Borneo and traversing the full width of the Indian Ocean to Zanzibar. Operations were then suspended for ten years due to the accession to power of an emperor who opposed oceanic exploration. Then, in 1433, the support of a new emperor allowed Cheng Ho to embark on his seventh and last voyage, which took him to Mecca and then far down the coast of Africa, almost to the Cape of Good Hope. However, immediately following this exploit, yet another shift in power at the imperial court caused the fleet to be recalled and all future voyages canceled.

Had this not occurred, it is highly probable that Cheng Ho would have rounded southern Africa in his next voyage and proceeded to discover Europe, thereby establishing China as the first global civilization.

Cheng Ho's political opponents, the Mandarin Confucian bureaucracy, claimed they opposed his exploration program on the grounds that the funds it required could be better spent at home on practical needs such as irrigation projects. Perhaps this was part of their reason. But it is also undoubtedly true that the Confucian bureaucrats, drawn from the conservative landlord class, found the ideas brought home from global exploration and encounters with other cultures disturbing to their self-contained worldview. In addition, oceanic expansion threatened to enhance the power of the bureaucrats' domestic anathema, the maritime merchant class of the coastal and river cities. That these latter considerations, rather than fiscal conservatism, were primary concerns is shown by the fact that not only was Cheng Ho's program de-funded and the fleet and shipyards demolished, but regulations were put in place to prevent private overseas ventures.

Thus, in 1433, 1449, and 1452, increasingly repressive laws were enacted forbidding Chinese to go abroad. By 1500, it was a capital offense to build a seagoing junk with more than two masts. In 1525, coastal officials were ordered to destroy all such ships and arrest anyone who sailed in them.

In 1434, just one year after the recall of Cheng Ho, Captain Gil Eannes, a sailor in the employ of Prince Henry of Portugal, succeeded in rounding Cape Bojador in West Africa. This feat of navigation was very modest by Cheng Ho's standards; yet, coming as it did after fifteen previous Portuguese failures, it constituted the first small but critical breakthrough in Henry the Navigator's program of global reconnaissance. If they had known how insignificant their heroic maritime efforts were compared to those of the Chinese, the Portuguese might have quit. But they didn't know, so they kept pushing. In 1498, Eannes' successor, Vasco da Gama, sailed into the Indian Ocean and encountered no Chinese fleet. It had ceased to exist. Visiting the same lucrative trading ports that had served Cheng Ho sixty-five years earlier, da Gama rapidly established European domination through the region. In 1511, Portuguese ships sailed into the Chinese port of Canton to establish, and soon dictate, trade relations. In 1557, the Portuguese seized the Chinese island of Macao, the first move in the accelerating process of Western humiliation and destruction of China that was to proceed for the next four centuries.

By 1793, the scepter of the world's oceans had been transferred among the Western powers, from Portugal to Spain to Holland and then to England. In that year, British ambassador Lord Macartney met with the Chinese emperor, who told him, "There is nothing you have that we lack. We have never set store on strange or indigenous objects, nor do we need any of your country's manufactures." Apparently the emperor had learned nothing. Britannia ruled the waves. The world and all its products, markets, ideas, innovations, and nations, including China, was at her disposal, to use, abuse, or ignore, as desired. Macartney must have smiled.

What the Chinese of Macartney's day lacked was not so much specific technologies, control of seas, or Indian ocean ports, as a social order capable of generating or assimilating new ideas. As a result of her decision to isolate herself, China threw away her massive initial advantages to rot in stagnation while the West leaped forward. In believing that she was the world, China had lost the world. As historian Daniel Boorstin has commented,[9] "When Europeans were sailing out with enthusiasm and high hopes, landbound China was sealing her borders. Within her physical and intellectual Great Wall, she avoided encounter with the unexpected. . . . Fully equipped with the technology, the intelligence, and the national resources to become discoverers, the Chinese doomed themselves to be the discovered."

CHAPTER 2

The Age of Dinosaurs

I speak of new cities and new people.
I tell you the past is a bucket of ashes.
I tell you yesterday is a world gone down,
A sun dropped into the West.
I tell you there is nothing in the world
only an ocean of tomorrows, a sky
of tomorrows.

—CARL SANDBURG, *"Prairie"*

I TALK OF the necessity for entering space, of humanity colonizing space. Yet how is that possible? When the Pilgrims emigrated to New England in the early 1600s, they funded the transportation themselves. It wasn't cheap: Upper-middle-class citizens liquidated their personal wealth simply to pay the price of passage. Skilled artisans often paid for their one-way passage with seven years of indentured labor. In modern-day American labor market equivalents, this puts the one-way price of the trip at about $300,000 per person. This is far beyond the discretionary spending level for casual travel, but for determined colonists willing to risk everything for a chance at a new life, it was, and is—perhaps barely—low enough. Nevertheless, by comparison with the costs of current-day space travel, it's dirt cheap.

A Space Shuttle launch carries a crew of perhaps six people and costs about $600 million. That's $100 million per person. Some may object that the Shuttle is a bad example, as even by current-day market standards it is vastly overpriced. Very well, launches on modern expendable boosters[1] cost about $10,000 per kilogram. [In this book I shall use kilograms and tonnes for units of mass: 1 kilogram (kg) equals 2.2 pounds; 1 tonne equals 1,000 kg or 2,200 pounds. The tonne, or metric ton, thus equals 1.1 ordinary or British tons.] If he or she were willing to ride on an expendable vehicle, a human passenger with minimal personal effects might represent only 100 kg for a launch cost of $1 million. But that just gets the passenger to low orbit. To go to an actual new world, say Mars, the passenger would also need to bring 400 kg of supplies. In addition, he or she would require a 500-kg share of the interplanetary spacecraft compartment mass. Thus, to enable the trip, 1,000 kg would have to be shipped from low Earth orbit (LEO) to Mars, which would require about 2,000 kg of rocket propellant and inert stage mass, for a total lift requirement of 3,000 kg. The resulting ticket price would be $30 million, roughly 100 times what the typical Pilgrim settler paid. It's hard to see how much interplanetary colonization could proceed at that rate.

Yet modern-day launch costs make no sense. The energy required to place a kilogram in orbit around Earth is 32 million joules, or 9 kilowatt-hours. At current U.S. electricity prices, this much energy could be bought off the grid for 45 cents—which would indicate that our 100-kg passenger should be launchable for about $45! This example is a bit of an oversimplification, as it ignores the energy required to orbit the spacecraft the passenger rides in, so let's choose another. If we assume that passengers make up 10 percent of an airliner's in-flight weight, and that the airplane has an aerodynamic lift-to-drag ratio of 10, then the amount of energy required for a 100-kg passenger to travel 5,000 km—New York to Paris, for example—would be 5,000 million joules, or about 50 percent *more* than the 3,200 million joules needed to send the passenger to orbit. As I write these lines, the price of a one-way transatlantic airline ticket is about $300; this analogy would suggest the possibility of trips to orbit with ticket prices in the $200 range.

This is still an oversimplification, however. Airplanes get most of their propellant from the air; rockets must carry all of theirs, a fact that carries a heavy performance penalty. There are air-breathing options for space lift, but they are complex and undeveloped, so let's assume that rockets provide the only viable launch technology. Current-day rockets, such as the kerosene/oxygen-fueled Atlas, can deliver about 1 percent of their takeoff mass to orbit—most (about 90 percent) of the remaining mass is propellant.

The cost of a kerosene/oxygen propellant mixture (at 3:1 oxygen/kerosene mixture ratio) is about $0.20/kg. Since the propellant consumed during launch has 90 times the mass of the payload delivered, the propellant cost of sending a mass to orbit is about $18/kg. Assuming a total system operating cost of six times the propellant cost (about double the total cost/fuel ratio of airlines), the resulting price of a rocket ride to orbit would be in the neighborhood of $100/kg, or $10,000 for a 100-kg passenger. There is no fundamental reason why space-launch prices in this range cannot be achieved.

Thus, we see that the reason it costs so much today to do anything in space has little to do with the laws of physics or engineering. What, then, is the reason for the astronomically high prices of contemporary space travel, and what can we do about it? To answer these questions, we need to take a look at the history, technology, and peculiar culture of the space-launch industry.

The first reason space flight costs so much today is that the primary launch systems have been developed in a totally cost-unconstrained environment. This has been true ever since Wernher von Braun used the imperatives of World War II to tap into the abundant coffers of the German Wehrmacht to obtain vast support which he used to build the technical empire that developed the V-2 missile. The V-2 was the first large liquid-fueled rocket and served as the technological progenitor of all subsequent space-launch systems. After World War II, members of von Braun's team joined with native talent in the United States and Soviet Union to exploit the comparably large imperatives of the cold war and mobilize unlimited funding to develop the superpowers' intercontinental ballistic missile (ICBM) weapons. Such rocket systems, including the Redstone, Thor-Delta, Atlas, and Titan missiles, subsequently launched the first satellites and astronauts. In a nuclear arms race, cost is no object, and so no serious effort was made to make these systems cheap.

If your launch costs are very high, then no expense can be spared in the effort to make your satellite payloads 100 percent reliable. Therefore, the higher the launch cost, the more expensive the satellite will be. As strange as it may sound, the more the satellite price increases, the less incentive there is to reduce launch costs. For example, if you are the U.S. Air Force and plan to launch a $1 billion reconnaissance satellite, it makes comparatively little difference if the launch vehicle costs $100 million or zero—at most, the mission cost is reduced by 10 percent. Furthermore, the higher the launch price, the fewer commercial launches there will be. This reduces the numbers of boosters produced, thereby also contributing to higher launch costs.

Beyond these considerations stands the government contracting system, known as "cost plus," which has been in place for some time now in the United States. According to the people who invented this system, it is essential that corporations be prevented from earning excessive profits on government contracts. Therefore, rather than negotiate a fixed price for a piece of hardware and allow the company to make a large profit or loss on the job depending on what its internal costs might be, regulators have demanded that the company document its internal costs in detail and then be allowed to charge a small fixed percentage fee (generally in the 10 percent range) above those costs as profit. This system has served to multiply the costs of government contracting tremendously, so much so that it has produced public scandals when news leaks out about the military paying $700 for a hammer or a toilet seat cover.

To see how this has worked, consider the case of the Lockheed Martin corporation, the largest aerospace contractor in the world. I was employed as a senior, and later staff, engineer at the prime facility of this company for seven years. Lockheed Martin almost never accepts hardware contracts on a fixed-cost basis. That is, the company rarely says to the U.S. government, "We will produce the ABC vehicle for you at a price of $*X*. If it costs us less than $*X* to make it, we will make a profit. If it costs us more, we will take a loss." Instead, most important contracts are negotiated along the following lines: "We will produce the ABC vehicle for a cost of about $*X*. We will then add a 10 percent fee to whatever it actually costs us to produce to provide the company with a modest profit." In other words, the more the ABC vehicle costs to produce, the more money the company makes. Hence, in addition to the vast numbers of accounting personnel that the cost-plus contracting system necessarily entails, the company is saturated with "planners," "marketeers," and "matrix managers," among swarms of other overhead personnel. Of the 9,000 people employed at the Lockheed Martin main plant in Denver (where the Atlas and Titan launch vehicles are made), only about 1,000 actually work in the factory. The fact that Lockheed Martin is keenly competitive with the other aerospace giants indicates that their overhead structures are similar.

In the context of this regime, government willingness to give such corporations cost-plus contracts for product improvement can actually serve as a disincentive for company investment in innovation. A number of years ago, I was part of a team that proposed a new upper stage for the Titan rocket, which would have increased the vehicle's performance by 50 percent. Creating the new upper stage would have required a company investment of about $150 million. (A single Titan launch sells for between $200 million and $400 million.) The corporation's management declined, saying, in effect, "If the Air Force wants us to improve the Titan, they will pay us to do it." As a result, the Titan was not improved and the company's commercial Titan line was shut down when all of its private-market business was taken by the slightly less obsolescent French Ariane. The company didn't mind much, however, as all of its cost-plus U.S. government launch contracts are protected by law from foreign competition, and it faces no U.S. competitors in the Titan's payload class.

The degree of technological stagnation that this system has caused in rocketry is extraordinary, not to say ironic, given that it has occurred in an industry which is generally taken as a symbol of progress. Yet the fact remains that the Delta, Atlas, and Titan rockets, which represent the primary U.S. launch vehicles today, all saw their first flights during the Eisenhower administration, and had evolved to essentially their current form well before the majority of people currently engaged in their manufacture or launch were born. There have been no new important rocket engines developed in the United States since the Space Shuttle Main Engines were first produced in the 1970s, and no major U.S. launch systems have been fielded since the first Shuttle flight in April 1981. This slow pace of progress has created a culture of extreme conservatism in subsystem design as well. For example, for years after its introduction, the Shuttle flew with 1970s-era computers using ferrite cores. Then, in 1997, in a wild leap of innovation, these were finally replaced—with IBM 386s! (IBM 386s became obsolete in the consumer computer market around 1992.)

In fact, there has been not only technological stagnation in the aerospace industry, but retrogression since the close of the Apollo era. Consider this: During the 1960s and early 1970s, the United States possessed capabilities including heavy-lift vehicles with a launch capability seven times greater than anything flying today; nuclear rocket engines with twice the exhaust velocity of any existing high-thrust engine; chemical engines with three times the thrust of any now flying; mercury ion engines; space nuclear power reactors; lightweight space suits; and many other capabilities that no longer exist. While the decades of retrenchment following the "crash program" Apollo period were supposed to be a period of deliberate technological progress that would allow us to reduce the cost of space launch so as to prepare the way for more sustainable space development, precisely the opposite has occurred. Even when adjusted for inflation, the cost of space launch in the United States today is actually *higher* than it was in 1972. This should be sobering to those today who would postpone substantial human space exploration initiatives to the Moon or Mars and wait for incremental progress to lower launch costs.

Another factor inflating launch costs is the fact that all existing launch vehicles are at least partly, and generally wholly, expendable. This comes as a result of the unique heritage of space-launch systems: Of all the methods of transportation known to human history, only launch vehicles are descended from ammunition. When John Glenn traveled to orbit in 1962, he rode atop an Atlas rocket. The Atlas was an ICBM directly derived from the German army's V-2, which itself was simply a replacement for the Paris Gun and other long-range artillery forbidden Germany by the victors of World War I.

One does not recover ammunition after it is fired. Thus, a $300-million Titan—a *weapon system* designed to deliver warheads to Soviet cities but later used as a *transportation system* to deliver Gemini astronauts to orbit, Viking to Mars, and Voyager to Neptune—can be used only once. Consider how expensive air travel would be today if Boeing 747s were scrapped after one flight, and a 747 costs only about $100 million! If the excessive overhead were squeezed out of the space industry and its immediate suppliers, the Titan could probably be built for $30 million (it's a much simpler machine than a 747), but even at that rate, the practice of expending each booster after a single use would still make space travel orders of magnitude more expensive than any other form of transportation.

So why not make launch systems reusable? Well, there are significant engineering reasons that favor expendability. If a vehicle is to be expended, it won't need landing gear, a deceleration system, or a reentry thermal protection system. Eliminating all these items reduces vehicle weight and therefore increases payload. Expending a rocket also makes it easier to adopt staging strategies, in which one rocket is launched from atop another, a practice that also increases payload or maximum range. In addition, expendability makes the launch system simpler overall, and drops the technological requirements on subsystems. For example, rocket engines only have to be designed to start once and endure a single burn. Nothing needs to be designed for servicing after use, and no special tools, procedures, or personnel to engage in such servicing need to be developed. If the launch rate is limited, many more launch vehicles will be needed to service the payload manifest (the ensemble of payloads available for launch) if the boosters are expendable than if they are reusable. Therefore, the economics of mass production will always favor expendables, and a single expendable booster that is part of a production line will have a much lower manufacturing cost than a single reusable vehicle with equivalent lift capability.

A reusable vehicle will require a ground support team to service it, and these people will have to be paid all year long regardless of how many times the craft is used. If the launch rate is too low, this payroll could conceivably exceed the costs of performing the same number of launches with expendables. In the case of the rather complex Space Shuttle (currently the only operational reusable system), the ground support team amounts to a virtual standing army, with an annual program cost of about $5 billion per year. Thus, at the current launch rate of about eight per year, a Shuttle launch costs close to $600 million. This is twice that of the pricey Titan IV–Centaur, which offers equivalent lift capability. But since the Shuttle is mostly reusable, if the Shuttle launch rate could be doubled to sixteen per year it could match the Titan costs, and if it were tripled to twenty-four per year, could significantly beat them.

The attempt to cope with these economic realities underlies much of the pathology associated with the Shuttle program for the past twenty-five years. For example, in selling the Shuttle program to Congress during the 1970s, NASA officials claimed that the Shuttle would fly forty times per year (one launch every nine days!). This prediction should have aroused skepticism on two grounds: (a) the technical difficulty in preparing a Shuttle for launch in so short a time and (b) the lack of a payload manifest large enough to justify such a launch rate. NASA leaders left (a) up to the engineers to solve as best they could, but attempted to solve (b) themselves through political action. Specifically, the NASA brass in the late 1970s and early 1980s obtained agreements from the White House to the effect that once the Shuttle became fully operational, *all* U.S. government payloads would be launched on the Shuttle. That is, NASA wanted the expendable Deltas, Atlases, and Titans phased out of existence so that the Shuttle could enjoy a bigger manifest and have its economics improve accordingly. The Air Force resisted this policy, as they feared that a Shuttle accident could cause a stand-down of the entire program, which would then make it impossible to launch vital military reconnaissance and communication satellites when required. It seems incredible today, but the NASA argument actually carried the day against the Air Force in Washington's corridors of power. During the 1980s, the expendable "mixed fleet" was in the process of being phased out. It was only after the *Challenger* disaster in January 1986 proved that Air Force concerns were fully justified that President Reagan reversed this decision.

The need to increase the launch manifest to justify Shuttle economics played a central role in the decision to initiate the Space Station program. In the early 1980s, NASA Deputy Administrator Hans Mark saw clearly that achieving a Shuttle launch rate of twenty-five per year would be impossible without the manifest created by the construction and supply needs of a permanently orbiting outpost, which he already supported as a facility for in-space scientific research (Mark did not believe the forty launches per year touted by earlier Shuttle advocates was feasible under any conditions). Based on this (probably accurate) assessment, Mark convinced first NASA Administrator James Beggs and then the Reagan White House of the need for a Space Station program.[2] The need to generate a large Shuttle manifest also helps to explain the bizarre nature of the engineering designs that have guided the Space Station program since its inception.

The right way to build a Space Station is to build a heavy-lift launch vehicle and use it to launch the station in a single piece. The United States launched the Skylab space station in this manner in 1973. Skylab, which contained more living space than the currently planned International Space Station (ISS), was built in one piece and launched in a single day. As a result, the entire Skylab program, end to end from 1968 to 1974, including development, build, launch, and operation, was conducted at a cost in today's money of about $4 billion, roughly one-eighth of the anticipated cost of the ISS. In contrast, the Space Station has gone through numerous designs (of which the current ISS is the latest), all of which called for over thirty Shuttle launches, each delivering an element that would be added into an extended ticky-tacky structure on orbit. Since no one really knows how to do this, such an approach has caused the program development cost and schedule to explode. In 1993, the recently appointed NASA Administrator Dan Goldin attempted to deal with this situation by ordering a total reassessment of the Space Station's design. Three teams, labeled A, B, and C, were assigned to develop complete designs for three distinct Space Station concepts. Teams A and B took two somewhat different approaches to the by-then-standard thirty-Shuttle-launch/orbit-assembly concept, whereas team C developed a Skylab-type design that would be launched in a single throw of a heavy-lift vehicle (a "Shuttle C" consisting of the Shuttle launch stack but without the reusable orbiter). The three approaches were then submitted to a blue ribbon panel organized by the Massachusetts Institute of Technology for competitive judgment. The M.I.T. panel ruled decisively in favor of option C (a fact that demonstrated only their common sense, not their brilliance, as C was much cheaper, simpler, safer, more reliable, and more capable and would have given the nation a heavy-lift launcher as a bonus). However, based on the need to create Shuttle launches as well as a desire to have a Space Station design that would allow modular additions by international partners, Vice President Al Gore and House Space Subcommittee chairman George Brown overruled the M.I.T. panel. By political fiat, these gentlemen forced NASA to accept option A, and the space agency has had to struggle with the task of building the Space Station on that basis ever since. The result has been a further set of cost and schedule overruns, the blame for which has been consistently placed on various NASA middle managers instead of those really responsible.

To sum up, realizing the advantages offered by making launch vehicles reusable is not as simple as it sounds. The combination of technical success/fiscal disaster that is the Shuttle program proves this. But can reusable launchers be designed that make economic sense? Absolutely.

The Shuttle is a fiscal disaster not because it is reusable, but because both its technical and programmatic bases are incorrect. The Shuttle is a partially reusable launch vehicle: Its lower stages are expendable or semi-salvageable while the upper stage (the orbiter) is reusable. As aesthetically pleasing as this configuration may appear to some, from an engineering point of view this is precisely the opposite of the correct way to design a partially reusable launch system. Instead, the lower stages should be reusable and the upper stage expendable. Why? Because the lower stages of a multi-staged booster are far more massive than the upper stage; so if only one or the other is to be reusable, you save much more money by reusing the lower stage. Furthermore, it is much easier to make the lower stage reusable, since it does not fly as high or as fast, and thus takes much less of a beating during reentry. Finally, the negative payload impact of adding those systems required for reusability is much less if they are put on the lower stage than the upper. In a typical two-stage-to-orbit system, for example, every kilogram of extra dry mass added to the lower stage reduces the payload delivered to orbit by about 0.1 kilogram, whereas a kilogram of extra dry mass on the upper stage causes a full kilogram of payload loss. The Shuttle is actually a 100-tonne-to-orbit booster, but because the upper stage is a reusable orbiter vehicle with a dry mass of 80 tonnes, only 20 tonnes of payload is actually delivered to orbit. From the amount of smoke, fire, and thrust the Shuttle produces on the launch pad, it should deliver five times the payload to orbit of a Titan IV, but because it must launch the orbiter to space as well as the payload, its net delivery capability only equals that of the Titan. There is no need for 60-odd tonnes of wings, landing gear, and thermal protection systems in Earth orbit, but the Shuttle drags them up there (at a cost of $10 million per tonne) anyway each time it flies. In short, the Space Shuttle is so inefficient because *it is built upside down.*

Additional inefficiencies in the Shuttle stem from the vehicle's birth and consequent design as a make-work project. The demands of the Apollo program gave rise to NASA as the large federal agency it is today. As Apollo drew to a close, the agency and its political allies required something for its large workforce of civil servants and private contractors to do. Unfortunately, rather than give this formidable army of technical talent a project worthy of it—the establishment of permanent human outposts on the Moon and Mars—the Nixon administration chose to maintain the agency in idle mode. The Space Shuttle program was the result. In other words, the de facto requirement of the Shuttle program was not that it accomplish anything, but that it keep a lot of people busy. In this latter quest it has succeeded admirably, but such success is exclusive to the goal of achieving low launch costs. No matter how reusable a system is, if you run an eight-launch-per-year medium-lift launch program that must nevertheless pay for the employment of 50,000 people, it's just not going to be cost effective.

So the keys to making reusable launch vehicles pay are to (a) have a high launch rate, (b) have a small ground staff, and (c) reuse the first stage. Now the only way to accomplish (a) and (b) simultaneously is to have a launch system that is extremely simple, thereby requiring the fewest possible maintenance and integration operations between launches. Since having multiple stages significantly increases vehicle complexity, this implies that the ideal reusable launch vehicle should have just one stage, which would accomplish objective (c) perforce, since the first stage would be the only stage. From this logic flows the conclusion that the best reusable system would be a single-stage-to-orbit (SSTO) vehicle.

Flying to orbit in a single-stage reusable system is extremely demanding technically. Using hydrogen/oxygen propellant, the highest performing realistic chemical rocket combination, an SSTO would have to keep its dry structural mass to just 10 percent of its takeoff weight simply to achieve orbit. The remaining 90 percent would all be propellant. The 10 percent allocated to dry mass would need to include the engines, the propellant tanks, the plumbing, and the overall structure, thermal protection covering, electronics and guidance systems, landing gear, wings or other aerodynamic surfaces, a payload compartment, and the payload. Accomplishing this is very hard to do. In fact it's never been done and some space-launch veterans still say it can't be done. I don't agree with that latter assessment; I think it is just another reflection of the stagnation in the space-launch industry. New materials and construction techniques are available today that can upset the conventional wisdom and make the 10 percent dry mass target needed for SSTO flight achievable. Even so, the decision to forgo the performance advantages offered by staging would still greatly reduce the payload capacity of an SSTO compared to either an expendable or even a reusable two-stage-to-orbit vehicle (TSTO) of the same size.

So if playing the difficult SSTO development game is to be worth the candle, it needs to be shown that such a system really can be maintained by a small launch crew and turned around quickly to perform launch after launch almost in the manner of an aircraft. It was to perform a demonstration of such SSTO operability that the Strategic Defense Initiative Organization (SDIO, now known as the Ballistic Missile Defense Organization, BMDO) initiated the DC-X program in 1991.

The DC-X (for *Delta Clipper*–Experimental) program was inspired by visionary engineers Max Hunter and Gary Hudson, who since the 1970s had been campaigning for the development of a wingless, conical SSTO vehicle. The vehicle would take off and land vertically on plumes of rocket exhaust just as, to quote Hunter, "God and Robert Heinlein would have wanted." To have any chance of successful implementation, the Strategic Defense Initiative needed a much cheaper and more responsive space-launch capability than that offered by current expendables. So SDIO technology development chief Simon "Pete" Worden hit on the idea of using some spare change to perform a "quick and dirty" demonstration of the basic feasibility of the Hunter/Hudson concept. Worden hoped that performing such a stunt would generate enough excitement within the political system to shake loose sufficient funds for a real SSTO development program. In carrying out these objectives. Worden and the DC-X program succeeded brilliantly.[3]

Worden appointed his protégé, Major Jess Sponable, to lead the program, and together they chose a McDonnell Douglas (Hunter's home base) team led by William Gaubatz to implement it. Operating on a relative shoestring budget of $60 million, they built the DC-X vehicle, a one-third–scale working model of a vertical-takeoff/vertical-lander SSTO. The DC-X didn't fly to orbit. In fact, it didn't even go supersonic, but it took off and landed again and again—eleven times in all—using hydrogen/oxygen rocket engines. In some of its flights, the vehicle demonstrated the essential attitude reversal maneuver that this type of system needs to land, an accomplishment that many had considered problematic. On one occasion, *it flew twice in the same day.* Best of all, its entire ground support team consisted of only twelve people.

I was present at the second DC-X flight in September 1993 at White Sands, and can speak from personal experience that the effect on the large crowd Worden had invited was electric. It flew to an altitude of only 200 meters that day, but the way it flew, the way it landed, and the obvious minimalist nature of the supporting infrastructure combined with the background knowledge of the program's very low cost sent a shock through all attending. On a technical level they had by no means proved that an SSTO was possible, but on an emotional level they had. The crazy thing worked!

Being a Star Wars type (his nickname among friends is Darth Vader), Worden is not popular among the political opponents of ballistic missile defense, and the Clinton-Gore administration was quick to punish him for his success by killing all future funding for the SDIO DC-X program. But by then his message had been sent. NASA, which had been very skeptical of the DC-X program, decided to ride the popular tide that Worden had unleashed by adopting the SSTO cause as its own. Thus, in 1994, NASA took over operation of the DC-X vehicle and then launched its own SSTO program, dubbed X-33.

Funded at $900 million, the X-33 program aimed to take big steps beyond DC-X. Instead of being one-third scale, it was two-thirds scale. While still suborbital, it would fly to space and reenter Earth's atmosphere at Mach 15 (60 percent of orbital reentry velocity). A competition was held for the contract, with Lockheed Martin, McDonnell Douglas, and Rockwell all submitting bids. Rockwell offered the most conservative concept: a Shuttle-shaped winged vehicle that would take off vertically and glide to an airplane-like landing. McDonnell Douglas proposed an enlarged version of DC-X. Lockheed Martin's concept proved the most radical, having a unique lifting body shape that combined wing and fuselage into a single structure, a novel metal thermal protection system, and a new unproved type of rocket engine known as an aerospike, which promised significantly improved performance over the standard bell-nozzle rocket engines chosen by the McDonnell Douglas and Rockwell teams. This system would take off vertically and land horizontally. Unveiling the concept, Lockheed Martin president Norm Augustine announced that if his company were given the $900 million award for the X-33 flight test vehicle they were prepared to follow the program up by spending $4.5 billion of the company's own money to develop it into a full-scale "VentureStar" launch system capable of SSTO flight.

In a very controversial decision in the summer of 1996, NASA selected the Lockheed Martin concept for X-33. Some people were upset because they saw the Lockheed Martin concept as technically risky, others because they felt that McDonnell Douglas really deserved the contract for its pioneering work with DC-X. In my opinion, neither of these arguments wash. The purpose of X vehicles is to innovate, so technical risk is necessary. And while Gaubatz's team had certainly demonstrated itself capable during DC-X, the Lockheed Martin Skunkworks had done so as well on many other programs. However, there may have been a more serious concern. By 1996, Lockheed Martin provided the lion's share of launch vehicles in the United States, including the Titan, Atlas, LMLV, and MSLS lines, and was also moving into position to control Shuttle launches through its United Space Alliance joint venture with Rockwell. So of all companies, Lockheed Martin has arguably the least interest in seeing a new, revolutionary, low-cost launch system come into being. Such a system would make worthless the entirety of the company's vast current launch business.

These concerns were underlined when, not long after the X-33 award, Norm Augustine retired from the company's leadership. The new management lost little time in clarifying Lockheed Martin's position on VentureStar. It now appeared that Lockheed Martin never actually promised to spend its own money to develop the VentureStar SSTO—it's just something that they *might* do if conditions warrant.

Norm Augustine is an exceptional visionary among aerospace executives, and I'm sure he made the VentureStar pledge with honest intentions. But he's out of the saddle now, and any decision to invest company money in an SSTO will have to be made by the usual standard-issue corporate suits. Based on both the logic of the company's interests and personal acquaintance with a substantial number of Lockheed Martin executives, I would estimate the chance that the corporation's management will decide of their own free will to invest billions of private funds to make their Titan and Atlas lines obsolete to be extremely small.

Once again, since the aerospace companies' profits are based on a cost-plus percentage fee of gross sales price, and the gross price of repeatedly producing many large expensive expendable launch vehicles is much greater than that involved in making a few reusable ones, the major current launch providers have no incentive or desire to introduce reusable systems, or cheaper expendables for that matter.

These are among the reasons why space launch is no cheaper today than it was thirty years ago, and why much-ballyhooed programs such as the NASA/Lockheed Martin X-33 are unlikely ever to produce cheap access to orbit. The only way that the major aerospace companies will ever introduce cheaper launch systems is if they are forced to do so, either by government imperative or by private competition. The former could come, but only in the context of a major national space initiative. NASA today is like a peacetime army filled with parade ground officers. It takes the imperatives and shock of war to force such an army to discard its McClellans and empower its Grants. It would take the moral equivalent of war imparted by the urgency of an Apollo-like drive for Mars to give the space agency the guts and authority required to whip the aerospace corporate bureaucracies into line—forcing them to liberate the talents of their powerful in-house technical organizations to solve the problem of cheap access to space. In their youth, the engineering divisions of the aerospace majors solved the problems of the Moon race, creating the entire range of today's spaceflight capabilities. They haven't done much since, but another wake-up call from above could mobilize them to contribute again.

As for competition, in the past it provided a bit of a prod, but today it is unlikely to be a driving force for innovation among the aerospace majors. In 1992, there were nine major aerospace corporations in the United States: Boeing, Martin Marietta, Northrop, Grumman, General Dynamics, McDonnell Douglas, Lockheed, Rockwell, and General Electric Astrospace. Then in a series of mergers, Martin Marietta took over Astrospace and General Dynamics Space Division, then merged with Lockheed. Boeing took over McDonnell Douglas and Rockwell, and Northrop merged with Grumman. So now, instead of nine competing aerospace majors, we have three, and this may fall to two if Lockheed Martin consummates its plan to take over Northrop Grumman. (At the time of this writing, the Justice Department, after letting all the big mergers go forward without hindrance, has finally raised objections to the Lockheed Martin/Northrop Grumman merger. This is a bit like locking the barn door after the horse has been stolen.) Boeing, because it has deep pockets, strong entrepreneurial instincts, and no launch system of its own to protect, previously threatened to destabilize the system by introducing something new. It had put considerable work into the design of a reusable rocketplane launch vehicle called RASV (Reusable Aerospace Vehicle) and actually initiated the development of the Sea Launch system, deploying low-cost Russian Zenit boosters from floating oceanic platforms. However, even though Sea Launch may well be put in operation, now that Boeing owns the McDonnell Douglas Delta line and has an interest in the Shuttle through its Rockwell division, it seems unlikely that any such ventures will be used to force a sharp drop in launch costs.

Thus, free of pressure from the government or from their peers, the aerospace giants laze away the twilight of the old Space Age. Like the dinosaurs of old, they dominate the landscape of their late Cretaceous. Having had a long and successful reign, they are the lords of all they survey and all they can remember. But, barely noticed by them, a new class of creatures has emerged, smaller, faster, more energetic, more alert, and more rapid to evolve. They are the space-launch entrepreneurs, and their appearance signals, at least to some, that the age of dinosaurs may be coming to a close.

FOCUS ON: FUNDAMENTALS OF ROCKETRY AND SPACEFLIGHT

A rocket engine derives its force through Newton's law of reaction. By expelling mass at high velocity to the rear of the engine, an equal amount of momentum is added in the forward direction of travel. The thrust (T) that such a system generates is simply the product of the propellant mass flow (m) and its exhaust velocity (C). That is:

$$\text{thrust} = (\text{propellant mass flow}) \times (\text{exhaust velocity}), \text{ or } T = mC \qquad (2.1)$$

The higher the exhaust velocity, the more thrust can be produced for a given propellant mass flow. Rocket engineers therefore use this characteristic to rate engines by their specific impulse, or "Isp," defined as the number of seconds a pound of propellant can be used to deliver a pound of thrust and stated in terms of seconds (e.g., an engine may have a specific impulse of 200 seconds or 200 s). The exhaust velocity can be found by multiplying the specific impulse by the Earth's gravitational acceleration. That is:

$$C = g(\text{Isp}) \qquad (2.2)$$

Thus, in metric units, one multiplies the Isp by 9.8 to obtain the exhaust velocity in meters per second. (If you prefer English units, multiply the Isp by 32 to get the exhaust velocity in feet per second.)

Chemical rockets can produce specific impulses ranging from 200 s (for hydrazine) to 350 s (for kerosene/oxygen) to 450 s (for hydrogen/oxygen).

These are low compared to the 900 s that can be produced by nuclear or solar thermal rockets or the 2,000 to 20,000 s specific impulses yielded by electrical propulsion systems. Nevertheless, since the propellant flow rate that can be put through a chemical rocket is essentially unlimited (because the propellant itself provides the energy needed to heat it), chemical rockets can produce much more thrust than the other systems. They are thus, perforce, the only systems that can be used to take off from the ground. (Nuclear thermal systems can be made fairly high thrust, and could in principle be used for ground takeoff as well. However, as a practical matter, the risk of radioactive release precludes doing this on Earth.) Thermal and electrical rockets offer the possibility of much greater propellant-use efficiency for orbit to orbit transfer propulsion, but due to the stagnation of the space program, such application remains to be demonstrated.

To get where you want to go in space, you generally need to change the velocity of the spacecraft from one speed and direction to another. Thus, velocity change, or ΔV ("delta-V"), measured in units of speed, such as meters per second (m/s), is the fundamental currency of astronautics. If you have a spacecraft with a given dry mass M (i.e., empty of propellant), and a certain amount of propellant, P, and a rocket engine with an exhaust velocity C, the following equation, known as the "Rocket Equation," shows how big a ΔV the system can generate:

$$(M + \mathrm{P})/M = e^{\Delta V/C} \qquad (2.3)$$

where e is 2.71828. So the quantity $(M + P)/M$, known as the vehicle's "mass ratio" (the ratio of the vehicle's weight when full of propellant to that of the vehicle empty), increases exponentially in proportion to $\Delta V/C$. If $\Delta V/C = 1$, then the mass ratio equals $e^1 = 2.72$. If $\Delta V/C = 2$, the mass ratio equals $e^2 = 7.4$. If $\Delta V/C = 3$, the mass ratio e^3 equals 20.1. If $\Delta V/C = 4$, the mass ratio e^4 equals 54.6. The exponential is a very strong function; a small increase in ΔV or decrease in C can cause a very big jump in the mass ratio. In fact, the situation is worse than this, because the dry mass M has to include not only the payload you are trying to push, but also the mass of the tanks required to hold the propellant and the engines big enough to push the spacecraft with its propellant, and both of these parasitic weights also increase in proportion to P. So as $\Delta V/C$ goes up, the mass of the spacecraft goes up faster than the exponential, so much so that depending on the lightness of the structural materials and the density of the propellants employed, somewhere between $\Delta V/C = 2$ and $\Delta V/C = 3$ the mass of a single-stage spacecraft will go to infinity! This is the reason why rocket engineers will kill to get ΔV down and C up.

Since the mass of the tanks, engines, and most other vehicle structures increases in proportion to the propellant load, for system performance estimation purposes they can be described by a single factor, the dry mass fraction, F, which when multiplied by the propellant mass gives the dry mass of the vehicle (not counting its payload). This dry mass fraction, F, is a function of system design and the lightness of the materials employed for its construction. So if $F = 0.1$, a ship carrying 90 tonnes of propellant would have a dry mass of 9 tonnes. If the mass ratio of the system had to be 10 to perform a certain ΔV, this would allow the craft to carry 1 tonne of payload. But if $F = 0.11$, the dry mass of the system would be 9.9 tonnes and the payload would fall by a factor of 10, to 0.1 tonne. If $F = 0.12$, the dry mass of the vehicle would be 10.8 tonnes, and the mission would be *impossible* even without any payload. So we can see that the payload delivery capacity of a rocket system can be *extremely* sensitive to the value of F, especially when the mass ratio of the system is high.

This problem can be partially mitigated by chopping the ΔV of the mission into a series of chunks and assigning the accomplishment of each to

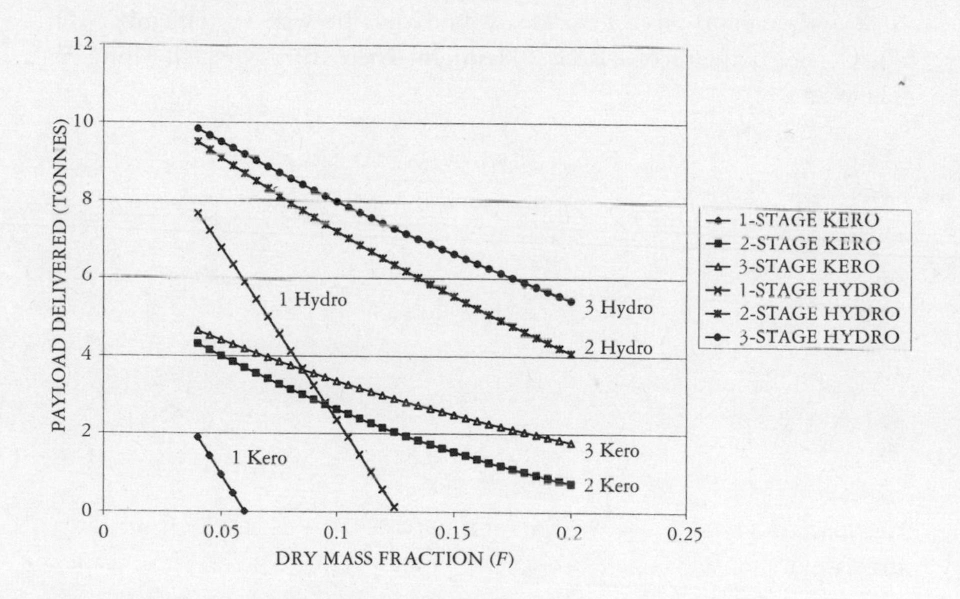

FIGURE 2.1 *Payload delivery capability of 100-tonne liftoff space-launch vehicles flying to LEO.*

a separate stage whose dry mass is dropped from the system after its propellant is expanded. This allows the mass ratio of each stage to be kept low and the mass of the system to be kept finite regardless of the values of C, ΔV, and F. Thus, while it seems inelegant, staging is frequently the only way that difficult (i.e., high ΔV) missions can be accomplished.

One such difficult mission of great interest to spaceflight engineers is travel from the surface of Earth to low Earth orbit (LEO). The ΔV for this mission is about 9,000 m/s. Figure 2.1 shows the payload delivery capability of a rocket vehicle with a gross takeoff mass of 100 tonnes attempting this mission using either one, two, or three stages. Vehicles using both kerosene/oxygen (Isp = 320 s, favored for convenience) and hydrogen/oxygen (Isp = 420 s, favored for performance) propulsion are considered (these Isp values are lower than the maximum that these propellants can generally deliver because the presence of the Earth's atmosphere during flight to orbit diminishes engine performance). Payload delivered is shown as a function of dry mass fraction, F.

Reducing the number of stages of a launch vehicle simplifies operations. But examining Figure 2.1, we can see that unless the dry mass fraction F can be kept very low, simplification to a single stage comes at an enormous cost in payload delivery capability.

Single-stage-to-orbit advocates would thus be wise to remember the motto of one of their icons, Robert Heinlein: There Ain't No Such Thing As A Free Lunch.

TANSTAAFL!

CHAPTER 3

The New Space Race

Among the map makers of each generation
are the risk takers, those who see the
opportunity, seize the moment and expand
man's vision of the future.

—RALPH WALDO EMERSON

On June 26, 1990, the aerospace industry met with startling news. The Motorola corporation announced that it planned to launch an integrated fleet or "constellation" of seventy-seven low-Earth–orbiting satellites to enable global communication through wireless telephone systems. Dubbed "Iridium" after chemical element number seventy-seven, the system would allow people in cities, on highways, on mountaintops, or at sea anywhere in the world to talk freely with handheld cell phones, or use mobile modems to communicate over the Internet. While later scaled down to sixty-six satellites, the name "Iridium" stuck, perhaps because it sounds attractive, perhaps because of its hidden irony. Iridium, after all, is the signature element of the enormous meteoric messenger from space whose impact 65 million years ago punctuated the period of the dinosaurs.

Seventy-seven satellites or sixty-six, the idea remained incredible. The mainstream aerospace industry refused to take the announcement seriously. The largest previous communication constellation had involved only eight satellites; seventy-seven was absurd. Motorola was either kidding or off its rocker. In exchange for a $50-million seed investment in the Iridium system, Motorola offered Martin Marietta Astronautics (the Titan manufacturer) an exclusive monopoly in launching the constellation. Martin Marietta Astronautics' president Peter Teets thought the idea was laughable. He turned them down flat.

But then the Loral Corporation announced that *it too* was planning a constellation of forty-eight satellites, and Orbital Sciences revealed plans for a group of twenty-four. A company called ECCO/CCI then appeared with its own constellation of forty-eight satellites. Ellipsat then produced an eighteen-satellite plan, and Starsys showed up with a further twenty-four.

This was getting pretty wild, but it was just the beginning. In 1994, financial heavyweights Bill Gates and Craig McCaw announced their plans for Teledesic, a constellation of 975 (now 288 larger) low-orbiting satellites to enable global mobile videophone and ultra-high-speed video Internet communication. Following this announcement, Motorola revealed plans to supplement Iridium with an additional seventy-two satellites in Iridium 2, then seventy-two more with Mstar, and then seventy-two more with Celestri, with each system more capable than the one before. A European group has since surfaced with plans for the seventy-two-satellite Satevod system, and rumors currently abound that plans are being developed for additional constellations financed from the Far East.

By the summer of 1997, Iridium had begun launching its first sixty-six satellites (on McDonnell Douglas Deltas, thanks to Mr. Teets's vision), and *U.S. News and World Report* published a cover story reporting a total of over 1,800 additional constellation spacecraft planned worldwide for launch within the next eight years. This satellite manifest is so large that it cannot be launched by the world's existing launch industry, at any price. Furthermore, the number of satellites to be launched is so great that not even Bill Gates can afford current launch-vehicle prices. The satellite constellation managers have thus started to explore alternatives to the existing major industry launch systems. Given the size of the constellation launch market—over $20 billion—financing has become available to aerospace start-up companies that might be able to meet its needs by developing new low-cost launch systems.

The first to be able to exploit the opportunity offered by this new situation was Walter Kistler. Kistler is a multimillionaire—rumor puts his net worth in the $60-million range—and owner and president of a medium-sized aerospace avionics company. He was extremely impressed by the DC-X flights of the fall of 1993, and these, combined with the Motorola Iridium announcement, convinced him that both the market and the technology existed to support the fully commercial (i.e., no government money) development of a single-stage-to-orbit (SSTO) reusable launch system. Many others had similar ideas at the same time, but compared to them Kistler had two important advantages: real business experience and real cash. The latter proved especially handy. Kistler initiated his launch company, Kistler Aerospace, with a $3-million personal investment. This clear bona fide gesture allowed him to attract other investors of similar caliber, and he rapidly built up a $20-million war chest. Since the idea of developing an SSTO on commercial money enthralls engineers, it was not difficult for Kistler to hire a team including many creative people from around the aerospace industry. By mid-1994, design work was underway.

While they may have been creative, the leaders of Kistler's first design team were, in my opinion, rather flaky. The DC-X experience seemed to indicate that a single-stage vertical takeoff/vertical landing (VTVL) system was the way to go, but the most straightforward incarnation of such a system concept was already being marketed by McDonnell Douglas as part of its X-33 proposal campaign. The Kistler team wanted something *different,* something jazzier, something that would allow them to put their own imprimatur on the basic VTVL story. What they came up with was something bizarre.

In an alleged stroke of genius, the Kistler team decided that the way to reduce dry weight on their vehicle would be to *eliminate its landing gear.* Instead, a separate "flying carpet" vehicle would be developed. Basically a platform driven through low-altitude air by a set of low-performing but presumably reliable hydrogen-peroxide rockets, the flying carpet would take off just after the Kistler launch vehicle had reentered the atmosphere, headed home for a crash landing. Meeting the hot post-reentry launch vehicle in midair, the flying carpet would *catch it* and then return it safely to base. You are welcome to draw your own opinion regarding the merits of this concept. In my view, and that of every competent aerospace systems engineer that I know, it was utterly preposterous. If the split-second timing required for the flying carpet midair-catch maneuver failed, not only the launch vehicle but the base and a significant fraction of its supporting personnel might all be lost as the residual hydrogen-oxygen propellant in the launch vehicle's tanks detonated on impact. It would be hard to design a worse range-safety problem—the Kistler concept was the subject of justifiable snickers throughout the aerospace community.

However, if you've got money, you can buy brains. In 1994, a great deal of money sitting around Hong Kong was looking for somewhere else to go and quickly, before the Communists showed up. A Mr. John Wang of that fair city decided that Kistler Aerospace was one of the many places his money should go. In the process of increasing Kistler's capitalization to well over $50 million, Wang gained a controlling position within the company. Hearing the snickers about the existing concept and having no ego investment in it, Wang was quick to make changes. He dispensed with the leadership of the original design team and hired veteran NASA Apollo manager George Mueller as president. Mueller in turn brought with him a team of former high-level NASA managers, including Aaron Cohen, former director of Johnson Space Center, to staff the company's upper management. This was a very smart move. Kistler made much of Mueller's Apollo connection, which conjured up visions of the development of the Saturn V heavy-lift booster, thereby mobilizing additional investment. In fact, Mueller was not really a booster man—the launch vehicles developed for Apollo were designed at the Marshall Space Flight Center under the direction of Wernher von Braun. Mueller and his team all hailed from the manned space-flight systems group at Headquarters and Johnson Space Center. But most people were unaware of the distinction, and regardless, Mueller really is a first-class manager with a documented record of good engineering judgment in developing large projects.

With the whiz kids out and the graybeards in, the Kistler vehicle was completely redesigned. Instead of a single stage powered by hydrogen-oxygen engines that landed with the help of a flying carpet, the new configuration was a conventional two-stage launch system using off-the-shelf Russian rocket engines burning kerosene and oxygen. Landing would be accomplished through the use of good old-fashioned parachutes and air bags. The Kistler team still claims that they can fly the beer-can–shaped first stage back to its landing site area using its rocket engines. It is very unlikely that they will really be able to obtain such system performance. In reality, the first stage will have to be recovered downrange. But once you accept that, there is no reason the system can't work. It's basically a fairly conservative expendable two-stage-to-orbit launch vehicle with a parachute recovery system tacked on. The relatively uncontrolled nature of the landing system would appear to make it likely that the Kistler vehicle will not be truly reusable in the sense that airplanes are, but perhaps each stage will average four or five flights before it suffers fatal mishap. This could still result in a substantial reduction in launch costs for payloads in the Kistler vehicle's lift class of 3,500 kg to LEO.

Furthermore, if the system does fly successfully, it could prove invaluable as an evolutionary step toward truly reusable multi-staged launch systems, including some with heavy-lift capability.

Another early entry into the constellation satellite gold-rush arena was Gary Hudson. Hudson had actually been pushing reusable launch vehicles similar to the DC-X VTVL design since the 1970s, and together with fellow visionary launch system engineer Max Hunter could rightfully claim credit as one of the spiritual fathers of that program. During the 1980s, Hudson actually tried to initiate commercial development of a VTVL SSTO vehicle called the *Phoenix,* but his Pacific American Launch Systems company could not raise capital. The successful flights of the DC-X in 1993 and the Motorola and Teledesic announcements clearly changed his prospects for the better. Together with his friend Bevan McKinney, Hudson decided to seek financing again. The name of his new company was HMX Engineering.

Hudson, however, had a problem similar to that faced by Kistler. McDonnell Douglas had stolen his thunder. Their projected *Delta Clipper* follow-on to DC-X was essentially identical to Hudson's basic *Phoenix* VTVL launch system. To attract investment, Hudson needed something different, since there was little chance that anyone would invest much with penniless HMX to try to create a launch system already under development at an aerospace giant.

Once again, the need to be different led to the bizarre. Hudson's invention was called Roton, and, if nothing else, it was something new under the sun.[1] Roton was a space-launch helicopter. Roton was supposed to work as follows: Rocket engines positioned at the ends of helicopter blades would spin around the base of the launch vehicle. Thrusting down and to the side, the rocket engines would directly impart an upward force but would also cause the helicopter blades to rotate, thereby adding aerodynamic lift to the system. The centrifugal force of the spinning blades would subject the propellants moving outward through the blades in pipes to hundreds of *g*'s of acceleration, thereby pressurizing the rocket engines without the use of pumps. The combination of aerodynamic lift and direct thrust would, supposedly, drive the vehicle upward with more efficiency than an ordinary rocket engine. Then, during stern-first reentry, aerodynamic forces would cause the blades to autorotate, thereby decelerating the vehicle without the use of a heat shield and allow the vehicle to perform a soft vertical landing without the use of rocket propellant.

Roton would not have worked. Shortly after the vehicle went supersonic during ascent, a rearward spreading shock wave would have intersected the rotor blades, making the outward part of them spin in supersonic air while the inner portion continued to spin in subsonic but shock-heated air. Effects similar to those experienced by propeller-powered airplanes attempting to penetrate the sound barrier in the late 1940s would have ensued, resulting in the loss of the launch vehicle. If that wasn't enough, the vehicle would certainly have been destroyed on reentry, since the spinning rotor really is no substitute for a heat shield. Finally, if by some miracle the vehicle survived reentry, it would be destroyed on landing since, with the rotor underneath the craft, it was aerodynamically unstable.

The design, however, did succeed in its true purpose, which was to bring in investment. In the fall of 1996, Hudson landed himself a major funder in the form of telecommunications multimillionaire Walt Anderson. Author Tom Clancy also invested. Anderson provided Hudson with an open checkbook, and to date an estimated $10 million has been spent in the development of the renamed Rotary Rocket Company. In the course of the engineering work paid for by these expenditures, the Rotary Rocket team has discovered many of the obvious failure modes of the Roton. As a result, the system has been completely redesigned, and now more closely resembles Hudson's old *Phoenix* VTVL concept. The rotor blades persist, but they lack rocket engines, are small, and sit atop the vehicle, where they can be used to provide an autorotation assist to deceleration and landing. The engines still spin, but on a short arm hidden behind a heat shield at the base. Kerosene/oxygen is the propellant mixture chosen, as it is cheap, dense, and easy to handle. As the spinning engines rapidly circle the base, an aerospike effect should be created that will generate aerodynamic flow-field effects to optimize the efficiency of the engines without the need for a large bell nozzle. This is a daring concept, as nothing like it has ever been demonstrated in a flight system, but theory suggests that it could work. Of course, the tough challenge of achieving the very light dry mass fraction required for single-stage-to-orbit flight still remains, and Hudson is attempting to tackle it head-on, even as he attempts to develop a novel propulsion system. No major aerospace company would ever dare to undertake so much high-risk technological innovation while attempting the development of a new flight system—and this in my view is Hudson's true value. Whether he succeeds or fails, he is going to break a lot of new ground. It will be interesting to see if he can pull it off.

In the summer of 1993 while attending an American Institute of Aeronautics and Astronautics (AIAA) technical conference in Monterey, California, I had a very interesting conversation with an Air Force flight test engineer named Captain Mitchell Burnside Clapp. I had known Clapp for a while, as he had participated in several Case for Mars conferences during the 1980s. Over a beer and a bar napkin, as engineers will, Clapp laid out his concept for a superior SSTO vehicle. According to Clapp, who was then serving on the staff of the SDIO's DC-X project, the ideal SSTO would not be a vertical takeoff system like the DC-X, but a winged rocketplane capable of horizontal takeoff and landing on an airfield. This idea in itself was not new; Boeing had been peddling such a concept, called RASV (Reusable AeroSpace Vehicle), for years. But RASV's numbers had never worked; the extra weight of the wings and landing gear required to get a fully loaded rocketplane off a runway made single-stage-to-orbit flight impossible. But here Clapp had an original twist. What if the rocketplane did not take off fully loaded? What if, instead, it took off nearly empty, and then received its propellant at altitude from a tanker aircraft, just as Air Force fighter planes do all the time? In that case, the wings and landing gear could be much smaller, as the rocketplane would not be heavy until it was flying fast with a lot of wind under its wings. Furthermore, the rocketplane would start its ascent to orbit fully fueled from a high-altitude, high-velocity condition. This would reduce both the aerodynamic and gravity losses involved in the flight, effectively giving the vehicle a head start compared to one that had to lift itself from the ground. With the help of aerial refilling of propellant, single-stage-to-orbit flight of horizontal takeoff/horizontal landing (HTHL) vehicles might finally become possible.

I thought that the idea had a lot of potential, and upon my return to work at Martin Marietta following the conference subjected it to extensive analysis. Examining its application parametrically across a range of technology assumptions and propellant combinations, I found that it offered strong performance advantages over conventional SSTO vehicles of either the VTVL or HTHL types. I then began working on ideas for the design of an X-vehicle that could demonstrate the concept.

Unknown to me, Clapp was working hard in a parallel direction. He managed to mobilize $100,000 in Air Force funds for a design study of his rocketplane concept. This vehicle design, called "Black Horse," became public around December 1993.

Black Horse was an SSTO rocketplane employing aerial refueling, or, more technically, aerial propellant transfer, since it was actually the oxidizer and not the fuel that was transferred in flight. It was equipped with seven hydrogen peroxide/kerosene rocket engines, two with short nozzles for low-altitude flight and five with large nozzles for high-altitude and space flight.

It was piloted, and had a payload to LEO of 2,300 kg. According to Clapp, it could be developed for about $350 million.

I examined Black Horse after it was published and found the design excessively optimistic. Clapp had based his design on a back-of-the-envelope estimate that the ΔV to orbit for this vehicle was about 8.2 kilometers per second (km/s)—in reality it was about 9 km/s. Even using Clapp's figure, the vehicle could have a dry mass fraction of only 8 percent. If the more accurate ΔV was used, the available dry mass dropped to 6 percent. From an engineering standpoint, neither would be achievable in practice. Furthermore, achieving even these structural mass fractions required developing hydrogen peroxide/kerosene engines with a specific impulse of 330 seconds, whereas the maximum that had ever been attained by rockets using this propellant combination was about 270 seconds. If a more realistic specific impulse of 300 seconds were assumed, the available dry mass fraction dropped to 4.7 percent, which is just hopeless. Furthermore, the aerial propellant transfer would have to be accomplished with the receiver aircraft driven by rocket power, which has never been done. Because of the gas-guzzling nature of rocket engines, most of the propellant transferred in the air would be consumed by the rocketplane while it was hanging on the tanker, making it questionable whether a tanker large enough to carry sufficient propellant to actually fill Black Horse's tanks could actually be obtained. The rocketplane's mass changed by a factor of 7 during the propellant transfer operation, and this would cause its flight control laws to change, thereby posing a severe challenge to the design of the aircraft flight control system. Finally, Black Horse had to use 98 percent pure hydrogen peroxide as its oxidizer. This would be dangerous. Hydrogen peroxide is an unstable molecule that would rather divide itself into water and oxygen. A small amount of impurity entering the propellant system could catalyze catastrophic decomposition of the hydrogen peroxide, causing the vehicle to explode.

In his presentations within the Air Force, Clapp managed to get by most of these points, and so for a while Black Horse became quite the rage. The Air Force, after all, is run by fighter pilots who don't know a lot about rocketry. But they know what they like, and for them the piloted Mach 25 rocketplane had enormous appeal.

However, I felt that if the aerial-refilled rocketplane was to have any chance of early development, it would first have to see light in a much less ambitious form. I therefore designed a smaller vehicle with much lower technology development requirements. In order to play off the Black Horse craze current at the time, I named my design "Black Colt," thereby suggesting it could be a developmental step toward a more fully mature "Black Horse."[2]

The main propulsion for Black Colt was provided by an NK-31 rocket engine burning kerosene and liquid oxygen, an off-the-shelf unit with a demonstrated specific impulse of 350 seconds and a thrust of 90,000 pounds. In addition, two F-125 fighter jet engines provided propulsion for takeoff, landing, and tanker operations. Instead of performing the full 9 km/s ΔV to achieve orbit, Black Colt would fly on a suborbital trajectory with a velocity at apogee of 4 km/s (it takes about a 5 km/s ΔV to achieve this). The remaining 4 km/s ΔV required to put a payload in orbit would be achieved by attaching a small solid rocket motor upper stage, such as a Thiokol Star 48, to the payload. In other words, Black Colt would use its jet engines to take off from an airport fully loaded with kerosene but with just enough liquid oxygen on board to pre-chill its tanks. It would then rendezvous with a tanker at an altitude of about 8,000 meters where it would receive its supply of liquid oxygen. After tanking had been completed, it would fire its rocket engine and fly to an altitude of 150 km (i.e., space) and half orbital velocity. Once there, it would open its payload bay doors and, in a vacuum and zero gravity environment, drift the payload/upper-stage combination out of its bay. When a sufficient degree of separation had been achieved between the payload and the rocketplane, the upper stage would fire, propelling the satellite to orbit. Simultaneously, the rocketplane would close its payload bay and then reenter the atmosphere at about Mach 12. The vehicle would then glide quite a distance shedding velocity until it went subsonic. At that point, the jet engines would be restarted and the vehicle would land at an airport in a conventional fashion.

The Black Colt concept had numerous advantages over Black Horse: High-performance engines to implement it were already available. And, because Black Colt's rocket engine's specific impulse was higher and the system's required ΔV was much lower, the vehicle's dry mass fraction could be as high as 23 percent, which makes building the whole system quite feasible. Tanking would be done while flying with jet engines, as currently practiced. The mass of the rocketplane during tanking would only change by a factor of 2, which simplified the control system development. Because Black Colt only went to half orbital velocity, it reentered at half the velocity of an SSTO and so its thermal protection system would take much less of a beating. The vehicle also had more utility (military or otherwise) than Black Horse or any other launch vehicle, as its jet engines would allow it to ferry itself around the world in the manner of an ordinary airplane. Jet engines also allowed the vehicle to perform conventional powered landings, a much safer procedure than the dead-stick glider landing required of Black Horse. Safety would also improve by using liquid oxygen over 98 percent hydrogen peroxide. Liquid oxygen has its hazards—it is cryogenic and will react spontaneously with many combustible materials. However, the amount of energy released in such reactions is proportional to the amount of combustible contaminant exposed. In contrast, introducing a tiny speck of the wrong kind of dirt can cause an entire 100-tonne tank of hydrogen peroxide to explode. Finally, liquid oxygen is about one-tenth the cost of hydrogen peroxide. For launch-vehicle applications, this does not matter—propellant costs are minor compared with the rest. But a rocketplane clearly has other potential applications, such as long-distance fast travel, and here propellant costs matter a lot.

Against all these advantages, Black Colt had one important disadvantage compared to Black Horse. It required an expendable upper stage. However, the cost of a Star 48 is only about $1.5 million, whereas the current commercial value of the 400-kg payload launch of the Black Colt system is around $10 million. Since it made engineering the entire system much more feasible using present-day technology, this cost hit was certainly worth taking.

In 1994 I was able to mobilize sufficient company R&D funds at Martin Marietta to undertake an internal design study of Black Colt. We determined that the vehicle was feasible and that the first flight test unit could be designed and produced for a cost of about $90 million. In relative terms, this is quite low. The economics of developing Black Colt were then improved further when NASA announced a competitive procurement for a vehicle called X-34, whose performance requirements Black Colt satisfied. NASA said it would be willing to provide up to $75 million for the system, with the winning private company paying the rest. NASA officials at the Marshall Space Flight Center were intrigued by Black Colt and dropped hints (including comments in the industry newspaper *Space News*) that they hoped we would bid for X-34. However, Martin Marietta management refused to bid. This angered me a great deal. Here we had a practical concept for a near-term reusable launch vehicle and NASA would foot almost the entire bill for its development, and yet—no bid. I protested this decision to several influential people within the company and got nowhere. Eventually, a director took me aside and said, “Look, Bob, it’s a very clever idea, but you've got to get the picture. We build Titans. You sell one of these to the Air Force and we're out of business."

His statement was more than a bit exaggerated, since Black Colt's payload capacity was only 400 kg, whereas a Titan can lift 20,000 kg to LEO. But his point was clear. Lowering launch costs was not perceived by management as being in the company's interests. I realized then that if Black Colt was to have any chance of becoming real, I would have to leave the reservation.

So in January 1996 I quit Martin to found two companies in my basement, Pioneer Astronautics and Pioneer Rocketplane. The former I started on my own as a vehicle for continuing to perform R&D work for NASA on advanced space systems and Mars exploration technology, an area where I have considerable interest. The latter I founded in partnership with Clapp (who by this time I had argued around to see the merit of the Black Colt–type design) and Chuck Lauer, a Michigan businessman who was fascinated by the commercial possibilities of the rocketplane. Since then both businesses have grown, with labs, offices, and a staff of about half a dozen people each. Between investors and NASA contracts, the Rocketplane venture has thus far raised about $3 million, which has allowed a fairly serious design effort, including wind tunnel work, to advance. The basic Black Colt concept has been redesigned to allow it to capture most of the projected satellite market. This was achieved by enlarging the vehicle and replacing the NK-31 rocket engine with an RD-120 (also a Russian kerosene/oxygen engine but with a thrust of 180,000 pounds and a specific impulse of 350 seconds) and the F-125's with more powerful Pratt and Whitney F-100's (the same engine used on the F-15 fighter). This enlarged Black Colt–type system, dubbed the *Pathfinder,* has an estimated payload delivery capacity of 1,400 kg, making it capable of launching Iridium or Teledesic satellites.

Since January 1998, Clapp and I have gone separate ways, he to lead Pioneer Rocketplane, and I Pioneer Astronautics. Clapp's task will not be easy. While a significant start, $3 million is a long way from the $100 million or so that is required to develop the *Pathfinder.* It is quite possible that by the time you read these lines, Pioneer Rocketplane will have ceased operations due to lack of sufficient capitalization. On the other hand, it may be trading at 17 on the NASDAQ, which would really suit me fine since I own 2.7 million shares. But whatever the fate of the Pioneer Rocketplane Company, I believe that the rocketplane concept that it is attempting to develop holds tremendous promise, because rocketplanes have numerous applications besides launch to orbit. Let's consider three of these: fast package delivery, military missions, and passenger travel.

ROCKETPLANE POSSIBILITIES

Fast Package Delivery

The world is moving toward an international market whose long distances demand faster transportation. Rocketplanes could meet this demand, enabling transportation of substantial payloads between transoceanic destinations with travel times of less than one hour.

In 1994, a group of leading aerospace contractors, consisting of Boeing, General Dynamics, Lockheed, Martin Marietta, McDonnell Douglas, and Rockwell, released a report, termed the Civil Space Transport Study (CSTS), that documented this need. In the course of the CSTS, numerous players in the current fast package delivery business (Federal Express, et al.) were interviewed and expressed little doubt that a strong market would exist for the kind of rapid global package delivery that a suborbital rocketplane could offer. In fact, in 1986, Fred Smith, the CEO of Federal Express, published a paper advocating the value of precisely such a service. As Mr. Smith put it at the time, "Based upon our research and investigation, such a hypersonic aircraft would be economically viable in our business; most importantly, our customers, more than any others, need this type of improved trans-oceanic speed."[3] Indeed, according to industry contacts interviewed by CSTS, a strong market would exist for such a service even if the price of delivery were as large as $2,000/kg. At a price of $400/kg, the study concluded that an annual market of about 1,000,000 kg of mail per year would be available in the year 2000, for a total revenue of $400 million per year ($400/kg surface-to-surface represents a price that even a primitive *Pathfinder*-based service could attain if used in a suborbital mode and assuming a 2,500-kg payload, $500,000 flight cost, and 100 percent margin for other overheads and profit). At these prices, packages would include electronic devices, biological reagents, human organs, precious stones, microcircuits, just-in-time parts needed at a factory, and important business documents. Based on industry trends, it is anticipated that this market would grow at a rate of 7 percent yearly. Assuming that $100 million of the $400 million per year represented profit, a rough estimate based on a 10 percent rate of investment return for the fast package providers would indicate that $1 billion in rocketplane sales would be possible.

An interesting analogy for the relationship between using rocketplanes for satellite launch and using them for fast package delivery exists in the history of humanity's use of the world's oceans. For thousands of years, commerce has been practiced upon the seas in two primary modes. In one case, individuals have tried to extract wealth directly from the sea, primarily by means of fishing. This activity compares with satellite use, which attempts to extract wealth directly from space by providing communications or remote sensing services from that vantage. However, as economically viable as fishing has sometimes been, for the past 3,000 years far more wealth has been generated on the oceans by maritime commerce, which has taken advantage of the fact that the sea is a low-drag medium allowing relatively easy transport of goods over long distances. Similarly, space is a global ocean with zero drag. The fast package delivery capability of rocketplanes takes advantage of this attribute to enable a form of economic activity—flying *through* space—that in the long run will probably dwarf the commerce carried on by satellites or other assets residing *in* Earth orbital space.

Military Applications

As a result of its study of Black Horse, the U.S. military has recently issued reports calling for the development of a transatmospheric vehicle that could perform ultra-high-speed missions, including reconnaissance, interception, and strike. The potential of rocketplanes to meet this demand compares favorably to other alternatives.

For example, with their capability for fast response, high speed, and long range, rocketplanes would be able to intercept an adversary's air transport deep within enemy airspace, thereby performing the same commerce interdiction role in the air that navies have historically performed at sea. Used as a strike vehicle, a rocketplane could release payloads from outside the atmosphere and then repeatedly bounce off the atmosphere well above fighter altitude to hit a series of targets in an unpredictable fashion. As a strategic weapon, the rocketplane combines the speed and range of an ICBM with the flexibility and recallability of a piloted bomber. If the United States were ever again to face a nuclear balance of terror analogous to that of the recent Cold War, the possession of rocketplanes by our armed forces would do much to eliminate any imperative by national leaders to "launch on warning." Instead, if early warning data were received indicating that an enemy strike were under way, the rocketplanes could be scrambled to safety in the air but not launched unless hostile action was confirmed.

Global Passenger Service

If rocketplanes should prove successful as vehicles for global fast package delivery, passenger versions could also evolve. These could fly from New York to Paris in less than one hour or New York to Sydney in less than two. It might be argued that the coming age of expanded global electronic communications will obviate the need for rapid personal travel. However, the evidence to date contradicts that argument. Starting with the telegraph, and continuing with the telephone, radio, TV, fax, and the Internet, the expansion of electronic communications has contributed to a vast expansion of global commerce and exponentially rising demand for intercontinental personal travel. Moreover, as electronic communications become faster, the more anachronistic today's current subsonic flight travel times will seem. A faster-moving society will require faster airplanes.

Economics is frequently raised as an objection to rocketplane passenger service. A long-distance subsonic aircraft might typically require a fuel load about equal to the weight of the airplane and its payload. In contrast, a long-distance rocketplane might need propellant equal to four times its dry mass. This seems like a formidable obstacle to rocketplane economics, until it is recognized that about 75 percent of the rocketplane's propellant load is liquid oxygen, which costs only about $0.12/kg, compared to about $0.40/kg for jet fuel. If this difference is taken into account, the propellant cost for a long-distance rocketplane ride is likely to be less than double that of a subsonic airliner. This cost will be counterbalanced by the fact that the same rocketplane could be used several times in the period it takes a long-distance subsonic airliner to fly a single trip, thereby lowering the rocketplane's per-flight interest costs. Provided that maintenance costs can be kept to levels comparable to those of existing aviation systems, there is no reason the rocketplane could not be made commercially competitive, especially when the desirability for its greater speed is taken into account.

One interesting feature of rocketplane passenger travel will occur several minutes into the middle of the flight when the vehicle will travel through space in a condition of free-fall. For those several minutes, the passengers will experience zero gravity and perhaps enjoy the novelty of a brief float around the cabin. If they look out their windows, they will see a black, starry sky—the universe in its full glory as viewed from space.

The ultimate importance of all this is that if the real Space Age is ever going to arrive, there will have to be a market for rocket vehicle technology that supports the manufacture of spacecraft components not in lots of ones or twos, but in hundreds or thousands. If travel to orbit is ever to be as cheap as air travel, we will need a worldwide launch infrastructure that supports not hundreds of flights per year, but hundreds of flights per day. The only market that can support that is global long-distance package and passenger delivery. For better or worse, the fact is that far more people want to fly to Tokyo than to orbit. To service that market, winged rocketplanes capable of horizontal takeoff and landing from existing airports will be necessary. In contrast, vertical takeoff vehicles would require the construction of literally trillions of dollars of new launch facilities near cities all over the world to assume such a role.

The use of aviation-style mass production methods in place of the expensive small-lot production techniques that dominate the space industry today will make possible the inexpensive development of new variants of rocketplanes and other space systems. Furthermore, it will also provide a driver for competitive privately sponsored R&D in propulsion and other technologies that simply does not exist today. These innovations in turn will accelerate technological progress and lower costs, creating the industrial infrastructure for a true Space Age in which the benefits of space grow exponentially and become available to all.

For the same reason that military and then postal aircraft preceded passenger aircraft, satellite-launch, military, and fast package rocketplanes will no doubt precede passenger rocketplanes. Nevertheless, the day can be foreseen when thousands of rocketplanes crisscross the globe daily, serving business and vacation travelers from New York to Sydney, or perhaps to orbit.

That's why I like rocketplanes.

And I'm not the only one. At about the same time that the Black Horse/Black Colt work was proceeding at the Air Force and Martin Marietta, Bob Kelly, a former TRW engineer, was developing a rocketplane concept of his own. While much larger than Black Colt or *Pathfinder,* Kelly's vehicle has a similar flight profile. The important difference is that instead of flying on jet power to 8,000 meters to meet a tanker and then firing its rocket engine, Kelly's rocketplane would be towed to altitude by an airliner. This leads to simpler vehicle design than the aerial-refilled rocketplane concept (since no jet engines are needed) but adds numerous operational difficulties associated with towing a very heavy rocketplane filled with high-explosive off a runway, tow-line disengagement, risk of vehicle loss if the rocket engine fails to start after separation, and the need for dead-stick (unpropelled) landings. Also, a vehicle that requires towing clearly has less operational utility than one that can ferry itself. Still, the simplicity of the design has its attractions, and as of this date Kelly Space Technologies has raised about as much money as Pioneer Rocketplane.

These are just a few of the aerospace launch-system start-up companies that have received significant funding as a result of the satellite boom. There are others: Space Access, Beal Aerospace, Microcosm, Advent, Universal Space Lines . . . the list goes on.

The aerospace launch start-ups described in this chapter are all immature and nearly all are critically underfunded. Yet they are growing in management and engineering maturity, and financial strength every day. Unquestionably, many will fail, to have their places soon taken by new contenders. But as shaky as they are, the benefit that their success would offer the satellite customers is enormous. Therefore, despite the inherent risk of dealing with new aerospace launch start-up ventures, the owners of these constellations have shown themselves willing to let contingent contracts for large numbers of satellite launches to companies whose launch systems do not yet exist. For example, in the final two months of 1996, Motorola gave a $90-million contract to Kelly Space Technologies for nine satellite launches, while Loral gave the new Kistler Aerospace corporation a contract for $100 million for ten launches. Teledesic has indicated that it is likely to follow suit. It is thus the clear intent of the communication satellite constellation companies to do everything they can to break the dinosaur's launch monopoly. Provided that they stick with that determination, at least one of the start-ups will succeed.

As a result, a new commercial space race has begun. The prize: tens of billions of dollars of launch contracts.

However, as exciting as this might be and as significant as it is for the future cost of satellite launch, the limits of such commercial enterprise for the creation of a spacefaring civilization need to be understood. The development of new, innovative, aggressive competing organizations is valuable, as is the reusable technology they are demonstrating, and some of the systems being explored by these organizations could lead to the creation of an industrial infrastructure that would significantly lower the cost of space vehicle hardware. But the actual launch systems under development are being designed not to support the colonization of space, but to launch satellites. Payload capacities are in the range of 1 to 5 tonnes to LEO, a far cry from the approximate 100-tonne lift capability desirable for supporting human exploration missions, let alone colonization. Transportation follows destination and application. So long as the only driver for commercial development of new launch systems is the delivery of satellites, commercial development of space-launch systems that can enable human settlement of space will be impossible.

The new entrepreneurial space race is crucial for the human spacefaring future because it is bringing into being the enterprising creatures that will seize the opportunity offered by a major opening in space to produce cheap transportation systems. The satellite rush may provide enough nourishment to give these creatures birth, but by itself is insufficient to grow them to the point where they can launch a spacefaring civilization. A larger opening, demanding greater launch capabilities, is needed.

If satellite launch is the only driver the private sector can provide to promote launch-system development, then market forces by themselves will prove insufficient to get humankind into space. But perhaps there are other forms of orbital private enterprise whose lure of profit may force the creation of the kinds of launch systems that humanity really needs to pioneer the new frontier. Such possibilities are the subject of the next chapter.

FOCUS ON: THE X-PRIZE

Most people reading this book recognize the name and accomplishments of American aviator Charles Lindbergh. I would venture that far fewer are familiar with the name of French hotelier Raymond Orteig, even though he played a central role in Lindbergh's successful nonstop Atlantic crossing. It was Orteig who offered up a $25,000 prize for the first nonstop flight between New York City and Paris, which Lindbergh claimed with his 33.5-hour flight on May 21, 1927.

The Orteig Prize was just one of more than 100 aviation prizes offered between 1905 and 1935 that challenged the minds and spirits of early aviators and, in doing so, helped give birth to today's multibillion-dollar air transport industry. Indeed, goal-specific prizes have played a critical role in numerous technological areas, and they've done so not just for decades but for centuries.

An early and famous prize offered in the 1700s sought to spur the development of accurate seagoing navigation. That prize, established in the English Parliament's Longitude Act of 1714, established an award of £20,000 payable to the individual who devised a method to reliably determine longitude at sea. The prize spurred clockmaker John Harrison to develop an accurate, seaworthy chronometer that could be used by ship's captains to precisely record the time when taking a fix on the sun's position. By comparing the shipboard fix and time with information recorded at a known location, a navigator could determine his longitude.

This century, physicist Richard Feynman helped open an entirely new field of molecular-scale engineering by offering $1,000 of his own money to the first individual to construct an electric motor with dimensions no larger than 1/64th of an inch on any side or to shrink written text by a factor of 25,000—small enough to inscribe an encyclopedia on a pinhead. Feynman laid the challenges down in a memorable 1959 speech entitled "There's Plenty of Room at the Bottom." Within a year the motor prize had been claimed. A quarter century passed before anyone could claim the text prize.

So the question arises: "If it has worked before . . ."

Not long ago a group of individuals led by space whiz kid and later entrepreneur Peter Diamandis asked themselves the same question about goal-specific prizes, this time with an eye to opening low Earth orbit and beyond to tourism. In May 1996, Diamandis and members of the St. Louis, Missouri, business community announced the establishment of the X-Prize Foundation and the X-Prize competition—$10 million to the first team that develops and flies a spaceship capable of launching three passengers to an altitude of 100 kilometers on two consecutive flights within a two-week time period.

Thus far, fourteen teams, including Pioneer Rocketplane, have registered in the competition. The entries run the gamut. Burt Rutan of Scaled Composites has entered the Proteus, which uses an aircraft as a first stage to carry the Proteus rocketship to launch altitude. Rick Fleeter and AeroAstro are proposing a conventional, vertically launched rocket that will glide back to its launch site under the wing and guidance of a parafoil system. James Ackermann and Harry Dace of Advent Launch Services have submitted their Advent CAC-1 Passenger Rocket (they're already selling seats), which uses neither runway nor launchpad but floats like a cork in the ocean before blastoff.

The curious thing about prizes is that the prize money, per se, is not what all the entrants are chasing. I mentioned that development of the *Pathfinder* rocketplane might cost $100 million or so, far beyond the $10 million offered by the X-Prize. Obviously something more than money is at work here: publicity possibilities, ego enhancement, societal validation, whatever. The point is that the offer of a seemingly limited amount of money can promote the infusion of large amounts of capital toward developing a technology. The $25,000 Orteig Prize spurred nine separate attempts at crossing the Atlantic, which, taken together, cost some $400,000. We can expect the X-Prize and future awards to spur a similar amount of investment, and, hopefully, help produce the huge returns in investment generated by goal-specific technology prizes of the past.

CHAPTER 4

Doing Business on Orbit

"Desire of gold, great sir?
That's to be gotten in the Western Ind."

–CHRISTOPHER MARLOWE,
Tamburlaine, 1587

TODAY, THE ONLY significant money-making activities conducted in space are those based on communications, broadcasting, and remote sensing of terrestrial weather and resources. All of these can be, and are, efficiently performed by robotic satellites whose masses seldom exceed 5 tonnes (and are generally closer to 1 tonne). As the art of miniaturizing electronics advances, the size of satellites required for these missions can be expected to shrink. Thus, while the market for these activities is rapidly expanding, there is little hope that they will lead to the development of low-cost launch vehicles adequate for supporting the human colonization of space. Put simply, new boosters that can inexpensively launch people and manned payloads are not a natural by-product of the satellite boom. Rather, they will only be developed to meet the needs of *human activity* in space.

Let us suppose, as now seems probable, that the satellite rush will succeed in generating a wave of highly competitive new companies sporting an array of low-cost reusable or semi-reusable boosters for 1- to 5-tonne-class payloads. It would be reasonable to assume that such a development would encourage entrepreneurs to float business plans for orbital projects, even if those projects required heavier lift capability. Having succeeded in creating themselves out of nothing to serve the satellite constellation launch market, the new launch companies could certainly be expected to be able and willing to expand their systems to meet the demands of weightier customers. But still, business plans for heavy-duty orbital activity would only get funded (and thus could provide an effective driver for the development of larger commercial launch systems) only if they could show that, with low launch costs assumed, they could return a profit. Well, could they?

A number of possibilities have been advanced for commercially viable human orbital activity. These include the development of orbital commercial research labs, industrial factories, space business parks, hotels, space asset servicing, and the construction of orbiting solar power generation stations.[1] The potential benefits offered by many of these concepts were used in the past as hype to justify developing the Space Shuttle and the Space Station. As a result of the high cost of Shuttle launches, very little of their promise has been realized. But assuming that the new wave actually does succeed in creating cheaper access to Earth orbit, might they not then become profitable? Let us consider each of these options in turn.

ORBITAL RESEARCH LABS

Of all the potential concepts for commercial human on-orbit activity, I think orbital research labs that take advantage of the unique zero gravity and high vacuum environments available in LEO have the best chance of producing a profit in the relatively near future. The product of such labs is knowledge, which is massless. Thus, precious little raw material is required, at least in principle, to produce marketable products of enormous cash value. This is so forcefully true that even the overpriced Space Shuttle has produced what might be considered a kind of profit during two 10-day missions in which zero gravity experiments helped researchers determine the structure of certain animal viruses, thereby enabling the development of veterinary vaccines worth several billion dollars to the economy. Further, a successful company, called SpaceHab, rents out the use of its lab module, which flies periodically on the Shuttle. Because of the high costs, commercial enterprise has yet to sign up for this service in a big way, leaving NASA as the primary customer. However, a dedicated orbiting research lab with a long-duration professional staff could offer much more than simply lab space as is now the case. With a lower-cost launch vehicle to support the operation, prices could conceivably drop to the point where investment in such research would be competitive with the return offered by terrestrial research facilities.

On orbit, the distorting influence of gravity is nearly absent, which creates conditions enabling the production and determination of the structure of various types of crystals and other compounds. In addition, low Earth orbit provides access to very high quality vacuum conditions that cannot be economically produced in Earthbound labs. The knowledge that flows from investigations conducted in these environments can allow the development of a range of products from disease cures to "brains" for new supercomputers so advanced that their proponents claim they would revolutionize life on Earth. Potential microgravity or high-vacuum research products with astronomical value exist in the form of vaccines, synthetic collagen (which could be used to construct corneas), targetable pharmaceuticals, structured proteins, crystal materials (for computer chips and quantum devices), ultrapure epitaxial film production, unique polymers and alloys, and electrophoresis applications. These products could lead to breakthrough applications in such high-growth areas as semiconductors, computers, instruments, biotechnology, and drug manufacturing, areas that today represent a business base of over $240 billion per year.

The ability of orbiting labs to effectively conduct this kind of research will be tested on the International Space Station, which will be fairly well equipped for such work. Indeed, microgravity research has been frequently used as one of the primary justifications of the Space Station program. Because the Space Station program has been misdesigned and chronically mismanaged, it has produced large cost overruns that have drained funds from other NASA endeavors, causing many to take shots at it. Some have advanced the notion that zero gravity research could better be done on unmanned satellites. Having run a research lab myself, I believe that such claims are completely wrong. Yes, isolated well-planned experiments can be flown on automated spacecraft and useful data returned. But to effectively perform an actual investigative research program into unknown intellectual territory requires real live human experimenters with constant access to their apparatus. An automated experiment can record data that are expected; only a human investigator can respond to surprises—and most big discoveries come as surprises.

The problem with the Space Station is not that it is a poor instrument for zero gravity research—it actually offers high value in this area. The problem is that it will accomplish this at too high a cost and that it has preoccupied NASA for too long. A simpler and more limited space station could have been developed (and be flying by now) that would have offered the same capability much cheaper and quicker. It need not take anything like twenty years and $30 billion to develop an orbiting research lab. And that's good news—it means that commercial orbiting labs could potentially be profitable.

ORBITAL INDUSTRIES

Producing patents, not products, is the best way for an orbiting lab to make money. The discovery of knowledge in space that enables industrial processes to be realized on Earth is clearly the highest payoff path for such space-based facilities. But what if that is not possible? What if the lab discovers a process that can be replicated only in the zero gravity environment of space? Could profitable mass production operations actually be initiated on-orbit?

The answer to this question depends upon a variety of factors, chief among them the cost of space launch, which today stands at roughly $10,000/kg to low Earth orbit. Let us assume that advances in launch technology allow this to drop by a factor of 5, to $2,000/kg (or roughly $1,000/pound), a goal that is NASA's current battle cry. Then, clearly, in order for space-based manufacturing to be profitable, the value of the goods produced per unit must exceed this figure. In fact, it must exceed it by a good deal, because in addition to transporting the raw materials, the launch system will also have to transport the orbital factory, its spares, consumables, and power system, the workforce and their consumables, and the propellants and other consumables necessary to keep the factory spacecraft functioning and stable in its proper orbit. In addition, the orbital factory business will have to support the salaries and fringe costs of the company's Earthbound and spacebound staffs, its offices, advertising, insurance, taxes, interest payments, and other overhead, and the standard wholesale/retail sales price markup; plus, given the high level of risk in such a business, the company must be able to pay large dividends to investors. So, if the launch cost is $2,000/kg, the orbital factory's product will have to have a *retail* sales price of at least $20,000/kg for there to be a net payoff sufficient to motivate investment. Roughly speaking, this is the price range of gold ($20,000/kg = $568/ounce). In addition, the product produced would have to be so superior to terrestrial alternatives that it would still sell well despite the fact that it would cost more. Taken together, these factors would tend to rule out almost all alloy or other materials production operations, but the production of advanced computer chips or unique pharmaceuticals would still qualify.

Finally, the sales volume of the product must be sufficiently high. If the basic costs of running the orbital lab are $40 million per year (an optimistically low number that would allow shipping just 20,000 kg/year of various supplies), the operation will still fail if only 1,000 kg of its $10,000/kg wholesale-priced product can be sold. To allow for all the business's costs, at least $80 million in gross revenue would be needed, or 8,000 kg of product. Let's say that the end use for the product is a drug or computer chip selling retail for $200 for a 10-gram unit. In that case, 800,000 units would have to be sold per year. Given a sufficiently desirable and unique product, sales numbers such as these are entirely feasible. But the prospective investors would have to be convinced of this in advance. This could be difficult, given that test marketing of a space-manufactured product is essentially impossible without an orbital facility in place. But perhaps the orbiting lab that discovered the product could be mobilized for such purposes.

Thus, given a sharp drop in launch costs in the relevant class of space lift, profitable orbital manufacturing facilities are possible. There would be a bit of a chicken-and-egg problem with respect to convincing investors of probable profitability. The deeper problem, however, is that the apparent size of the launch market offered by such operations (especially in their initial phases) may not be sufficient to motivate private investment to create the enabling low-cost 10-to-20-tonne-to-LEO launch systems required.

ORBITAL HOTELS

On-orbit hotels are another form of commerce that could conceivably become viable. Tourism, after all, is one of the largest businesses on Earth. Why should it not be in space? The potential viability of space tourism has received a great deal of attention, not least because it offers space development advocates the possibility of traveling to orbit themselves.

Why would anyone want to take a vacation on orbit? Well, for those who have had too much of the Aegean Islands, Aspen, and Tahiti, it offers something truly different. Still not convinced? How about the attractions of zero gravity, which, it has been argued, will be of special interest to honeymooners and other fun-loving couples (although there is a hilarious folk song popular in the science fiction community in which exactly the opposite proves to be the case[2]). This experience could be expected to be enhanced, at least for some people, if the bedroom suite module had a huge transparent window facing downward to give the couple a spectacular view of the blue rotating Earth (and vice versa). For those with other tastes, the module window could face outward toward the endless sea of space with its myriad of unblurred stars glistening like a million jewels on black velvet. In between bouts in the bedroom, the couple could enjoy unique zero gravity sports such as tennis, racquetball, basketball, soccer, gymnastics, or martial arts carried out in a large module suitably designed to accommodate numbers of people rapidly bouncing off the walls. For a modest extra charge, guests could take classes in extra-vehicular activities and become certified to wear space suits and go EVA. An astronaut certification suitable for framing would also be provided. Those of a more sedentary bent could while away the hours between bedtimes engaged in astronomy or Earth studies in the hotel's observatory. To increase the variety offered by the hotel's primary attraction, a matchmaking service could also be provided. This would be especially valuable, since in addition to being fun-loving and adventurous, most of the people you would meet at the hotel would undoubtedly be rich.

Okay, so there's a product here, and while it may not tap the most noble aspirations of human nature, it certainly has been known to sell. The question is, how much are people willing to pay for a hot date?

According to some marketing surveys, quite a bit. For example, a 1995 study of the North American market by Patrick Collins, a visiting researcher at Japan's National Aerospace Laboratory and Tokyo University, found that about 60 percent of the American population would want to take a vacation in orbit and that, of these, 18 percent said they would be willing to pay for it with six months' salary.[3]

They'd better be well paid. Even at our assumed reduced price of $2,000/kg, the cost of transporting each guest to the hotel with minimal luggage would run about $200,000. Of course, we would also have to lift the hotel crew and the guest and crew consumables, as well as the cabin containing the passengers. If we throw in all the other costs for the business, a good ballpark estimate for the price of a two-week-long vacation package would be at least $500,000 per person, or $1 million per couple. So the target market for the space tourism industry would not be mere millionaires but people with annual incomes exceeding $1 million per year.[4] To raise the same $80 million in gross revenue required by our orbital factory example, 160 guests per year would be needed, or an average of about six guests at a time, assuming each stays for two weeks. In principle, six people could be housed in one module, but that would defeat the purposes of the hotel's primary attraction. So we would need three bedroom modules for the guests, plus one for the crew, plus a lobby/dining room, an observatory, a sports arena, and a utilitarian service module, all attached to a sufficient photovoltaic power array and spacecraft propulsion, control, and communication bus. This is sounding pretty massive, with large initial costs to develop and launch the entire complex. Maybe a guest ticket price of $1 million or $2 million each might be needed. Would the high rollers shell out these kinds of bucks for the experience? I don't know and doubt that anyone does. Those who argue that they would can point to the fact that 160 guests per year is only a fraction of 1 percent of the worldwide population in the relevant income bracket. I'm not sure that investors would find such an argument convincing.

If these prospects weren't difficult enough, a crowd of skunks capable of spoiling the space tourism picnic is standing by in the form of the Federal Aviation Administration and a host of other regulatory agencies, which will demand that the vehicles for transporting passengers to orbit as well as their spaceborne accommodations be certified for safety. This will no doubt be very expensive and time consuming, and may well prove impossible in advance of the development of such items. But without the assurance in advance of certification to allow operation, little investment in space hotels or their supporting transportation systems is likely to occur.

SPACE BUSINESS PARKS

So both space tourism and space industry face a serious chicken-and-egg problem. One possible way out of this conundrum is offered by the concept of a mixed-use space business park. This idea is a favorite of Chuck Lauer, one of my partners in the Pioneer Rocketplane venture, and he has set up a company of his own called Orbital Properties to pursue it. The appeal of this concept for Chuck is obvious—he makes most of his living as a real estate developer creating shopping centers and business parks here on Earth.

Whether it will prove equally appealing to private investors is less clear; to the best of my knowledge neither Orbital Properties nor any other space real estate development plan has ever received significant private investment. But that could change, if the price of space launch were to drop.

The idea of the space business park is *not* to define the business, just create the infrastructure. If you build it, they will come—or so the theory goes. In other words, you build a large spacecraft with a truss, a power array, attitude control systems, and some pressurized modules and then announce that you have space on-orbit for rent. Perhaps your first customer might be an orbital research outfit. That would be logical; as we have seen, of all the orbital businesses we have discussed so far, research has the best chance of producing a profit under present-day technological assumptions. In the course of its investigations, the research company may discover a potential unique product that, contrary to their initial hopes, can only be produced on orbit. So, willy-nilly, they are forced to rent additional modules from you for factory space. If necessary, you expand your space business park with additional pressurized modules to meet this demand. The operation of the orbital factory might create sufficient demand to develop new launch systems with sufficient economy, capacity, and reliability to make space tourism a reasonable proposition. At *that* point, the space hotel entrepreneur will find his funding, and you add on the deluxe bedroom modules with the reflecting mirror walls.

So the space business park scheme has the advantage of being evolutionary, with an initial form of space-based activity (research) that could be viable even under current launch prices, and which, moreover, will have a basis of experience in the operation of the International Space Station. But if the ISS exists, why should anyone rent lab space in your business park? One reason might be security. Even if the ISS subsidizes its lab rental to the point where it is free, a private pharmaceutical company might well prefer to pay premium prices for orbital research space in a place where it can keep the results of its investigations secret.

The lack of definition inherent in the concept of the mixed-use space business park is both its greatest strength and greatest weakness. It offers flexibility, which allows the business to avoid being trapped by fixed ideas of space enterprise that may prove to be crackpot. But its business plan had better contain more than "If we build it, they will come." Solid advance commitments from a well-funded set of orbital research projects or other initial customers will be required to get the business park off the ground. That could take some time to develop.

SPACE ASSET SERVICING

Space-based assets are very expensive to both build and launch, and everyone who owns one wants to get as much use out of it as possible. A service that could extend the useful life of satellites and other orbital equipment by providing refueling, repairs, or necessary upgrades might find itself with many customers willing to pay out very big bucks.

The advantages offered by possessing an orbital servicing capability were forcefully illustrated by NASA Space Shuttle astronauts in their repair and refurbishment operations of the Hubble Space Telescope. Hubble was launched in April 1990 and it was soon discovered that the telescope's mirror was defective, giving the device blurred vision. This was enormously disappointing, not to mention embarrassing. Hubble had cost over $2 billion to develop over a period of almost two decades, and carried with it the hopes of a generation of professional space astronomers and millions of others interested in their findings. Had it worked properly, the instrument offered the promise of a breakthrough in humanity's ability to view the universe. Instead, it had faulty optics. Had this occurred with nearly any other satellite, the situation would have been a total disaster. Fortunately, however, the Hubble had been designed for on-orbit servicing by Shuttle astronauts. So a set of corrective lenses was designed and in 1993 installed on Hubble by the crew of the Shuttle *Endeavour.* The orbiting observatory obtained crystal clear vision, allowing Hubble to deliver on all that it had promised with fantastic discoveries in astronomy and cosmology. In a subsequent Shuttle mission in February 1997, the telescope's instruments were upgraded, endowing it with technology unavailable at the time it was launched. As a result of these servicing operations, the Hubble failure was turned into an unqualified success—an epic achievement not only for NASA but in the history of science.

A cynic might remark that the Hubble repair mission cost $600 million, which is an awful lot to pay for a house call. Indeed it did, but even the most narrow-minded government accountant could see that it was worth it. The observatory cost $2 billion to develop and $600 million to launch. Compared to the cost of replacing it, the $600 million repair mission was a bargain.

The above cost-benefit calculation indicates something about the orbital repair business that makes it rather different from the orbital labs, factories, hotel, and business park opportunities discussed thus far. Orbital servicing does not require, or even benefit very much from, low launch costs. High launch costs may make a servicing mission expensive to perform, but by the same token it also increases the value of the repair job, since the alternative option of launching a replacement satellite also must pay the same high launch fee. In addition, if launch costs are high, the original satellite itself will be more expensive, and therefore more worthy of expensive repair operations, since it will be designed to higher standards of reliability than would be required if the cost of replacing it were low. On the other hand, low launch costs would lower the financial threshold required for entrepreneurs to get into the space servicing business, and would also provide a larger number of potential customers, since more satellites would be flying. Taken together, the advantages and disadvantages offered to the orbital servicing business by lowering launch costs pretty much result in a wash. The launch cost invariance of the orbital servicing business is a two-edged sword. On the positive side, it means that orbital servicing could be profitable without a miraculous drop in the cost of space lift. On the negative side, it suggests that unless they are vertically integrated with their own space-launch companies, orbital servicing businesses will not provide much drive toward the development of low-cost access to space.

Unlike Hubble, most satellites are not designed for orbital refurbishment. Accommodating the requirements of such operations adds cost, mass, and complexity to a spacecraft, and no commercial satellite builder will do so until there is an operating space servicing business to make it worthwhile. There is, however, one major spacecraft that will need plenty of servicing in the near future: the International Space Station. Just as peddlers, blacksmiths, and tailors found it profitable to hang out near cavalry forts in the Old West, so one could imagine a keen entrepreneurial group setting up a little supply depot or workshop of their own, co-orbiting with the Space Station. "Houston, you have a problem. We have a solution. Now, if you're willing to talk business . . ."

The above scenario may sound a bit far-fetched, for two reasons. The first is that NASA and its Russian counterparts are not accustomed to taking advantage of commercial services supplied to them in space—they have never done so and they have no plans to do so. This is undoubtedly because there have been few instances in the past where (a) NASA had a space asset capable of being serviced, and (b) there was a commercial company capable of providing a vital service. (An important exception to this general rule occurred in the late 1970s when Martin Marietta developed plans to boost the Skylab space station to a higher orbit using an upper stage delivered by

Martin's Titan launch vehicle. NASA declined the offer, preferring to wait until the Shuttle was available to service Skylab. Unfortunately, the Shuttle was delayed in becoming operational and Skylab's orbit decayed, terminating with a crash into Australia.) The second reason relates to the first: On-orbit servicing is technically challenging and quite risky. Two spacecraft must be brought in physical contact with each other. If this is not done with the greatest of finesse, you run the risk of having one spacecraft ram the other, just as happened to the Mir space station during a Progress module resupply practice mission in the summer of 1997. Now consider: The Hubble Space Telescope cost $2 billion; the International Space Station will cost $20 billion. NASA managers are understandably jittery about the idea of just anybody flying up and attempting to dock their Radio Shack–wired cheapo spacecraft with one of these assets.

One solution that might get past this problem would be the development, by NASA, of a proximity operations vehicle (a "proxops" stage) that would hang around the station and grab resupply payloads delivered to the general vicinity by commercial suppliers. The NASA-run proxops vehicle would then gently ferry the items (which would all have standardized packaging interfaces analogous to the containers used on merchant ships) to the station. If this were done, both the regular and emergency servicing supply manifest of the Space Station could be opened to commercial enterprise and competition. Developing the proxops vehicle would cost money, but it would be worth it since the cost of both routine resupply and emergency service would drop drastically. Moreover, by opening its launch manifest (which is actually greater than that of Teledesic) to competition in this way, the Station could serve as a useful driver for investment in and development of low-cost private launch systems.

Thus, with the help of a government-run proxops vehicle, commercial resupply launch-on-demand service could be provided for the Space Station, but that is not what is ordinarily meant by space servicing. Well, what about our previous example of a private co-orbiting space station, equipped with a good machine shop, loaded with critical parts, and crewed by handyman astronaut-mechanics selected for their ability to fix wiring and plumbing? Provided that NASA was willing to accept support from such an operation, a problem that probably could be worked politically, it's possible that it could be profitable. The investment and technical threshold required to get into such a business would be rather large, but the funds might be mobilized by a well-qualified company, provided that NASA awarded it an exclusive service contract in advance. This, however, would likely mean that only one such operation could succeed.

Satellites can fail for any number of reasons. A simple form of comsat servicing would be to fly an inspection spacecraft to rendezvous with defunct satellites to determine the cause of failure. This could be of considerable interest to constellation operators and space insurance companies, as the data obtained could be used to reduce failure rates. Of course, one of the most common sources of satellite termination (and the one guaranteed to occur if the others are avoided) is exhaustion of onboard propellants needed for attitude control and stationkeeping. If a way could be found to refuel the satellites on-orbit, great savings could be obtained. The task is technically challenging, because two automated spacecraft bristling with solar panels, antenna dishes, and the like must rendezvous, dock, and mate without entangling or damaging each other. Since most communication satellites are in higher orbits than manned Earth-to-orbit vehicles such as the Shuttle or future SSTOs can reach, robotic or teleoperated spacecraft must service such satellites. (Even if the piloted vehicle could be refueled in low orbit, allowing it to reach the comsats, the propellant consumption required to move a relatively large piloted ship up to the satellite orbit to transfer the small amount of propellant used by the much less massive comsat would make such operations uneconomical.) That said, there is nothing in principle impossible about developing such a robotic tanker, and when plans for such a system are sufficiently mature to convince satellite constellation operators of its feasibility, it should be possible to get them to install the necessary interfaces on their vehicles, since the payoff to them will be large.

Thus orbital servicing of communication and other commercial satellites actually makes good business sense, and I predict that within a decade or so such operations will be implemented. The effect, however, will be to *reduce* the number of satellite launches required to maintain the constellations. Thus, while it offers a viable form of space commerce, on-orbit servicing will *not* spur the development of launch systems providing cheap access to orbit. Quite the contrary—orbital servicing provides an *alternative* that will make constellation operators *less* sensitive to the need for low-cost space launch.

SOLAR POWER SATELLITES

Many who see orbital commerce as the driving force for the development of a spacefaring civilization look to the generation of electricity for use on Earth by large Solar Power Satellite (SPS) systems.[5] Orbital labs, factories, and space servicing operations may offer fairly good near-term prospects for commercial viability, but the launch capability and volume required to support them is modest. In contrast, were it possible to generate electric power in space for terrestrial consumption at competitive rates, the market would be nearly unlimited. Vast numbers of huge SPS systems would then be built, and their construction and operation would require a huge fleet of reusable medium- and heavy-lift launch vehicles. Truly cheap access to space with booster systems of every payload capacity would be rapidly developed, and the doorway to the final frontier thrown wide open.

In space, solar energy is available twenty-four hours a day, unmasked by the dulling effect of the Earth's atmosphere. Moreover, while most terrestrial solar arrays are fixed in orientation, an orbital solar array can readily track the Sun. Avoiding the atmosphere increases the effective solar incidence by a factor of about 1.5, while the ability to track multiplies the average power produced by the orbiting array by a factor of 4. Thus, when both of these advantages are taken into account, an orbital solar array can produce a time-averaged output that is about six times greater per unit area than its counterpart fixed in orientation on the ground in an equatorial desert. The SPS unit beams it solar-produced power via microwaves to Earth where it is received by a "rectenna." The microwave energy is then converted to high-voltage alternating power for consumption on the consumer grid. The beaming process could be made about 50 percent efficient, so half the power is lost, reducing the orbital array's advantage over its terrestrial counterpart to a factor of 3. However, as a countervailing advantage, the groundside rectenna is smaller than a solar array, and cheaper, and can be put nearly anywhere in the world, including places where solar power is frequently unavailable due to weather. So, once an SPS is in operation over the appropriate hemisphere, a relatively cheap rectenna could be installed nearly anywhere in that hemisphere to obtain power. This could make enormous quantities of electricity available in remote areas of the Third World and avoid the need for installing expensive power generation equipment in countries where political instability might make such installations insecure.

But what would the price of such power be? Let's consider first the case assuming current launch costs and solar cell technology. Solar incidence at the Earth's distance from the Sun is about 1,300 watts (W) per square meter, and silicon solar panels with weights of about 4 kg per square meter and 15 percent efficiency are currently available. Such panels therefore have a power/mass ratio of about 49 W/kg. Of course, half the power is lost during transmission to Earth, and the weight of the SPS spacecraft, including all supporting structures, mechanisms, attitude control systems, and the microwave transmissions system, could be expected to be at least double the weight of the solar panels themselves. Thus, net power produced by the SPS spacecraft would be closer to 12 W/kg. Also, the SPS spacecraft could not be in low Earth orbit. If it were, it would zip around the Earth once every ninety minutes and be unable to provide constant, or even frequent, service to a rectenna station on Earth. Instead, the SPS would have to be in a slow-moving high orbit, with the best choice being geosynchronous Earth orbit (GEO), 35,000 km up. At that altitude, the SPS would orbit the Earth once every twenty-four hours, and since the Earth turns at the same rate, this would allow the satellite to hover over a fixed position on the Earth's equator. (For this reason, the GEO orbit, which was discovered by Arthur C. Clarke in the 1940s, is the most popular today for communication satellites that are not part of large constellations.) While the orbit of the SPS would be equatorial, its high altitude would give it a good line-of-sight for transmission over most of the hemisphere. The cost of delivering payloads to GEO, however, is about four times that of LEO, running the range of $40,000/kg. So just the *launch cost* of the SPS would be about $3,300/W, or $3.3 *trillion* for a 1,000-megawatt (MW) unit suitable for providing the power needs of a city the size of Denver. And that's just the SPS launch cost. If we add in the costs of assembly (the 1,000-MW SPS would be over 5 square kilometers in size and would weigh 41 million kilograms), maintenance, insurance, spacecraft hardware, construction, real estate costs for the rectenna and its power conditioning system, salaries, taxes, and so on, the cost of the total SPS would undoubtedly run at least $6 trillion. That's about 3,000 times the cost of a nuclear power plant producing the same amount of power, or about 6,000 times the cost of a natural gas fired unit. At these installation prices, the fact that the SPS requires no fuel would make very little difference. Just the interest cost on $6 trillion would be about $600 billion per year; if we add in maintenance and depreciation over twenty years, the cost would be at least $1 trillion per year (the entire yearly budget of the U.S. government is about $1.5 trillion). That boils down to a user price of $114 per kilowatt-hour (assuming that nothing is added for profit), over 2,000 times the $0.05 per kilowatt-hour currently prevailing in the United States.

So in order for SPS to become competitive, the price of space lift needs to drop by a factor of more than 2,000 to $4.30/kg to LEO or $17/kg to GEO. That's *impossible.* The reason I say it is impossible (and not just very improbable) is because the propellant cost *alone* needed for rocket-driven space launch runs a factor of 4 larger than that, and total system operating expenses for a reusable booster would drive launch prices up a further multiple of 6—and the two combine to knock the SPS business plan off base by a factor of 24 even in a future world of minimum-cost, totally reusable SSTO rocket vehicles. Conceivably you could reduce launch propellant costs by a factor of 2 by using an (as-yet-unproved) advanced air-breathing propulsion system like a supersonic combustion ramjet, and another factor of 2 in launch weight could be saved by using gigantic ion engines (100,000 times the size of any yet built) to propel the SPS from LEO to GEO, and maybe in the future the weight of the solar panels could be cut in half relative to the 4 kg per square meter that we have assumed. Twenty-four cut in half three times equals 3. In other words, even under these assumptions combining an *ideal* air-breathing reusable launch system with an *ideal* orbit transfer propulsion system and ultra-lightweight construction, the SPS would still come in at triple the cost of ground-based forms of power generation.

The late Princeton professor Gerard O'Neill was a space solar power advocate who realized that, if the materials required to construct an SPS had to be lifted from the Earth, the system could never pay. He therefore recommended that the required solar panels, structural elements, and everything else needed for the SPS be manufactured out of lunar material. Because of the Moon's low gravity and vacuum environment, it would be possible to launch these materials into space using electromagnetic catapults called "mass drivers." The catapulted debris would accumulate in the stable gravitational loci of the Earth-Moon system known as LaGrange points. Residents of large orbiting space colonies (built out of billions of tons of lunar material) would gather the materials and use them as raw material for SPS components, which would be manufactured and delivered to SPS orbits for final satellite construction. The immense profits derived from providing space solar power liberated from the costs of Earth-to-orbit launch would both fund the construction of the lunar base and orbiting space colonies and provide a luxurious lifestyle for millions of people within the orbital settlements. This grandiose vision has attracted numerous adherents, including Congressman Dana Rohrabacher (R-CA), the chairman of the House Space Subcommittee, and his aide Mr. Jim Muncy.

It is certainly true that if you want to put very large quantities of mass into GEO, it is easier to do so from a lunar base or near-Earth asteroids than it is to launch from Earth. But the size and complexity of the O'Neill operation, complete with lunar base, catapults, orbital ferries and propellant depots, billion-tonne rotating LaGrange-point colonies (each complete with greenhouse farms, life support systems, hospitals, schools, public libraries, shopping centers, lakes, parks, and houses with white picket fences), orbital ore refineries, factories, and construction docks, boggles the mind. Just for starters, even if O'Neill's lunar catapult could deliver mass to GEO at 1/10,000th of current launch prices, the cost just to transport unrefined raw mass for his billion-tonne colony would be $4 trillion, or $4 million per resident. A reasonable guess might be that if refining, processing, manufacturing, and construction costs are added, the cost of building the floating colony would be at least ten times greater, or $40 trillion. It's impossible to estimate the total cost of the complete system (which includes several space colonies, a lunar colony, catapult, interorbital tugs and their refueling and servicing stations, orbital refineries and factories, and more), but it is clear that even if a long-term return could be shown, the up-front cost of developing all this infrastructure would put it well outside the budget not only of venture capital, but of the world's leading financial institutions.

Moreover, there's no reason to believe that such a system ever would produce a return. I've noted that space-based solar arrays hold a power generation per unit area advantage over ground-based arrays of a factor of 3. But this is only because current ground-based arrays do not track the Sun. If ground-based systems were fitted with mechanisms to make them track—a much easier feat than putting an array in orbit—this advantage would be cut in half. This would leave the SPS with just 50 percent more power generation per unit area than a ground-based array, which is a margin far too thin to ever compensate for the greatly added cost and complexity of a space-based power system.

Finally, as if all this were not bad enough, the SPS operators would have one hell of an environmental impact statement to write, as the system requires sending thousands of megawatts of microwaves through Earth's atmosphere. While from an engineering standpoint such a system could probably be designed to be safe, the chance that an environmental lawsuit could stop operation of a trillion-dollar SPS system after it has been constructed (as has happened with multibillion-dollar nuclear power plants) would chill any possibility of private investment in such an enterprise.

In short, solar power satellites and O'Neill colonies, based on their business plans, are completely implausible for the foreseeable future.

A more modest proposal than the SPS is the power relay satellite (PRS). The PRS generates no power; it is just an orbiting reflecting antenna (or "bent horn") that allows electricity generated at some point on Earth where power is cheap (say, the Pacific Northwest of the United States) to be transmitted to another location (say, central Africa) where power is unavailable. Because it does not need to generate power, a PRS would have a mass an order of magnitude less per unit of power transmitted than an SPS. Also, because the customer is in the Third World and presumably has no alternatives, the PRS can deliver less power and charge prices for it above current advanced-sector utility rates. So instead of a 1,000-MW SPS weighing 41 million kilograms, we only need to build a 100-MW PRS with a mass of 410,000 kg. If launch prices to GEO can be dropped by a factor of 50 to $800/kg, such a satellite would cost $320 million to launch, and perhaps $800 million overall. In contrast to the astronomical investment requirements of the SPS, funds in such quantities can at least be found in major banks and investment firms. Interest payments would be $80 million per year, however, and if we add in depreciation and other costs, we obtain an end price to the user of about $0.20 per kilowatt-hour. This is four times what the average U.S. citizen pays for electricity, but people in regions with no alternatives might be willing to pay it.

Thus, provided a 50-fold drop in launch costs is achieved, power relay satellites could become a reasonable business proposition. Unlike the factor of 2,000 launch cost reduction required for SPS systems, such a 50-fold drop is possible and will probably happen sometime in the middle of the next century as a result of the imperatives of interplanetary colonization. But don't expect to see power relay satellites until then.

CAN BUSINESS PLANS OPEN SPACE?

Gerard O'Neill was a visionary humanist who inspired many people, and it gives me no great pleasure to debunk his ideas. But it has to be done, because he was leading in the wrong direction. The mistaken belief that his program—or any of its many SPS-based variants—can give rise to a spacefaring civilization has caused severe disorientation and misguided efforts among many people seeking that goal. Putting aside the ludicrous specifics of the O'Neill/SPS plans, the core notion that the final frontier can be opened on the basis of entrepreneurial business plans is wrong. Fundamentally it serves as a cop-out for those who shirk the responsibility of initiating a national space program that really *does* open new worlds. "The private sector will do it" is their standard dodge. Well, it won't, at least not by itself. The private sector can be expected to fund the development of new small-launch vehicles to lift private commercial satellites. It may also finance the creation of limited human operations on-orbit, including, as we have seen, orbital labs, manufacturing facilities, and space servicing. This is possible because the technology, method of operation, and a good deal of the market necessary for these launch and orbital operations have already been paid for by substantial government funding over the past four decades. There is nothing wrong with this; in fact, it follows a near-universal historical pattern of terrestrial frontiers being first opened either by governments or by social groups motivated by transcendent purposes and only afterward developed by private commerce. People can be courageous, but money is timid; it prefers to reproduce itself in tried and proved ways. If your only fundamental goal is to make money, there are far more reliable ways to do so than to venture into the unknown. Thus, on Earth, developing new frontiers for profit has occurred only after such regions have been explored and pioneered at considerable risk and cost by individuals possessing rather different motives.

Government space initiatives over the past forty years have tamed near-Earth space to the point where it is now a potential arena for private enterprise. This is an extremely positive development; while they will not build the first orbital research lab, private companies will create the next generation of such facilities and will do so for orders of magnitude less cost than the International Space Station. Such labs, eventually supplemented by orbital manufacturing stations, will make available an array of products that may revolutionize medicine and computer technology. The combination of low-cost space access with orbital servicing operations will also allow the development of global communication systems whose capabilities will impact society in ways that exceed the imagination of most people today. For example, such augmented communication constellations could enable low-cost wristwatch-sized communication devices that would be able to access on a real-time interactive basis all the storehouses of human knowledge from anywhere in the world. In addition, these devices would enable their users to transmit very high volumes of data—including voice, video, and music—either to each other or to the system's central libraries. The practical value of such systems is obvious, but their implications go far beyond the practical into the social and historical. We will see human society thoroughly linked together, resulting in deep cultural fusions and a radical generalization of the dissemination of human knowledge. In a real sense, the establishment of the full range of global communication services that orbital industry will enable represents the final step establishing humanity as a Type I civilization.

That said, the fundamental problem facing the human race today—the creation of a true spacefaring Type II civilization—will not be solved by developing orbital private enterprise in geocentric space. True, such operations will serve as a "school for sailors," training the people and honing the skills and organizations for future space ventures in a way analogous to the manner in which coastal fisheries helped to provide the men to handle the ships of the great nautical explorers of the past. But human beings will never settle Earth orbit, because there is nothing there to settle. We need to reach beyond. When we do so, we will not be led, but be followed by the entrepreneurs. Their aid in providing low-cost cargo delivery and other services to help initial outposts grow into settlements will be vital. But the trail will have to be blazed by those who live for Hope and not for cash.

TYPE II

Creating a Spacefaring Civilization

From that most holy wave I now returned
to Beatrice; remade, as new trees are
renewed when they bring forth new boughs, I was
pure, and prepared to climb unto the stars.

—DANTE ALIGHIERI
Purgatorio
Canto XXXIII, Lines 142–145;
translated by Allen Mandelbaum

CHAPTER 5

The View from the Moon

If God had not meant for mankind to colonize space,
he wouldn't have given us the Moon.

—KRAFFT EHRICKE

THE PURPOSE OF the human venture into space is to renew our species through accepting the challenge to create new civilizations on new worlds. Economic advantage will motivate certain enterprises that play a role in our outward migration, but humanity does not need to move into space for economic reasons. Enormous amounts of knowledge and technology will be generated through the encounter with unexplored worlds, and this will be of inestimable value. But fundamentally, humanity's entry into space is not about profits, or even knowledge—it's about social reproduction. We go to grow, to reproduce, to bring into being peoples and possibilities that today exist only as potentials. By accepting this challenge, we will benefit ourselves in many ways, but as in all truly meaningful activities, the primary beneficiaries will not be ourselves, but our posterity. We are planting orchards: For us is the sense of accomplishment and the delight in watching the seedlings grow. The fruit is for our children.

But where to plant? What world shall we make our first new home beyond the Earth? For some, the answer is the Moon.

Earth's moon is a world with a surface area the size of Africa, thus somewhat justifying lunar colonization visionary Krafft Ehricke's designation of it as our "eighth continent."[1] As a destination for space colonization, the

Moon has the undeniable advantage that it is the closest of any major or minor planetary bodies, reachable with existing chemical propulsion in a three-day flight. It is also clear that we have the capability to establish permanent bases on the Moon—after all, we had piloted lunar vehicles before we had VCRs, hand calculators, microwave ovens, or push-button telephones. The lunar surface contains vast amounts of oxygen, silicon, iron, titanium, magnesium, calcium, and aluminum, tightly bound into rocks as oxides, but there nevertheless. Data substantiating the existence of these resources are given in Table 5.1, which shows the results of chemical analysis of various Apollo lunar samples.[2]

These resources give the Moon an enormous advantage as a destination for colonization over geocentric orbital space, where there is nothing at all to work with. These materials could be used to produce part of the consumables, rocket propellants, power systems, and building or shielding materials to support lunar settlements or related activities. The Moon also possesses scarce, but in principle obtainable, supplies of helium-3, an isotope otherwise naturally nonexistent in the inner solar system. Helium-3 offers a number of potentially important advantages as a fuel for thermonuclear fusion reactors, and thus perhaps could provide a future lunar colony with a cash export commodity. The Moon has a vacuum environment and a gravitational pull only one-sixth as large as the Earth, thus making it much easier to launch spacecraft from than the home planet. This has caused many to speculate that the Moon might serve as the optimal port of departure for interplanetary expeditions to points beyond. In addition, the airless environment of the Moon provides a unique near-Earth location in which large-scale vacuum processing of various locally available materials can be

• TABLE 5.1 •

Chemical Analysis of Typical Apollo Lunar Samples

COMPOUND	APOLLO 11 BASALT	APOLLO 14 BRECCIA	APOLLO 17 REGOLITH
SiO_2	40.46	48.09	44.47
TiO_2	10.41	1.51	2.84
Al_2O_3	10.08	16.72	18.93
FeO	19.22	9.53	10.29
MgO	7.01	10.18	9.95
CaO	11.54	10.67	12.29
Na_2O	0.38	0.73	0.43

undertaken. These advantages have caused some to postulate ambitious plans for lunar settlement based upon commercial development.

There are problems, however. As an examination of Table 5.1 indicates, while the Moon's rocks and soils possess ample supplies of oxygen and several important metals, they are entirely lacking in such vital substances as organics, hydrates, carbonates, nitrates, sulfates, phosphates, and salts. The key primary biogenic elements of hydrogen, carbon, and nitrogen are present on the Moon, but in general only in extremely rare quantities (~50 parts per million, or ppm) in surface materials impregnated by the solar wind. The leading secondary biogenic elements, sulfur and phosphorus, are also on the rare side, with typical concentrations in lunar soils ranging from 500 to 1,000 ppm. Among the leading secondary industrial elements, potassium, manganese, and chromium are reasonably common (~2,000 ppm), but nickel is rather scarce (~200 ppm) and cobalt even scarcer (~30 ppm), while copper, zinc, lead, fluorine, and chlorine are extremely hard to come by (~5 to 10 ppm each). Helium is present in solar-wind-impregnated regolith in concentrations of about 10 ppm, while argon and neon can be found in concentrations of about 1 ppm each. The potentially commercially valuable helium-3, however, constitutes only 1 part in 2,500 of the total helium supply (the vast majority of which is ordinary helium-4), or just 4 parts per billion of the surface regolith.

There are other problems as well. The Moon has virtually no atmosphere, so its surface is naked to solar flares. As a result, human colonies and their agricultural greenhouses would have to be placed either underground or, if on the surface, beneath glass domes with walls about 10 cm thick. This would make the creation of large amounts of habitable volume and arable land comparatively difficult. Supergreenhousing of surface domes to very hot temperatures during the day would be a serious concern. Moreover, terrestrial plants are not adapted to growth in the Moon's two-week light/two-week dark day/night cycle. Short of massive genetic engineering of a sort yet undemonstrated, crops would have to be grown using artificial light. This is a much bigger problem than it sounds. Plants are enormous consumers of light energy, typically using about 3,000 kW/acre, or 750 MW/km^2. If this does not seem too much consider this: The amount of sunlight that illuminates the fields of the state of Rhode Island (not usually thought of as an agricultural giant) is around 2,000,000 MW (or 2 terawatts), which is comparable to the total electric power currently generated by all of human civilization. In other words, growing plants in significant quantities using artificial lighting sources is wholly impractical, yet this is the only way in which agriculture can be conducted on the Moon. Combine this with the fact that the carbon, hydrogen, and nitrogen elements required to make plants are nearly absent from the Moon and it soon becomes apparent that lunar colonization faces severe difficulties. On the Moon, manure would be harder to come by than chromium.

Of course, it is true that no society need be entirely self-sufficient. If a lunar colony could make a substantial portion of what it needs, and produce a commercially useful export, could it not use the cash generated by the exports to import that which is locally unavailable? Well sure, and certainly any interplanetary colony, or terrestrial one for that matter, will always have to go through a period in which advanced manufactured goods are imported. But when the required imports are things as fundamental (and massive!) as food, fabrics, plastics, and a large fraction of the range of basic elements of life (organic chemistry, metallurgy, and industrial processing), the demand for the colony's cash export had better be enormous.

ENERGY FROM THE MOON

In recent years, two schemes have been proposed for supporting lunar colonization through the large-scale export of a cash commodity to Earth. In both cases the product is energy. On the positive side, energy is an item with a guaranteed large and growing terrestrial market. On the negative side, it is a product without unique qualities. In the terrestrial electric power marketplace, a kilowatt is a kilowatt is a kilowatt, and consumers don't know or care whether the juice in their power lines comes from the Moon or burning garbage at the city dump. Power can be sold in huge quantities, but only if it can be produced at or below the going rate. The two lunar-power–based development schemes embrace this challenge, but in radically different ways.

One of the concepts, whose foremost champion is Dr. David Criswell of the University of Houston–Clear Lake, is to mass-produce solar power arrays on the Moon out of lunar materials, deploy the arrays on the lunar surface, and beam the power to Earth. It will be observed that this idea is rather similar to Gerard O'Neill's solar power satellite plan, but it has a number of advantages. Foremost among these, it eliminates the need to build a billion-tonne orbiting colony out of materials transported across space, as well as the need to transport millions of tonnes of solar power satellite material from the Moon to Earth orbit for assembly into gigantic power generation and transmission stations on orbit. These simplifications would tend to make Criswell's approach to generating power several orders of magnitude cheaper than O'Neill's, thereby confirming the wisdom of utilizing the resources of a planet that already exists instead of trying to build one yourself. However, there are a number of problems. In the first place, Criswell's concept requires beaming energy across 400,000 km of space, rather than the "mere" 36,000 km required by O'Neill's systems. This itself is not a show stopper, but it means that either the transmitting antenna needs to be ten times as big, or a transmitter frequency ten times as great needs to be employed, if the transmitted energy is to be focused on a receiver of a given size. For example, in order for an O'Neill-type SPS positioned in geosynchronous orbit 36,000 km above the Earth to focus its power transmitted at a frequency of 3 GHz (10-cm wavelength) on a 1-km-diameter rectenna receiver on the ground, the SPS would need a transmitting antenna 36 km in diameter. That's big, but Criswell's would need to be a lot bigger—400 km in diameter to achieve the same focus. If the systems were designed to operate at a higher frequency, say 30 GHz (1-cm wavelength), the transmitting antennas would scale down in proportion, but at the higher frequency at least half the transmitted energy would be absorbed by Earth's atmosphere.

Producing all the silicon needed for large-scale solar arrays won't be easy either. As we can see in Table 5.1, it's certainly true that SiO_2 is abundant on the lunar surface, but reducing this material to metallic silicon requires reacting it with carbon, which the Moon essentially lacks. The required reaction is:

$$SiO_2 + 2C \rightarrow Si + 2CO \qquad (5.1)$$

While the carbon monoxide waste stream so generated can be processed to allow the carbon used to be recycled and used again, in reality there is always loss in such cycling chemical engineering systems. Thus large supplies of hard-to-get carbon will be needed.

Another problem is that, fixed on the Moon's surface, Criswell's solar arrays would experience precisely the same average power generation losses due to day-night cycling (albeit at slower, longer intervals) and non-optimal Sun angles as those encountered by solar panels positioned on Earth's surface. Compared to Earth-based photovoltaics, the lunar array would only obtain the advantage of, at best, a factor of 2 increase in power generation due to its perpetual clear weather. This advantage would be wiped out by the losses encountered in transmitting the power, even at low frequency where atmospheric losses are minimized. Thus, it is hard to see how, with all the additional costs involved in manufacturing and positioning the solar panels and their huge transmitting system on the Moon, a lunar array could be economically competitive against Earth-based photovoltaics (which themselves have yet to become competitive against fossil fuels, hydroelectric, nuclear, wind, or geothermal power).

The other proposal, that of Professors Jerry Kulcinski and John Santarius of the University of Wisconsin, is considerably more interesting. These gentlemen propose to mine the lunar regolith for its helium-3 and then export this unique substance to Earth for consumption in terrestrial fusion reactors. Now, one obvious and frequently noted flaw in this plan is that fusion reactors do not exist. However, that fact is simply an artifact of the mistaken priorities of the innocent gentlemen in Washington, D.C., and similar places who have been controlling scientific research and development's purse strings for the past few years. Lack of funding, not any insuperable technical barriers, currently blocks the achievement of controlled fusion. The total budget for fusion research in the United States currently stands at about $250 million per year—less than half the cost of a Shuttle launch, or, in real dollars, about one-third of what it was in 1980. Under these circumstances, the fact that the fusion program has continued to progress and now is on the brink of ignition is little short of remarkable.

All atomic nuclei are positively charged and therefore repel each other. To overcome this repulsion and get nuclei to fuse, they must be made to move very fast while being held in a confined area where they will have a high probability of colliding at high speed. Superheating fusion fuel to temperatures of about 100 million degrees Celsius (°C) gets the nuclei racing about at enormous speed. This is much too hot to confine the fuel using a solid chamber wall—any known or conceivable solid material would vaporize instantly if brought to such a temperature. However, at temperatures above 100 thousand degrees Celsius, gases transition into a fourth state of matter, known as a plasma, in which the electrons and nuclei of atoms move independently of each other. (In school we are taught that there are three states of matter: solid, liquid, and gas. These dominate on Earth, where plasma exists only in transient forms in flames and lightning. However, most matter in the universe is plasma, which constitutes the substance of the Sun and all the stars.) Because the particles of plasma are electrically charged, their motion can be affected by magnetic fields. Thus, various kinds of magnetic traps (tokamaks, stellarators, magnetic mirrors, etc.)

have been designed that can contain fusion plasmas without ever letting them touch the chamber wall.

At least that is how it is supposed to work in principle. In practice, all magnetic fusion confinement traps are leaky, allowing the plasma to gradually escape by diffusion. When the plasma particles escape, they quickly hit the wall and are cooled to its (by fusion standards) very low temperature, thereby causing the plasma to lose energy. However, if the plasma is producing energy through fusion reactions faster than it is losing it through leakage, it can keep itself hot and maintain itself as a standing, energy-producing fusion fire for as long as additional fuel is fed into the system. The denser and hotter a plasma is, the faster it will produce fusion reactions, whereas the longer individual particles remain trapped, the slower will be the rate of energy leakage. Thus, the critical parameter affecting the performance of fusion systems is the product of the plasma density (in particles per cubic meter), the average particle confinement time (in seconds), and the temperature, measured in kilo-electron volts (keV) achieved in a given machine. The progress that the world's fusion programs have had in raising this parameter, known as the Lawson parameter, is shown in Figure 5.1. To

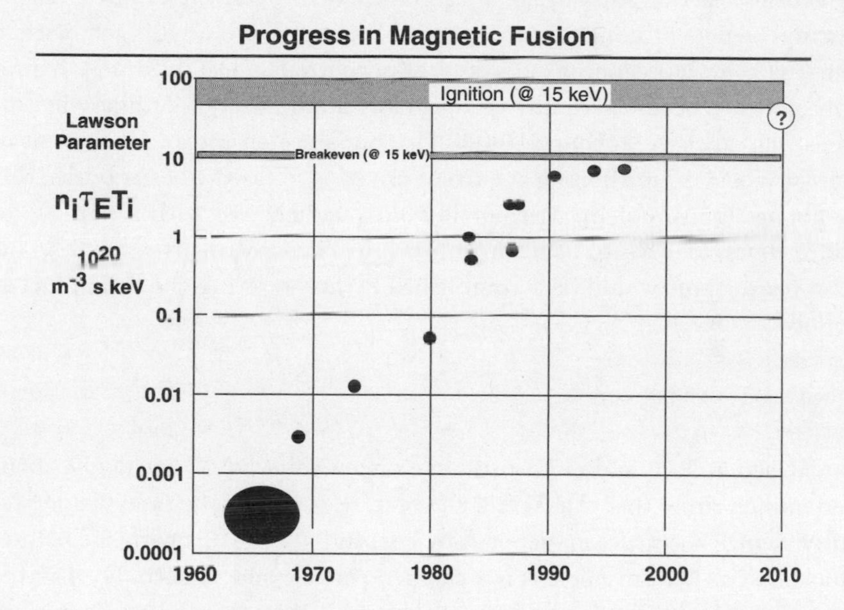

FIGURE 5.1 *Progress in magnetic fusion. (Courtesy Dale Meade/Princeton Plasma Physics Lab)*

produce energy at a rate equal to the external power being used to heat the plasma (via microwave heaters or other means), a deuterium-tritium fusion reactor must have a Lawson parameter of 1×10^{21} keV-particle-seconds/m^3 (or keV-s/m^3 for short). Such a condition is known as "breakeven" and was finally reached at the European JET tokamak in 1997. A deuterium-tritium plasma with a Lawson parameter of 4×10^{21} keV-s/m^3 would produce energy at a sufficient rate that no external heating power would be needed. Once started up, such a plasma would heat itself. This condition is known as "ignition" and is the next, and final, physics milestone that needs to be achieved before actual energy-producing fusion reactors can be engineered.

As can be seen in Figure 5.1, the world's fusion programs have made an enormous amount of progress over the past thirty years, raising the achieved Lawson parameter by a factor of over 10,000 to reach breakeven. Another factor of 3, which can certainly be accomplished if funds are provided to build the next generation of experimental tokamaks, would take us to ignition.

Fusion can certainly be developed, and when it is, it will eliminate the specter of energy shortages for millennia to come. However, not all fusion reactors are created equal.

Currently, the world's fusion programs are focused on achieving the easiest fusion reaction, that between deuterium (hydrogen with a nucleus consisting of one proton and one neutron) and tritium (hydrogen with a nucleus containing one proton and two neutrons). Deuterium is nonradioactive and occurs naturally on Earth as 1 atom in 6,000 ordinary hydrogens. It's expensive (about $10,000/kg), but since an enormous amount of energy (about $5 million/kg at current prices) is released when it burns, this is not really a problem. Tritium is mildly radioactive with a half-life of 12.33 years, so it has to be manufactured. In a deuterium-tritium (D-T) fusion reactor, this would be accomplished by first reacting the fusion fuel as follows:

$$D + T \rightarrow He4 + n \qquad (5.2)$$

Reaction 5.2 yields 17.6 million electron volts (MeV) of energy, about ten million times that of a typical chemical reaction. Of the total yield, 14.1 MeV is with the neutron (denoted by "n") and 3.5 MeV is with the helium nucleus. The helium nucleus is a charged particle and so is confined in the device's magnetic field, and as it collides with the surrounding deuterium and tritium particles, its energy will heat the plasma. The neutron, however, is uncharged. Unaffected by the magnetic confinement field, it will zip right out of the reaction chamber and crash into the reactor's first wall, damaging the wall's metal structure somewhat in the process, and then plow on until it is eventually trapped in a "blanket" of solid material positioned behind the wall. The blanket will thus capture most of the neutron's energy and in the process it will be heated to several hundred degrees Celsius. At this temperature it can act as a heat source for high-temperature steam pipes, which can then be routed to a turbine to produce electricity. The blanket itself is loaded wth lithium, which has the capacity to absorb the neutron, producing helium and a tritium nucleus or two in the process. The tritium so produced can later be separated out of the blanket materials and used to fuel the reactor. Thus a D-T reactor can breed its own fuel.

However, not all of the neutrons will be absorbed by the lithium. Some will be absorbed by the steel or other structural elements composing the reactor first wall, blanket cooling pipes, etc. In the process, the reactor's metal structure will become radioactive. Thus, while the D-T fusion reaction itself produces no radioactive wastes, radioactive materials are generated in the reactor metal structure by neutron absorption. Depending upon the alloys chosen for the reactor structure, a D-T fusion reactor would thus generate about 0.1 to 1 percent of the radioactive waste as a nuclear fission reactor producing the same amount of power. Fusion advocates can point to this as a big improvement over fission, and it is. But the question of whether this will be good enough to satisfy today's and tomorrow's environmental lobbies remains open.

Another problem caused by the D-T reactor's neutron release is the damage caused to the reactor's first wall by the fast-flying neutrons. This damage will accumulate over time, and probably make it necessary to replace the system's first wall every five to ten years. Since the first wall will be radioactive, this is likely to be an expensive and time-consuming operation, one that will impose a significant negative impact on the economics of fusion power.

So, the key to realizing the promise of cheap fusion free of radioactive waste is to find an alternative to the D-T reaction, one that does not produce neutrons. Such an alternative is potentially offered by the reaction of deuterium with helium-3. This occurs as follows:

$$D + He3 \rightarrow He4 + H1 \qquad (5.3)$$

This reaction produces about 18 MeV of energy and no neutrons. This means that in a D-He3 reactor, virtually no radioactive steel is generated and the first wall will last much longer, since it will be almost free from neutron bombardment. (I say "virtually no" and "almost free" because even in a D-He3 reactor some side D-D reactions will occur between deuteriums that will produce a few neutrons.) In addition, no lithium blanket or steam pipes are needed. Instead, the energy produced by the reactor, since it is all in the form of charged particles, could be converted directly to electricity by magnetohydrodynamic means at more than twice the efficiency possible in any steam-turbine generator system.

There are two problems, however. In the first place, the D-He3 reaction is harder to ignite than D-T, requiring a Lawson parameter of about 1×10^{22} keV-s/m^3. That should not be fundamental; it just means that D-He3 machines need be a little bigger or more efficient at confinement than D-T devices. If we can do one, then in a few more years we can do the other. The bigger problem is that helium-3 does not exist on Earth. It does, however, exist on the Moon.

The solar wind contains small quantities of helium-3, and over ages of geologic time has implanted the surface layers of the lunar regolith with about 4 parts per billion of this unique isotope.

Four parts per billion is not much—for almost any substance this would represent much too low a concentration to be economic for any sort of industrial recovery. At 4 ppb you need to process 250,000 tonnes of raw material to obtain 1 kg of product. It certainly would not be worthwhile trying to refine gold from such dilute feedstock. But helium-3 is much more valuable than gold. A kilogram of gold, at today's prices, is worth about $15,000. A kilogram of helium-3, on the other hand, if burned in a fusion reactor using a 60 percent efficient MHD conversion system, would produce 100 million kilowatt-hours (kWh) of electricity. At a typical current rate of $0.06/kWh, this represents a gross product value of $6 million/kg. This means that if a utility were willing to spend one-sixth of its gross revenue on fuel, helium-3 could sell for about $1 million/kg. This is so high that, even at current space transportation rates, it would be economical to ship such material from the Moon.

However, you still need to process 250,000 tonnes of raw lunar regolith to get it. That would be mean "farming" an area 1 km on a side and 10 cm deep and taking all that material to a "shake-and-bake" system where the soil would be heated to about 700°C, causing the helium-3 and all other embedded volatiles (including the several thousand-fold more common nonprecious helium-4) to outgas. If the processor could handle 5 tonnes a minute, this would take about 35 days, working around the clock. Isotope separation would then need to be performed to divide the 1 kg of helium-3 from the 2,500 kg of common helium-4. This would be done by bringing the helium to very low temperatures where it will fractionate, since the different isotopes have different boiling points.

In the process of producing the kilogram of helium-3, enough soil would be processed to also yield about 10 tonnes each of solar-wind–implanted nitrogen, hydrogen, and carbon. While a thin basis for settlement, these byproducts could go a long way toward relieving some of the logistics needs of a lunar mining base.

Processing 250,000 tonnes of Moon dirt to make $1 million of cash product ($4/tonne) may seem like a hard way to earn a living. Indeed, unless the operation is heavily automated, it is unlikely to be economical. However, because of the fundamental simplicity of the shake-and-bake processing system, such automation might be possible. One can envision groups of remotely operated bulldozers rapidly plowing tonnes of regolith onto a continuously moving conveyor belt, which dumps the material into an oven. Two ovens would be employed at a given site, one sealed for baking and the other open for filling. When the first oven had caused its soil to outgas its volatile content, a trap door would open and the "dried-up" waste dirt would be dumped out the back onto a waste conveyor, while the front would open for a refill of fresh soil. Meanwhile, the second oven would seal itself up and start baking its load of regolith. And so it would go, with each oven alternating roles in turn. Once the gases have been separated from the soil, they can be handled by fairly standard sorts of chemical engineering fluid-processing techniques that are highly susceptible to automation.

But there are other problems. The equipment required to process 5 tonnes per minute of lunar soil would have non-negligible mass. An optimistic low-end guess for the ovens, the isotope separation system, the conveyor belts, and a small fleet of bulldozers and trucks would be on the order of 100 tonnes. Now at current launch costs of $10,000/kg to low Earth orbit, it would cost $1 billion to launch 100 tonnes to orbit, but transportation to the Moon using hydrogen/oxygen rocket propulsion increases the total amount to be lifted by a factor of 5. Thus, it would cost $5 billion to ship enough equipment to the Moon to produce about $1 million per month or $12 million annual revenue. At this rate it would take about 400 years (!) for the equipment to pay back its shipping cost. But as we have seen, in the long term it is within the realm of engineering feasibility to reduce launch costs by a factor of 100 relative to current rates. If such $100/kg to LEO launch prices were available, the cost of shipping the helium-3 mining gear could be recovered in about four years, which would be reasonable.

So, unlike the space solar power beaming schemes, the helium-3 business plan might actually be workable, but not anytime soon. It requires the development of both economical controlled fusion power plants and very cheap space-launch and translunar transportation systems, all of which are well in the future. The cost uncertainties in all aspects of the plan make it clear that no one is going to invest big bucks to create a spacefaring civilization in order to be able to mine lunar helium-3. Rather, lunar helium-3 will become available as a resource to humanity as a result of developing a mature Type II spacefaring civilization for other reasons. Helium-3 won't provide the magnet that will draw us into space, but mastery of space will give us helium-3.

This is important for two reasons. In the first place, lunar helium-3 represents a large resource, enough to power human civilization at its current level of energy consumption for about a thousand years. Second, and more important, however, D-He3 fusion represents not simply another source of energy but a new kind of energy. Aside from antimatter, which does not exist naturally in the habitable portions of the universe, D-He3 fuel has the highest energy/mass ratio of any substance known. If used as the fuel for a fusion rocket, the D-He3 reaction could produce exhaust velocities as high as 5 percent the speed of light. Since rockets can generally be designed to achieve speeds of about twice their exhaust velocities, this means that a D-He3 rocket could reach 10 percent of lightspeed, making travel to nearby stars possible on time scales of four to six decades. This may seem long, but it is less than a human lifetime, and in any case it compares quite favorably with the millennia required for starflight using more conventional systems.

A tremendous amount of engineering advance will be required before such high-performance fusion rockets become a practical reality. The point, however, is that the D-He3 rocket is one of very few systems based on currently known physics that is capable of enabling interstellar flight. That is a promise not to be taken lightly. It is the reason controlled fusion must be pursued despite all apparent obstacles, setbacks, and attractive alternatives. The stars are worth more than kilowatts.

CLEMENTINE AND LUNAR PROSPECTOR

I said that hydrogen is generally present on the Moon only in the ~50 ppm quantities impregnated in the lunar regolith over eons of geologic time by the solar wind. In general, that is true. However, data received as a result of two recent spacecraft missions suggest that in certain isolated and unique places on the Moon the case may be otherwise.

Down through the ages, the Moon, like the Earth, has occasionally been the target of incoming comets. Comets are mostly composed of ice. Upon impact, the comet would be completely vaporized, if not ionized, and because of its high temperature and the weak lunar gravity, the vast majority of the resulting vapor would immediately escape back into space. However, in the complex dynamics of collisions between water vapor molecules and other material thrown up by the impact, some of the water vapor molecules might collide with other particles, bounce back toward the Moon, and, by chance, land in the shade. Because the Moon has no atmosphere to transfer heat around, it's very cold in the shade, and water vapor impacting on such shadowed ground might get frozen and held in place. Of course, as soon as the Sun came up, the shadow effect would be lost and such frost would be vaporized once again. Carrying on this way with repeated vaporization and frost deposition is a rather risky lifestyle for a water vapor molecule on the Moon. The lunar surface is unshielded from solar ultraviolet radiation, and sooner or later a water molecule that hangs around too long in the open will be hit by an ultraviolet (UV) photon strong enough to dissociate it into hydrogen and oxygen atoms, both of which will immediately and permanently escape the Moon. But what if, in its random travels, a water vapor molecule happens to land in a place where the Sun never shines? Then it could be frozen in place for good. Such permanently shaded locations could in principle exist in areas of depressed topography near the lunar poles. Based on such logic, Jet Propulsion Lab scientist Bruce Murray hypothesized in the early 1960s that ice deposits might exist in such anomalous polar lunar craters.

The theory seemed so far-fetched, however, that no one took it seriously until the early 1990s, when astronomical observations revealed ice deposits sitting in precisely such locations near the poles of Mercury. Mercury is the closest planet to the Sun—it's hot there. No one every expected to find ice on Mercury. But there it was, in permanently shadowed, and thus permanently cold polar craters on what otherwise is certainly the hottest and driest planet in the solar system. If ice could exist on the poles of Mercury, there was no reason to believe that it could not exist in similar locations on the Moon.

Unfortunately, it is impossible to observe the Moon's poles from the Earth—the angle is wrong and the view is blocked. In order to look for lunar polar ice deposits, a spacecraft mission is necessary.

In the early 1990s, Strategic Defense Initiative (SDI) technology development director Colonel Pete Worden (of DC-X fame) was developing a satellite to carry advanced sensors capable of detecting and characterizing enemy missiles during launch. Intrigued by the polar ice issue and other questions of planetary science and space resources, Worden proposed that the SDI sensors could best be tested in lunar orbit. Thus, in January 1994, the SDI spacecraft, dubbed *Clementine* (after the miner's daughter), was sent hurtling to the Moon. In the following months, *Clementine* positioned itself in a polar lunar orbit and surveyed the entire surface with its advanced sensors, demonstrating their utility for the SDI mission. In the process, *Clementine* produced by far the best topographic and multi-spectral map of the Moon yet. Unfortunately, not even the most advanced sensors can see into permanently shadowed craters—by definition, there is no light to see with in such locations. However, Worden's team, led by Lawrence Livermore Laboratory scientist Stu Nozette, had another trick up their sleeves. They bounced radio transmissions from the spacecraft off the lunar poles when the spacecraft was positioned well for the returning signal to be received on Earth. By doing this, they created a kind of bistatic radar. Radar signals returning from ice have a slightly different quality than those bounced off of dry soil. Examining the radar reflections from the Moon's south pole, Nozette's team announced that the results were consistent with ice.

The *Clementine* mission created a sensation, not least because it was done for $80 million, making it by far the cheapest planetary mission the United States has ever flown. It was a major embarrassment to NASA and the Jet Propulsion Lab, which had recently spent $1 billion on the *Mars Observer* spacecraft only to see it fail completely. As a form of sour grapes, the NASA brass made much of the fact that *Clementine* was unable to continue beyond the Moon with a hoped-for extended mission to a near-Earth asteroid, and anti-SDI people within the administration acted forcefully to prevent Worden from repeating his success by blocking funds for *Clementine II.* But sour grapes alone wouldn't do. NASA was forced to rethink its planetary mission program into a "faster, better, cheaper" model of low-cost ($150 million or less) Discovery-class missions based on *Clementine.* But the *Clementine* radar results, while consistent with lunar polar ice, were not conclusive. There were plenty of alternative explanations for the data. So for its first openly competed Discovery mission, NASA chose *Lunar Prospector,* a spacecraft designed to settle the polar ice issue once and for all.

Lunar Prospector, also built and flown for about $80 million, is actually a simpler and less capable spacecraft than *Clementine.* It has no laser altimeter/rangefinder, multispectral sensors, or even a camera. But it does have a gamma-ray spectrometer for use in estimating the concentrations of most elements found on the lunar surface and, most important, carries a neutron spectrometer—a sensitive instrument for detecting hydrogen in the lunar soil.

Lunar Prospector was launched in January 1998. By March, the mission's principal investigator, Dr. Alan Binder, was ready to announce results. According to Binder and his team, *Lunar Prospector*'s neutron spectrometer *had* detected water, in concentrations of about 0.5 percent in both of the Moon's polar regions. The actual neutron spectrometer measurement indicated hydrogen in concentrations of about 0.05 percent (500 ppm). Binder and his team inferred that the detected hydrogen was in the form of water (which weighs nine times as much as the hydrogen it contains), an assumption that is supported by most, though not all, of the planetary science community.

Soil containing 0.5 percent water is a lot wetter than any previously known to exist on the Moon, but it's still drier than the Sahara, Martian desert dirt, or dry concrete for that matter. However, Binder believes that the water he detected might not be in the form of 0.5 percent dilute permafrost spread over the whole pole, but instead might exist as small crater ponds of pure or nearly pure ice scattered across the polar region. Such a result would be more consistent with the *Clementine* radar findings (which hardly would have noticed dilute permafrost). The 0.5 percent ice signal would then result from the fact that these frozen water concentrations cover about 0.5 percent of the polar area under study. If that were the case, it would make the water detected by *Lunar Prospector* a much more readily exploitable resource.

In its current high orbit, it is impossible for *Lunar Prospector* to distinguish between the two alternatives of dilute permafrost across broad polar areas or scattered local ice concentrations. As the mission proceeds *Lunar Prospector*'s controllers plan to lower the orbit, which should increase the resolution of the measurements as the spacecraft flies ever closer to the surface. Even that might not be good enough—at lowest flight altitude the spacecraft will still be tens of kilometers above the ground. If the ice concentrations have dimensions of a few kilometers across or less, the spacecraft will still not be capable of distinguishing them from dilute permafrost.

What's really needed is a spacecraft that can actually observe directly. Some have proposed landing a robotic rover and walking it into a dark polar crater. This might work, although such a project faces a significant engineering challenge, as the temperature within a permanently shadowed lunar crater is estimated to be about –230°C. The rover would certainly have to be either nuclear- or radioisotope-powered, and its launch would thus arouse extensive protests from "radical activists" who (rather than confronting the old-hat problems of poverty, war, and injustice dealt with by their predecessors) have chosen to devote their lives to stopping nuclear-powered space exploration. In addition, the rover might walk into the wrong crater and, given the limited range of robotic planetary rovers, might not be able to make it to a second crater for another look. A better approach might be to send another spacecraft into lunar orbit, this one equipped with a suite of high-powered active sensors. In other words, what's needed is a spacecraft with enough power to flash a light into the permanently shadowed craters and photograph their contents. As an important secondary instrument, it should also carry a well-designed ground-penetrating radar (GPR) capable of operating at several different frequencies. Low-frequency GPR can go very deep but has low resolution. High-frequency GPR provides more detail but does not penetrate as far. The mission should carry both. Such a spacecraft is not beyond our capability and could probably be built and flown for about the same cost as *Lunar Prospector* or *Clementine.*

MOON BASING

For reasons discussed earlier in this chapter, confirming that water exists in useful quantities on the lunar surface would still not make the Moon a viable target for colonization. It would, however, greatly simplify the logistics of establishing and amplify the value of operating a lunar base.

In addition to being one of the staffs of life, water can be broken down into hydrogen and oxygen, an excellent rocket propellant combination. If water is present on the Moon, then rocket vehicles transporting people and payloads from Earth to the lunar base need not take their return propellant with them. Instead, they would be able to refuel with propellant manufactured at the lunar base. Even more important, a local supply of rocket propellant available at the base would allow crews to sortie from the base repeatedly to visit numerous locations all over the Moon in rocket-propelled, long-range ballistic hopper vehicles. This would allow them to service a widely dispersed set of scientific assets spread across the Moon. Thus, the Moon's Arctic or Antarctic could become the location where we would site the base that would support Moon-wide arrays of optical telescopes that could offer a spectacular view of the universe.

The possible discovery of water on the Moon gives new life to an idea that has been discussed in both science fiction and the astronautical engineering literature for some time—that of using a lunar base as a staging point for missions to worlds beyond. The idea is that since the Moon has only one-sixth Earth's gravity and no atmosphere, it's possible to reach any destination in space much easier from the Moon than it is from the Earth's surface. Thus, if indeed rocket propellant can be made available on the lunar surface, the Moon could well turn into an excellent refueling station and port of call for interplanetary traffic. This proposal was advanced even before water was detected on the Moon—it was always known that the Moon contained plenty of oxygen in the form of metal oxides in its rocks, and techniques have been demonstrated for getting it out. In particular, the mineral ilmenite ($FeTiO_3$), which occurs in about 10 percent concentrations in some lunar soils, can be reduced by hitting it with hydrogen at 1,000°C. The reaction involved is

$$FeTiO_3 + H_2 \rightarrow Fe + TiO_2 + H_2O \qquad (5.4)$$

The water produced is then electrolyzed to produce hydrogen, which is recycled back into the reactor, and oxygen, which is the net useful product of the system. The high temperatures required, the necessity for hydrogen feedstock to replace leakage, and the need to mine and refine ilmenite make this system somewhat difficult. However, the reactor itself has worked on test stands at the Carbotek Corporation, and the balance of the system's complexities should be manageable at a mature lunar base. Of course, the only useful propellant product is oxygen—if this was the best the lunar base could do, visiting spacecraft would still need to bring their own fuel (hydrogen, methane, or kerosene) to burn in the oxygen. But since for rocket vehicles the oxygen generally constitutes at least 75 percent of the total fuel/oxygen propellant combination, a supply of lunar oxygen alone could still be quite useful. But if lunar water is available, then both oxygen and hydrogen can be provided, and the chemical process required to produce them becomes much simpler (only electrolysis is required) as well.

So, the idea of the Moon as a refueling station is interesting. It has its possibilities, but also its limitations. Using lunar propellants as a means of refueling Moon base spacecraft for their return to Earth or for hopping around the Moon makes perfect sense. Surprisingly, however, using a Moon base to refuel spacecraft on their way from Earth to Mars offers no benefits at all. This is because the spacecraft, its equipment, and the large majority of its provisions must come from Earth, and the rocket ΔV required to go from LEO to Mars (4.2 km/s) is less than that required to go from LEO to the surface of the Moon (6 km/s). So even if there was rocket propellant, already made, sitting at a lunar base right now and available for free, it would make no sense for a Mars-bound spacecraft to fly there to get it! It would be easier and cheaper to fly to Mars directly. If the Moon base operated a reusable lunar-surface-to-orbit shuttle and thus was able to provide propellant not only on the lunar surface but in lunar orbit as well, the situation would change, but not by enough. The ΔV to go from LEO to lunar orbit is 4.2 km/s, the same as that required to fly from LEO to Mars, so lunar orbital refueling of Mars-bound ships would still be pointless, especially since lunar-produced propellants delivered to lunar orbit would most certainly not be free. A Moon base refueling station would be of no assistance for Mars missions.

However, if the destination chosen is well beyond Mars, the balance of benefits shifts. For example, the ΔV to go from LEO to Ceres, a planetoid in the heart of the main asteroid belt, is 9.6 km/s, which is greater than that required to go to either lunar orbit or the lunar surface. So, if lunar propellants could be made available in these locations cheaply enough (compared to simply lifting the required 9.6 km/s worth of propellant from Earth to LEO), Moon-based refueling could be advantageous. As the destination chosen is moved farther out, the required mission ΔV's grow, and so do the potential benefits offered by lunar refueling. Of course, it would never pay to set up a lunar refueling station for the benefit of one or two outer solar system missions; the basic infrastructure would cost too much. But if there were regular interplanetary traffic, say to support mining operations in the main asteroid belt (see Chapter 7), a lunar refueling station might find itself with a vital supporting role.

THE LUNAR OBSERVATORY

Why, then, should we be interested in returning to the Moon? The O'Neill colony mining base and translunar solar power beaming concepts are all fantastical, and the helium-3 and space commerce refueling business plans are at least fifty years premature. The Moon lacks sufficient resources for true colonization. So if we can't make money there, and we can't create settlements there, why should we go?

For science.

It has been known for some time that the Moon would be a superb location for astronomical observatories. It has no obscuring atmosphere and the fact that it rotates only once every 28 days affords a telescope 28 times as much time to gather light from a distant object as would be possible on Earth (or over 400 times as long as generally is possible in LEO). In addition to being atmosphere-free, the Moon is also seismically dead and thus provides a rock-steady platform for mounting telescopes. This is an essential attribute required to create optical arrays in which groups of telescopes all focus on a single object and coordinate the signals they receive via computer. While the implementation of such arrays requires knowing the distance between telescopes to an accuracy of less than a millionth of a meter (and is thus nearly impossible on the seismically vibrating Earth or in free-floating space), their advantage to astronomy is extraordinary—while the power of a single telescope to resolve detail is proportional to its diameter, the resolving power of an array of telescopes is proportional to the diameter of the array. So while the Hubble Space Telescope has a diameter of 2.5 meters, on the Moon an array of telescopes could be stationed across the Moon's diameter, some 1,700 kilometers, and achieve a resolution nearly a million times better than Hubble.

Such observatories would allow us to map Earth-sized planets around nearby stars, perhaps out to a distance of 100 light-years or so, within which some 10,000 solar systems are likely to be found. With the aid of the lunar optical array, we would be able to learn as much about these other solar systems as we knew about our own before the advent of interplanetary space probes in the 1960s. Perhaps more important, the optical array would allow us to penetrate deep into the cosmos to examine objects dating from the time of the "Big Bang" that is generally believed to have initiated our universe.

Optical astronomy would not be the only branch of humanity's oldest science to benefit enormously from a lunar base. The Moon's lack of atmosphere makes it an ideal platform for conducting cosmic-ray, gamma-ray, x-ray, and ultraviolet astronomy. These techniques, which are difficult to impossible to perform through Earth's thick atmosphere, are key to examining high-energy processes in the universe. The low temperatures available in the Moon's permanently shadowed craters make them ideal locations for stationing infrared telescopes. The Moon's far side is the only place in the solar system that is shielded from terrestrial civilization's massive radio chatter, and so is the best place for positioning radio telescopes. In addition, because the Moon has no ionosphere, a radio telescope positioned on the surface of the Moon can pick up low-frequency (30 MHz or less) radio waves from space that the Earth's ionosphere completely masks from reception by ground-based instruments here. Each of these windows in the electromagnetic spectrum offers unique advantages and opportunities to discover new physical phenomena, and all of them can best be studied from the Moon.

This is of much more than academic interest. Historically, astronomy has led in the development of new physical law. The laws of gravity, electromagnetism, relativity, and nuclear fusion were all discovered through astronomical observation. It is strange to think that the scientific basis for the technologies that dominate world commerce and power politics today, including accurate global navigation, ballistic missiles, electronic communications, and nuclear and thermonuclear weapons, all stem from the apparently impractical studies of astronomers. Yet this should be expected, since the universe provides a much bigger lab than any we can build. By allowing us to probe far deeper into time and space than ever before possible, a system of lunar observatories may well allow us to discover new laws of physics, including those of the creation process itself.

What the Moon lacks in amenities, it makes up for with the view.

But while the benefits of astronomy provide good and sufficient reason to go to the Moon, they are insufficiently sensuous to move the soul of a civilization. Humanity needs a challenge from space that will inspire it to transform itself into a multi-planet species. Such a broad and deeply felt challenge can only come from the lure of a true new frontier, a new world with the full potential for the development of new, and hopefully better, branches of human civilization. Such a challenge can only come from Mars.

FOCUS ON: THE LUNAR BEANSTALK

The Moon may be a lousy place to try to grow plants, but it's an excellent place to build a beanstalk. Confused? Okay, I'll start at the beginning.

Back in 1960, the Soviet newspaper *Komsomolskaya Pravda* published an interview with engineer Y. N. Artsutanov containing a description of a novel means of Earth-to-orbit transportation. In the Artsutanov scheme, a satellite placed in geostationary orbit would simultaneously extend cables down toward the Earth and in the opposite direction, keeping its center of mass, and thus its orbit, constant. This procedure would continue some 36,000 km until the lower cable reached the surface of the Earth, where it could be anchored and used to support elevator cabs. These cabs, in turn, could then be used to transport payloads up to the satellite where they could then be released into geostationary orbit. If the payloads continued farther out along the cable, they would have greater than orbital velocity, and could be released on trajectories that would take them to the Moon, Mars, Jupiter, or beyond. In other words, Artsutanov had designed a skyhook, or beanstalk, which just as in the fairy tale of Jack and the Giant could allow someone to literally climb from the surface of the Earth to the land beyond the sky.

This concept was published in both Russian and English only to be widely ignored. It was eventually rediscovered in 1975 by Jerome Pearson of the Wright Patterson Air Force Base.[3] Pearson published a series of papers on the concept going into far greater detail than earlier authors, including derivations for system mass, tapered tether designs, and allowable rates for moving payloads along the tether without exciting dangerous vibrational modes. Subsequently the geostationary beanstalk concept was widely publicized by Arthur C. Clarke, who made it a central feature of his novel *The Fountains of Paradise.*

The beanstalk concept as envisaged by Artsutanov, Pearson, and Clarke was a wonderful idea that offered a complete and easy solution to the problem of cheap Earth-to-space transportation. It had just one problem: It was impossible. It was impossible because if one places a load at the bottom of a geostationary tether, the bit of tether holding it must be thick enough to support that load. The next bit of tether must be thick enough to support not only the load, but the bit of tether supporting the load. Thus as it proceeds to 36,000 km from the ground to geostationary orbit, the tether must get thicker and thicker, and its diameter and weight will grow exponentially. Depending upon the strength-to-weight ratio of the tether material assumed, the cross-sectional area of the tether at the satellite would be 10 to 20 orders of magnitude greater than its area at its base, with similar incredible ratios holding between the tether mass and the mass of the payload it is required to lift. Unless fantastical materials, such as 36,000-km-long single-crystal graphite fibers with incredible strength-to-weight ratios, were assumed, a beanstalk designed to lift 1 tonne would itself have to weigh quadrillions of tonnes. With real materials, the beanstalk just wouldn't work.

Well, not on Earth anyway, but on the Moon things could be quite different. On the Moon, the beanstalk would have to lift loads only against one-sixth of Earth's gravity, and the length of the tether required could be a lot shorter too. The effect of both of these factors on the tether design equations is exponential, and the net result is that practical beanstalks really could be built on the Moon using current state-of-the-art materials like Spectra, a high-strength plastic. The lunar beanstalks would still have to weigh about 100 times the mass of the payload they would lift or lower, but since they could be used again and again to transfer payloads back and forth between the lunar surface and lunar orbit, an investment in such infrastructure could pay off well to support the operations of a mature lunar base.

There are two ways a lunar beanstalk could be designed. One way would be as a stationary system, with its center of mass at the "L1" LaGrange point in space where the gravity of the Earth and the Moon balance. Such a stationary beanstalk would need cable cars to deliver payloads up and down, just as specified in Artsutanov's original article. Another way, however, would be to position the center of mass of the beanstalk in low equatorial orbit, and have the tether rotate at just the right speed so that its backward moving tips have zero velocity with respect to the ground. Using such a system, you would just wait on the ground for the tether tip to come by, at which time you could step into the seat positioned at its tip and ride it up to orbit like a Ferris wheel! Six such "tether stops" could be positioned at different positions spread across the tether's ground track, and a tether tip would swing by each stop every couple of hours. If you wanted to travel rapidly across the Moon, you could get on at one stop and get off at another. With an equatorial orbiting beanstalk there would be only six stops and they would all be along the equator. If you put the rotating beanstalk in a polar orbit, you could create a lot more stops scattered all over the Moon, but service to any particular stop would be correspondingly less frequent.

I can see the advertising slogans now: "Travel by Roto-Beanstalk! It's the Next Best Thing to Beaming There!" Such slogans, however, would conceal the real difficulties involved. Lunar orbits are unstable and ground contact elevations vary, so it will no doubt take all the ingenuity and round-the-clock hard work of a dedicated corps of roto-beanstalk engineers to make the system work. Their motto: "Keep 'em rolling!"

CHAPTER 6

Mars: The New World

We hold it in our power to begin the world anew. . . .

—TOM PAINE, 1776

BEYOND THE MOON lies Mars, the decisive step in humanity's outward migration into space. Mars is hundreds of times farther away than the Moon, but it offers a much greater prize. Indeed, uniquely among the extraterrestrial bodies of our solar system, Mars is endowed with all the resources needed to support not only life but the development of a technological civilization. In contrast to the comparative desert of Earth's moon, Mars possesses oceans of water frozen into its soil as permafrost, as well as vast quantities of carbon, nitrogen, hydrogen, and oxygen, all in forms readily accessible to those clever enough to use them. In addition, Mars has experienced the same sorts of volcanic and hydrologic processes that produced a multitude of mineral ores on Earth.[1] Virtually every element of significant interest to industry is known to exist on the Red Planet. With its 24-hour day-night cycle and an atmosphere thick enough to shield its surface against solar flares, Mars is the only extraterrestrial planet that will readily allow large-scale greenhouses lit by natural sunlight.

Mars can be settled. For our generation and many that will follow, Mars is the New World.

MARS DIRECT

Some have said that a human mission to Mars is a venture for the far future, a task for "the next generation." Such a point of view has no basis in fact. On the contrary, as I explained in detail in my book *The Case for Mars,* the United States has in hand, today, all the technologies required for undertaking an aggressive, continuing program of human Mars exploration, with the first piloted mission reaching the Red Planet within a decade.[2] We do not need to build giant spaceships embodying futuristic technologies in order to go to Mars. We can reach the Red Planet with relatively small spacecraft launched directly to Mars by boosters embodying the same technology that carried astronauts to the Moon more than a quarter century ago. The key to success comes from following a "travel light and live off the land" strategy similar to that which has well served terrestrial explorers for centuries. The plan to approach the Red Planet in this way is called "Mars Direct."

Here's how the Mars Direct plan works. At an early launch opportunity, for example 2005, a single heavy-lift booster with a capability equal to that of the Saturn V used during the Apollo program is launched off Cape Canaveral and uses its upper stage to throw a 40-tonne unmanned payload onto a trajectory to Mars. Arriving at Mars eight months later, the spacecraft uses friction between its aeroshield and Mars' atmosphere to brake itself into orbit around the planet and then lands with the help of a parachute. This payload is the Earth Return Vehicle (ERV). It flies out to Mars with its two methane/oxygen-driven rocket propulsion stages unfueled. It also carries 6 tonnes of liquid hydrogen cargo, a 100-kW nuclear reactor mounted in the back of a methane/oxygen-driven light truck, a small set of compressors and an automated chemical processing unit, and a few small scientific rovers.

As soon as the craft lands successfully, the truck is telerobotically driven a few hundred meters away from the site, and the reactor is deployed to provide power to the compressors and chemical processing unit. The hydrogen brought from Earth can be quickly reacted with the Martian atmosphere, which is 95 percent carbon dioxide gas (CO_2), to produce methane and water, thus eliminating the need for long-term storage of cryogenic hydrogen on the planet's surface. The methane so produced is liquefied and stored, while the water is electrolyzed to produce oxygen, which is stored, and hydrogen, which is recycled through the methanator. Ultimately, these two reactions (methanation and water electrolysis) produce 24 tonnes of methane and 48 tonnes of oxygen. Since this is not enough oxygen to burn the methane at its optimal mixture ratio, an additional 36 tonnes of oxygen is produced via direct dissociation of Martian CO_2. The entire process takes ten months, at the conclusion of which a total of 108 tonnes of methane/oxygen bipropellant will have been generated. This represents a leverage of 18:1 of Martian propellant produced compared to the hydrogen brought from Earth needed to create it. Ninety-six tonnes of the bipropellant will be used to fuel the ERV, while 12 tonnes are available to support the use of high-powered, chemically fueled long-range ground vehicles. Large additional stockpiles of oxygen can also be produced, both for breathing and for turning into water by combination with hydrogen brought from Earth. Since water is 89 percent oxygen (by weight), and since the larger part of most foodstuffs is water, this greatly reduces the amount of life support consumables that need to be hauled from Earth.

The propellant production having been successfully completed, in 2007 two more boosters lift off the Cape and throw their 40-tonne payloads toward Mars. One of the payloads is an unmanned fuel factory/ERV just like the one launched in 2005; the other is a habitation module carrying a crew of four, a mixture of whole food and dehydrated provisions sufficient for three years, and a pressurized methane/oxygen-powered ground rover. On the way out to Mars, artificial gravity can be provided to the crew by extending a tether between the habitat and the burned-out booster upper stage and spinning the assembly. Upon arrival, the manned craft drops the tether, aerobrakes, and lands at the 2005 landing site where a fully fueled ERV and fully characterized and beaconed landing site await it. With the help of such navigational aids, the crew should be able to land right on the spot, but if the landing is off course by tens or even hundreds of kilometers, the crew can still achieve the surface rendezvous by driving over in their rover. If they are off by thousands of kilometers, the second ERV provides a backup. However, assuming the crew lands and rendezvous as planned at site number one, the second ERV will land several hundred kilometers away to start making propellant for the 2009 mission, which in turn will fly out with an additional ERV to open up Mars landing site number three. Thus, every other year two heavy-lift boosters are launched, one to land a crew and the other to prepare a site for the next mission, for an average launch rate of just one booster per year to pursue a continuing program of Mars exploration. This is only about 12 percent of the U.S. heavy-lift launch capability and is clearly affordable. In effect, this "live off the land" approach removes the manned Mars mission from the realm of mega-fantasy and reduces it in practice to a task of comparable difficulty to that faced in launching the Apollo missions to the Moon.

The crew will stay on the surface for 1.5 years, taking advantage of the mobility afforded by the high-powered chemically driven ground vehicles to accomplish a great deal of surface exploration. With a 12-tonne surface fuel stockpile, they have the capability for over 24,000 kilometers' worth of traverse before they leave, giving them the kind of mobility necessary to conduct a serious search for evidence of past or present life on Mars—an investigation that is key to revealing whether life is a phenomenon unique to Earth or general throughout the universe. Since no one has been left in orbit, the entire crew will have available to them the natural gravity and protection against cosmic rays and solar radiation afforded by the Martian environment, and thus there will not be the strong driver for a quick return to Earth that plagued previous Mars mission plans based on orbiting mother ships with small landing parties. At the conclusion of their stay, the crew returns to Earth in a direct flight from the Martian surface in the ERV. As the series of missions progresses, a string of small bases is left behind on the Martian surface, opening up broad stretches of territory to human cognizance.

Such is the basic Mars Direct plan. In 1990, when it was first put forward, it was viewed as too radical for NASA to consider seriously, but over the past couple of years, with the encouragement of former NASA Associate Administrator for Exploration Mike Griffin and current NASA Administrator Dan Goldin, the group at Johnson Space Center in charge of designing human Mars missions decided to take a good hard look at it. They produced a detailed study of a Design Reference Mission based on the Mars Direct plan but scaled up about a factor of 2 in expedition size compared to the original concept. They then produced a cost estimate for what a Mars exploration program based upon this expanded Mars Direct would cost. Their result: $50 billion, with the estimate produced by the same costing group that assigned a $400-billion price tag to the traditional cumbersome orbital assembly of mega-spacecraft approach to human Mars exploration embodied in NASA's 1989 "90 Day Report" (whose sticker shock caused the congressional abortion of President George Bush's Space Exploration Initiative). If scaled back to the original lean Mars Direct plan described here, the program could probably be accomplished for $20 to $30 billion. This is a sum that the United States, or Europe, or Japan, could easily afford. It's a small price to pay for a new world.

In essence, by taking advantage of the most obvious local resource available on Mars—its atmosphere—the plan allows us to accomplish a manned Mars mission with what amounts to a lunar-class transportation system. By eliminating any requirement to introduce a new order of technology and complexity of operations beyond those needed for lunar transportation to accomplish piloted Mars missions, the plan can reduce costs by an order of magnitude and advance the schedule for the human exploration of Mars by a generation.

Exploring Mars requires no miraculous new technologies, no orbiting spaceports, and no gigantic interplanetary space cruisers. We can establish our first small outpost on Mars within a decade. We and not some future generation can have the eternal honor of being the first pioneers of this new world for humanity. All that's needed is present-day technology, some nineteenth-century industrial chemistry, a solid dose of common sense, and a little bit of moxie.

COLONIZING MARS

The question of colonizing Mars is not fundamentally one of transportation. If we were to use the same heavy-lift boosters used in the Mars Direct plan to launch people to Mars on one-way trips, firing them off at the same rate we currently launch the Space Shuttle, the United States today could populate Mars at a rate comparable to that at which the British colonized North America in the 1600s—and at lower expense relative to our resources. No, the problem of colonizing Mars is not that of moving large numbers to the Red Planet but of the ability to use Martian resources to support an expanding population once they are there. The technologies required to do this will be developed at the first Mars base, which will thus act as the beachhead for the wave of immigrants to follow. Initial Mars Direct exploration missions approach Mars in a manner analogous to terrestrial hunter-gatherers and utilize only its most readily available resource, the atmosphere, to meet the basic needs of fuel and oxygen. In contrast, a permanently staffed base will approach Mars from the standpoint of agricultural and industrial society. It will develop techniques for extracting water out of the soil, for conducting increasingly large-scale greenhouse agriculture, for making ceramics, metals, glasses, and plastics out of local materials, and for constructing large pressurized structures for human habitation and industrial and agricultural activity.[3,4]

Over time, the base could transform itself into a small town. The high cost of transportation between Earth and Mars will engender a strong financial incentive to find astronauts willing to extend their surface stay beyond the basic 1.5-year tour of duty to four years, six years, and more. Experiments have already been done showing that plants can be grown in greenhouses filled with CO_2 at Martian pressures—the Martian settlers will thus be able to set up large inflatable greenhouses to provide the food required to feed an expanding resident population. Mobile microwave units will be used to extract water from Mars' abundant permafrost, supporting such agriculture and making possible the manufacture of large amounts of brick and concrete, the key materials required for building large, pressurized structures. While the base will start as an interconnected network of Mars Direct–style "tuna can" habitats, by its second decade the settlers could live in brick and concrete pressurized domains the size of shopping malls. Not too long afterward, the expanding local industrial activity will make possible a vast expansion in living space by manufacturing large supplies of high-strength plastics like Kevlar and Spectra that will allow the creation of inflatable domes encompassing Sun-lit pressurized areas up to 100 meters in diameter. Each new reactor landed will add to the power supply, as will locally produced solar panels and windmills. However, because Mars has been volcanically active in the geologically recent past, it is also highly probable that underground hydrothermal reservoirs exist on the Red Planet. Once such reservoirs are found, they can be used to supply the settlers with abundant supplies of both water and geothermal power. As more people steadily arrive and stay longer before they leave, the population of the town will increase. In the course of things, children will be born and families raised on Mars, the first true colonists of a new branch of human civilization.

We don't need any fundamentally new or even cheaper forms of interplanetary transportation to send the first teams of human explorers to Mars. However, meeting the logistical demands of a Mars base will create a market that will bring into being low-cost, commercially developed systems for interplanetary transport. Combined with the base's own activities in developing the means to use Martian resources to allow humans to be self-sufficient on the Red Planet, such transportation systems will make it possible for the actual colonization and economic development of Mars to begin.

While the initial exploration and base-building activities on Mars can be supported by government largess, a true colony must eventually become economically self-supporting. Mars has a tremendous advantage compared to the Moon and asteroids in this respect, because unlike these other destinations the Red Planet contains all the necessary elements to support both life and technological civilization, making self-sufficiency possible in food and all basic, bulk, and simple manufactured goods. However, for quite a while, some high-technology imports from Earth will be required, which will have to be purchased with cash. The Mars colony will be able to obtain such earnings by exporting both ideas and materials. Just as the labor shortage prevalent in colonial and nineteenth-century America drove the creation of Yankee ingenuity's flood of inventions, so the conditions of extreme labor shortage combined with a technological culture will drive Martian ingenuity to produce wave after wave of invention in energy production, automation and robotics, biotechnology, and other areas. These inventions, licensed on Earth, could finance Mars even as they revolutionize and advance terrestrial living standards as forcefully as nineteenth-century American invention changed Europe and ultimately the rest of the world as well.

In addition to inventions, Mars may also be able to export minerals. Like the Earth, Mars has had a complex geologic history, sufficient to form rich mineral ores. Unlike the Earth, however, Mars has not had people on it for the past 5,000 years scavenging all the readily available rich mineral deposits to be found on its surface. Rich, untapped mineral deposits of gold, silver, uranium, platinum, palladium, and other precious metals may all exist on the Martian surface. Even at this early date in its exploration, however, Mars is already known to possess a vital resource that could someday represent a commercial export. Deuterium, the heavy isotope of hydrogen currently valued at $10,000 per kilogram, is five times more common on Mars than it is on Earth. Deuterium has its applications today, but it is also the basic fuel for fusion reactors, and in the future when such systems come into play as a major foundation of Earth's energy economy the market for deuterium will expand greatly. Still another source of revenue is represented by the sale of real estate—developed or undeveloped—to incoming colonists, development companies, and speculators.

But let us be clear: I am *not* suggesting that human colonies be set up on Mars in order to mine gold or deuterium or that business plans for the commercial colonization of Mars on such a basis would have much chance of success. That would be silliness of the same sort as the space hotel and solar power satellite quackery discussed in chapter 4. What I am saying is that the resources exist on Mars to establish viable human societies and that, once there, the means exist whereby such populations would be able to generate the revenue they need to grow. Deuterium will be produced as a byproduct of the hydrogen cycling necessary for life support and materials processing at a Mars settlement. We shall not go to Mars in order to set up inventors' colonies, but a Mars settlement would be a group of technologically adept people in a frontier environment where they would be both forced to and free to innovate, and as such, it would be a pressure cooker for invention. The new Martians will invent to meet their own needs; the inventiveness that will flow from the culture they will be forced to create will produce patents; the patents licensed on Earth will make the Martians rich. There is no source of cash profit on Mars today. There will be once smart people are living there.

Martian colonists will be able to use rocket hoppers using locally produced propellants to lift such resources from the Martian surface to Mars' moon Phobos, where an electromagnetic catapult can be placed capable of firing the cargo off to Earth for export. Larger or more complex cargoes could be shipped out from Phobos at low cost using robotic solar sail-powered spacecraft. Alternatively, on Mars it should also be possible to build a "skyhook" or stationary beanstalk consisting of a cable whose center of mass is located at a distance, from which it will orbit the planet in synchrony with Mars' daily rotation. To an observer on the Martian surface, such cables will appear to stand motionless, allowing payloads to be delivered to space via cable car. Because of strength-of-materials limits, such systems cannot be built on Earth, but, as on the Moon, in Mars' three-eighths gravity they may well be feasible. If so, they would further enhance the ability of Mars colonists to transport goods cheaply to Earth and to access the resources present throughout the rest of the solar system as well. As discussed in *The Case for Mars,* the rocket propulsion requirements to reach the main asteroid belt from Mars are much lower than from Earth. High-technology goods needed to support asteroid mining may have to come from Earth for some time. But since food, clothing, and other necessities can be produced on Mars with much greater ease than would be possible anywhere farther out, Mars could become the central base and port of call for exploration and commerce heading out to the asteroid belt, the outer solar system, and beyond.

Currently available scientific evidence shows that Mars was once a warm and wet planet, a place friendly to life. Atmospheric resources in the form of vast reserves of carbon dioxide and water adsorbed or frozen into the soil still exist. Once sufficient human industrial potential is developed on Mars, human colonists might begin to employ it to return the planet to the warm-wet climate of its distant past. By producing halocarbon supergreenhouse gases on Mars at a rate similar to what we are now doing on Earth (but being sure to use only perfluorocarbons, or PFCs, instead of chlorinated flurocarbons, or CFCs, so as not to destroy the planet's ozone layer), and willfully dumping these climate-altering substances into the atmosphere, Mars colonists could, over a period of several decades, warm the planet by as much as 10°C. This warming would have the effect of causing massive quantities of carbon dioxide to outgas from the soil. Since CO_2 is a greenhouse gas, this would raise the temperature of the planet still more. As the temperature rises, the vapor pressure of water in the Martian atmosphere would also rise, and since water vapor is also a very strong greenhouse gas, this would raise the temperature of the planet still more, forcing even more CO_2 out of the soil, and so on. As a result of these positive-feedback mechanisms, a runaway greenhouse effect could be created on a planet-wide scale on Mars, with the net result being to raise the average temperature on the planet by over 50°C within half a century. At the same time, the atmospheric pressure would rise from its current level of about 1 percent that of Earth to about 35 percent (i.e., 5 psi pressure). Now an air pressure of 5 psi may not sound like much, but it's what we used on the Skylab space station in the early 1970s. Provided that the atmosphere is enriched to be 60 percent oxygen and 40 percent nitrogen (instead of the 20 percent oxygen/80 percent nitrogen that prevails on Earth), such gas is perfectly breathable. So, while humans could not breathe the 5 psi CO_2 atmosphere that would prevail on such an altered Mars, they no longer would need to wear space suits—simple breathing gear providing oxygen-enriched gas would suffice. Available habitation volume would expand dramatically as well, since the availability of a 5-psi external pressure would allow very large inflatable domes featuring an internal breathable 5-psi atmosphere (3 psi oxygen/2 psi nitrogen) to be readily erected. Moreover, as the outside environment would be warm enough for liquid water, the Martian permafrost would start to melt, and plants could propagate across first the tropical and then the temperate regions of Mars. Over a period of a thousand years or so, such plants could put enough oxygen in Mars' atmosphere to make it breathable by humans and higher animals. Eventually, the day would come when the breathing gear and city domes would no longer be necessary.

This feat, the "terraforming" of Mars from its current lifeless or near-lifeless state to a living, breathing world supporting multitudes of diverse and novel life forms and ecologies, will be one of the greatest and noblest enterprises of the human spirit. No one will be able to contemplate it and not feel prouder to be human.

Life in the initial Mars settlements will be harder than life on Earth for most people, but life in the first North American colonies was much harder than life in Europe as well. People will go to Mars for many of the same reasons they went to colonial America: Because they want to make a mark or a new start, or because they are members of groups who are persecuted on Earth, or because they are members of groups who want to create a society according to their own principles. Many kinds of people will go, with many kinds of skills, but all who go will be people willing to take a chance to do something important with their lives. Out of such people are great projects made and great causes won. Aided by ever-advancing technology, such people can transform a planet and bring a dead world to life.

THE QUESTION OF LIFE ON MARS

By exploring Mars, we will come face-to-face with one of the deepest and most important philosophical questions that humans have puzzled over since time immemorial, the question of the prevalence and significance of life in the universe. Is the universe alive or is it dead? Mars holds the answer.

Mars has been a prime suspect for some time in the search for life beyond the Earth. In the seventeenth century, the Italian astronomer Giovanni Cassini measured the Martian day at 24 hours, 40 minutes (just 2.5 minutes longer than current data indicate), and noted Earth-like ice caps on Mars's poles. During the eighteenth century, the British astronomer William Herschel observed that Mars rotates on its axis in a manner not dissimilar from Earth (he observed a tilt of 30 degrees; 24 degrees is the modern value) and therefore should have Earth-like seasons. In the nineteenth century, cloud movement was detected on Mars, proving that it had an atmosphere. Combined with the fact that the rate of solar heating on Mars' equator could be readily calculated to equal that experienced in Norway, all these observations suggested a fair degree of similarity to terrestrial conditions and thus the possibility of life.

The Boston Brahmin astronomer Percival Lowell thought he observed a network of canals spanning the surface of the Red Planet, implying the existence of advanced civilization on our neighbor world, and wrote a series of popular and extremely influential books and magazine articles on the subject early in the twentieth century. Lowell's canals were an optical illusion, but by the time this was proved by later observers the idea of Mars as an abode for life, and possibly intelligent life, had become deeply rooted in the public mind.

The hope of finding life on Mars took a series of blows in the 1960s, when NASA's Mariner 4, 6, and 7 probes flew by the planet and photographed Moon-like craters in its southern hemisphere. However, in 1971, the tide turned again when the Mariner 9 orbiter photographed the entire planet and discovered not canals, but networks of dry riverbeds, primarily in the northern hemisphere. Mars may be cold and desiccated today, but Mariner 9 data argued forcefully that at least in the distant past it was warm and wet enough for liquid water.

In 1976, NASA sent the Viking mission, consisting of two landers and two orbiters, to Mars. The orbiters confirmed Mariner 9's dry river networks with even better photographs. The robotic landers tried to search the surface for life, but obtained inconclusive results. Both landers carried biology packages containing three experiments that searched for life by collecting soil samples, culturing and incubating them, and detecting changes in gas concentrations within the experimental packages. All experiments detected these changes when Martian soil was inserted. This could certainly be viewed as a positive result. However, when directly examined by a gas chromatograph–mass spectrometer (GCMS), Martian soil at both landing sites showed not a trace of any organic material. This suggested to most of the Viking scientists that the gas release from the culture media was caused not by life, but by chemical reactions between the media and the soil. A minority, however, pointed out that the results of the GCMS did not preclude the existence of very scarce bacterial spores in the soil, which then multiplied rapidly in the favorable conditions available in the culture medium to produce a measurable gas release signal. Given its limited repertoire of instruments and experiments, Viking could provide no additional data capable of resolving the dispute.

However, in August 1996, a team of NASA and university scientists unveiled strong evidence for the existence of ancient bacteria within a 3.6-billion-year-old rock sample that had been ejected from Mars 16 million years ago by meteoric impact. The evidence includes actual organic molecules; magnetite, pyrrhotite, greigite, and other minerals that are typically the residues of bacterial activity; carbonate materials that strongly imply that the rock was immersed in a life-friendly aqueous environment on ancient Mars; and images of microstructures that resemble fossilized bacteria. While attacked by skeptics, the NASA-university team, consisting of Dave McKay, Everett Gibson, Kathie Thomas-Keprta, Richard Zare, and Chris Romanek, has offered powerful rebuttals to defend their sensational findings.

There is no doubt at all that the rock examined by McKay-Gibson et al. (known as Alan Hills 84001, or ALH84001, for the Antarctic location and year in which it was found) comes from Mars—its isotopic composition data provide a precise match with that found on the Martian surface by the Viking landers. Alternative nonbiological explanations can be offered for each of the team's data sets: Polycyclic aromatic hydrocarbons (PAHs), magnetite, and fossil-like rock structures can all be formed without life being present, and carbonates can be formed at very high temperatures without liquid water. But the thing that makes the data so compelling is that in ALH84001, all of these phenomena are found within microns of each other, making life the simplest explanation for their combined appearance. Indeed, the nonbiological explanation for some of the data is hard to support in the presence of the rest. For example, it's hard to see how the PAHs and greigite found in ALH84001 could have survived the high-temperature (450°C) shock heating processes suggested for nonaqueous carbonate formation.

In the period shortly after their August 1996 press conference, the putative microfossils displayed by the McKay-Gibson team were attacked as being too small to represent fossilized bacteria. It was even argued that there exists a fundamental minimum to the size required for complex biological structures and that it is impossible for any living microbes to exist on such a tiny (0.4 micron) scale. However, since that time actual living bacteria known as nanobacteria have been found on Earth in precisely this size range. While this does not prove that the structures observed in ALH84001 actually are microfossils, it certainly destroys the argument that they *can't* be.

The Alan Hills meteorite results have thus held up fairly well. The only way the Mars-life evidence will ever be fully confirmed, however, is via a direct discovery of fossils or extant life on the Martian surface itself. The importance of such a find can hardly be overstated, as it will put one of the last nails into the coffin of the Ptolemaic Earth-centered universe. To be completely clear, finding bacteria or fossils of extinct microbes on Mars is not important because it shows that Mars has or once had bugs. In itself, such a finding would merit little more than a footnote in the encyclopedia of science. No, the reason such a discovery would be of world-historic importance is that it would demonstrate that the processes leading to the development of life are not particular to Earth and in fact are highly probable wherever and whenever appropriate physical and chemical conditions exist. Currently, the only example we have for life is that found on Earth, so we have no way of knowing whether life's appearance on our planet was a sure thing based on natural self-complexification of organic chemistry or a one-in-a-trillion shot based on divine intervention or the nearly equally miraculous, extremely improbable chance formation of the superbly designed self-replicating DNA molecule. Finding life or its remains on Mars would decide this issue.

Astronomers have recently discovered some twenty extra-solar planetary systems, and as a result we now know that the processes that lead to planet formation around stars are non-exceptional. All theories of planetary system formation based on such unlikely events as collisions between stars have thus been shown to be false. Instead, some form of nebular hypothesis involving the formation of planetary systems as integral with the process of star formation must be true, which strongly implies that most stars have planets. There are 400 billion luminous stars in our galaxy alone, and every one of them has an appropriate zone, near or far depending upon the brightness of the star, where the right temperatures for liquid water and thus the development of life obtain. Therefore, if we can show that the processes that lead to life's appearance are also non-exceptional, it means that life is everywhere. Furthermore, the entire history of life on Earth shows a continuous tendency on life's part to evolve from simple forms to ever more complex and energetic forms capable of greater degrees of activity and intelligence. Therefore if life is everywhere, intelligent life is nearly everywhere. If we find fossil bacteria on Mars, it means that we are not alone. Except for finding extant life or the actual direct detection of extraterrestrial intelligence, no discovery may ever mean as much in telling us who we are.

We may go to Mars as Americans, Russians, and Japanese; if we find life we will come back as Humans.

THE NEED FOR NATIONAL AND INTERNATIONAL INITIATIVES

The establishment of human beings on Mars is the decisive step in the transformation of humanity from a single planet to a multi-planet species.

As such, it is the most important task facing our generation. Unfortunately, it won't be done by the invisible hand of market forces playing out their own logic. If a new branch of human civilization is going to be established on Mars, it is going to have to be accomplished by institutions, organizations, or movements willing and capable of undertaking great projects for reasons that transcend near-term cash profits. In the modern world, the foremost examples of such institutions are national governments.

It is currently fashionable to deride the capability of governments. Government projects are generally inefficient, and it is almost always the case that private companies can get a job done much more cheaply and quickly than government agencies attempting equivalent work. Unfortunately, private companies won't undertake any project unless there's the promise of profit, which is why governments are generally needed to mobilize a society's collective will to accomplish unprofitable but necessary tasks ranging from national defense to harbor dredging. For this reason, substantial government support has always been necessary in the risky initial exploration and settlement of new and unknown lands. From the practical point of view of opening Mars to mankind, the need to mobilize government in its traditional role in support of exploration cannot be ignored. Modern governments possess enormous resources, which today are largely dissipated on a plethora of projects with little long-term significance. If deployed correctly, a tiny percentage of those resources would be more than adequate to fund an aggressive program that could rapidly open the Red Planet to humanity. Sadly, that is not currently being done.

Instead, government-funded Mars exploration has been limited to two relatively low budget robotic efforts: one Soviet, the other American. The Soviet (later Russian) program has been an unmitigated failure. Out of more than a dozen orbiter and lander spacecraft launched toward Mars by the Soviets, not one has succeeded in accomplishing its mission. The American effort, led largely by the Jet Propulsion Lab, has had a much better record. Out of the thirteen American Mars spacecraft launched since 1965, no less than ten, including three fly-by spacecraft, four orbiters, and three landers, have been successful. In addition to the Mariner and Viking spacecraft already discussed, these have included the *Pathfinder* lander and the Mars Global Surveyor (MGS) orbiter, both of which reached Mars in 1997. The *Pathfinder* mission was a spectacular success. Accomplished with a cut-rate budget of $175 million, a young team, and a risky, uncontrolled bounce-and-tumble air-bag landing system, *Pathfinder* landed smack in the middle of an ancient water runoff channel on Mars on July 4 and successfully deployed a small (10-kg) robotic rover named *Sojourner* (after the anti-slavery heroine Sojourner Truth). For the next two months, *Sojourner* wandered about the landing site examining rocks with its camera and alpha-proton–x-ray spectrometer chemical analysis instrument. In the process it found rounded pebbles similar to those created by water erosion in terrestrial streambeds, as well as conglomerate rocks, which are also evidence of the past presence of liquid water. These travels, which were followed with enthusiastic attention in much of the world's press, ended only when a power failure on the *Pathfinder* lander terminated *Sojourner*'s radio communication link with Earth. (Some think that for days or weeks afterward, little *Sojourner* continued to circle around the silent lander listening for new instructions, like a forlorn dog hoping that its slain master would awake. That's what its programming would have dictated. Maybe some day astronauts will visit the site and, by examining the tracks, be able to tell us the true story of the last days of the orphaned rover.) Meanwhile, Mars Global Surveyor slipped into Mars orbit and, despite a number of technical glitches with its solar panels, began to photograph Mars with a camera offering roughly seven times better resolution (about 1.4 meters per pixel) than the best previous (Viking) photos (roughly 10 meters per pixel.) The MGS images have revealed the dry beds of what were once without question meandering river channels, thereby proving beyond doubt that liquid water not only once existed on Mars but it cycled there for significant periods of geologic time. In addition, MGS's laser altimeter discovered a large topographically depressed and relatively uncratered area in Mars' arctic region that is flatter than any feature on Earth except the ocean bottom. Apparently, Mars not only once had rivers, it had oceans too.

Like *Pathfinder,* MGS was also a cut-rate spacecraft, built and flown for a budget of $150 million. The success of these two units appeared to confirm the wisdom of NASA Administrator Dan Goldin's call for designing interplanetary missions along small-scale, "faster, better, cheaper" lines. In 1999, NASA sent two more inexpensive spacecraft to Mars. One of these, an orbiter carrying an advanced infrared instrument for studying Mars' atmospheric motions, was lost while attempting to enter Mars orbit in September 1999. The other, a small lander, is scheduled to land on Mars' south polar cap in December 1999.

However the limits of this approach are becoming all too clear. In May 1998, NASA had to announce that it was canceling plans to fly the larger (80-kg) well-instrumented *Athena* rover to Mars in 2001 because the "faster, better, cheaper" robotic Mars exploration program budget of $250 million per year (2 percent of NASA's total budget) could not afford it and the small landers dictated by the program's small budget could not carry it (especially so, after the administration pulled $60 million from the 2001 Mars mission budget to help pay for a space station overrun). Instead, the 2001 lander would have no rover at all, not even a Viking-style robotic arm. A small orbiter, similar to the 1998 unit, is still scheduled to fly in 2001, equipped with a gamma-ray spectrometer (GRS) for assessing surface chemical composition. This promises some important remote sensing data in 2001, but had the program office been able to spend an extra $10 million to buy a slightly more powerful Delta II launch vehicle, the 2001 orbiter would have been able to carry both its GRS and the infrared weather detector, thereby allowing it to back up and retrieve the science lost with the 1999 climate orbiter. The JPL robotic Mars exploration office claims that it will be able to use the pennies it will save by gutting the 2001 lander mission to allow *Athena* to be launched to Mars in 2003, followed by a Mars Sample Return (MSR) mission in 2005. I can believe the former claim but not the latter. At $250 million per year, the JPL Mars exploration program simply does not have enough money to fund two robotic missions to Mars every two years and finance the major new technology development required for the MSR mission at the same time.

Given President Clinton's promise of August 1996 that he would "put the full intellectual and technological might of the United States behind the search for life on Mars," this situation can only be described as ridiculous. Within its limits, the robotic Mars exploration program has been extremely productive, yet it is starving for funds while the rest of the federal government feasts on the national budget surplus. The robotic Mars exploration program needs to have its budget doubled if it is to progress at a reasonable rate, and this should be done forthwith. In 1998, NASA spent $600 million on a Space Shuttle mission whose primary accomplishment was to kill two dozen rats (that's about $25 million per rat) in zero gravity with a special spring-loaded miniature guillotine (ordinary weight-driven guillotines won't work in zero gravity). To make such expenditures while denying adequate support for capable Mars probes shows an incredible misallocation of priorities.

That said, the real limitations of robotic Mars exploration need to be understood. Photographic surveys from orbit or balloons, seismology, meteorology, and limited geochemical contact science can all be accomplished on Mars with robotic devices. But searching for fossils, let alone extant life, requires skills and versatility of an entirely different order. On Earth, fossil hunting requires hiking long distances through unimproved terrain and climbing up difficult slopes. It requires doing heavy work—digging trenches and wielding a pickax. It requires delicate work as well, such as carefully splitting open the thin layers of fossil shales edgewise to look for traces of past life pasted between the stone pages. It also requires very complex modes of adaptive perception. All these abilities are far beyond the capability of robotic rovers. The *Sojourner* rover had no manipulative abilities at all, could move only 2 or 3 meters per day, and was incapable of climbing over a rock 20 centimeters (cm) tall. Its "eyes" were positioned less than 30 cm off the ground and could produce a few grainy black-and-white images a day. Even the greatly improved *Athena* rover would have been able to travel just 100 meters or so per day and would be stopped dead by a boulder field, sand dune, or rocky hillside that would be child's play for any normal pedestrian between the ages of five and seventy. America's Rocky Mountain states abound in dinosaur fossils. Yet it is a fair bet that thousands of *Sojourner* and *Athena* rovers could be parachuted into the Rockies without ever finding a dinosaur fossil—at least not before the arrival of the next ice age, at which time they would be crushed by the advancing glaciers, which they would not be able to outrun. If brought to a paleontological dig on Earth, these high-tech devices would at best find use as tables upon which coffee cups could be placed. No, if you are serious about hunting fossils, real live human rockhounds are needed.

Looking for extant life on Mars, the demands become even greater. If there is any life on Mars today, it is most probably in liquid groundwater several kilometers below the surface. Getting to it will require drilling. Without question, this means human crews, as the complex operations required to set up a drilling rig in the field totally rule out robotic systems. (If you doubt that, just ask anyone who has done oil drilling on Earth. It will be a lot harder on Mars.) And we do want to look for extant life on Mars. Finding fossils will tell us whether or not we are alone. Finding living organisms will tell us a lot more. Right now we have only one example of life—the terrestrial kind—and so we have no way of knowing which features of life that we observe are peculiar to Earth and which form the general basis of life more broadly. We are like linguists whose only experience with language is our own; we might know what Ainu is, but we don't know what language is, and therefore we haven't got a clue as to what kinds of poetry might be possible. Similarly, today we know what *Earth life* is, but we don't know what *Life* is. It's worth finding out. At the very least, finding out what Life is could lead to all sorts of advances in genetic engineering, biotechnology, and medicine. It could very well lead to whole new sciences. If we want this knowledge, humans will have to go to Mars.

But the most important question about the Red Planet is not "Was there life on Mars" or even "Is there life on Mars?" It's "Will there be life on Mars?" That is, can humans go and establish a new branch of human civilization on a new world? Can humanity become a multi-planet species? Can we make the leap from Type I to Type II? Have we reached the "end of history" or the birth of a new age bursting with hope and infinite possibilities? This, Mars' most important question, can only be answered by human pioneers.

Robots are useful auxiliaries to Mars exploration, but fundamentally they are beside the point. We need a humans-to-Mars program. We don't have one.

The best chance in a generation to get a government-led humans-to-Mars program launched will probably occur in the spring and summer of 2001—the first season of the first term of the first U.S. administration of the new millennium. It will be an obvious time for new beginnings of grand projects, and a humans-to-Mars program launched by a new president in the spring of 2001 could make its first piloted landing by 2008, before the end of the administration's second term. The symbolism of the need for new starts in the new millennium will, if anything, be even stronger in Europe.

For five hundred years after Rome fell, no one built anything of consequence in western Europe. No one built any stone structures more than two stories high. Metaphorically speaking, no one planted any trees, because no one believed there was a future. Then with the turn of the millennium, people looked around and saw they had survived the barbarian invasions, the Huns and the Moors, the Vikings, and the threat of the world's supernatural destruction on the millennium itself. The realization hit them: "We made it!" And knowing there would be a future, they started building cathedrals.

Similarly, humanity, and most especially Europe, has just survived what may have been the most dangerous century in human history. The twentieth century has been filled with chaos—two world wars, the Great Depression, fascism, communism, the Cold War, and at least one near-miss at nuclear war. In retrospect it seems incredibly fortunate that civilization muddled through. But it did, and here we are on the brink of a new millennium with institutions being put in place that should preclude another general European war. In 1959, when the movie *On the Beach* was made, it was felt to be a realistic projection that humanity would extinguish itself in thermonuclear war—by 1964. We're past that kind of pessimism now.

There's going to be a future. Unless world events take a drastic unexpected turn for the worse in the next few years, when the millennium turns there is going to be a strongly—and rightly—felt sense that it is time to build cathedrals once again.

What better cathedral, what more profound statement of our faith in the importance of the future could there be than the establishment of mankind's first outpost on a new world?

The year 2001 will be one of unique opportunity. We need to seize the time. But how?

THE MARS SOCIETY

In 1996 I published *The Case for Mars,* advocating the launching of a humans-to-Mars initiative. Over the next two years, I received over 4,000 letters and e-mails from people worldwide who read the book, with the large majority asking the same question in so many words, "How do we make this happen?" On July 4, 1997, NASA landed *Pathfinder* on Mars, and the NASA Mars Web site was visited 100 million times. One hundred million hits in one day! That is more than the number of people who vote in the United States. It is more than the number of people who are actively for or against abortion, gun control, a balanced budget, and national health care, combined. In the six months following the landing, another 700 million hits occurred. Even assuming a fair rate of repeats, this suggests that more people care about opening the Martian frontier than care about all the political fetish issues that the Beltway pundits and image makers advise on and use to mold their client politicians' public profiles. This is as it should be; a hundred years from now, no one will either know or care what our gun control, abortion, balanced budget, or national health care laws were. But what we do to start the spacefaring stage of human history by opening Mars will be of supreme importance.

In a way, it is surprising that the politicians are not more aggressively pursuing a humans-to-Mars program. Taking the lead in such an initiative would offer a statesman the chance for true immortality. On the other hand, who today can remember any of the many big shots who said no to Columbus? Who today would remember Isabella had she not sponsored the Italian navigator? Can you even name the previous queen, or the one that followed? In her time, Isabella probably thought herself significant because she was a powerful monarch. In reality, her only enduring importance stems from her support for the exploratory vision of a wandering son of a Genoese weaver. Similarly, by avoiding the challenge of launching humanity's career on Mars, the current generation of politicians are condemning themselves to ultimate obscurity. But just as "the cones of silence" in the old comic TV series *Get Smart* prevented the transmission of sound to anyone within, so the gangs of Beltway makeup men have created analogous barriers around their clients through which important ideas penetrate only with the greatest of difficulty. A determined effort will be needed to break through the cones.

The people miss the frontier more than words (alone) can ever express. They need to express their feelings with votes, funds, letters, lobbying, and every other type of political pressure. Organization is needed.

In *The Case for Mars* I suggested that people deal with this problem by joining existing space activist organizations, such as the Planetary Society, the National Space Society, or the Space Frontier Foundation. However, since that time it has become apparent that none of these organizations has the determination, resolve, or focus required to launch and sustain an effort to pressure the system into initiating a humans-to-Mars program. Therefore, in the spring of 1998, I joined with other prominent advocates of that goal, including Mars Underground founders Chris McKay and Carol Stoker, former NASA Associate Administrator for Exploration Mike Griffin, my *Case for Mars* co-author Richard Wagner, and noted science fiction authors Kim Stanley Robinson and Greg Benford, in launching the Mars Society, an international organization dedicated to furthering the exploration and settlement of Mars by both public and private means. My wife, Maggie, also signed on and proved invaluable in getting things going. This organization became active in May 1998 and almost immediately scored a significant success in June by launching an e-mail barrage to the Senate Appropriations Committee, causing it to restore $20 million (of the $60 million looted by the administration) to the 2001 Mars mission. While this was not enough to get the *Athena* rover back on the mission, it was sufficient to allow the smaller *Marie Curie* rover (*Sojourner*-sized, but with some of *Athena*'s superior instrumentation) to fly. Then, in August 1998, we held our Founding Convention in Boulder, Colorado. This conference was a smashing success with over seven hundred gung-ho attendees from around the world, including representatives from the United States, Canada, Jamaica, Mexico, Brazil, Argentina, the United Kingdom, Ireland, France, Germany, Switzerland, Holland, Sweden, Spain, Italy, Austria, the Czech Republic, Poland, Rumania, Greece, the Ukraine, Russia, China, Japan, Indonesia,

Australia, New Zealand, Israel, Egypt, and Mozambique. Every NASA center, every national lab, and nearly all the top universities in the United States were represented. The conference received extensive coverage in leading world media, including the *New York Times,* the *Washington Post,* the *Boston Globe,* ABC-Discovery News, the CBC, as well as *Discover, Reason,* and *Popular Mechanics* magazines. Following the conference, a series of initiatives were launched, including establishing over seventy chapters in most of the countries represented.

As mentioned above, the Mars Society intends to go beyond supporting government Mars programs—we intend to launch privately funded exploration of the Red Planet as well. This might seem impossible, because a business plan for such a venture conducted on a for-profit basis today would be weak. But there is a successful alternative model: the ocean exploration program of Jacques Cousteau. The idea is to raise enough money to do something real, however small, and on the basis of that success be able to raise funds for ever more ambitious exploration ventures. Consider those 700 million hits on the NASA Web site following the *Pathfinder* landing. At $1 each, this would be enough to fund four *Pathfinder* missions. At $10 each, that's $7 billion, enough to send humans to Mars, if done in the private sector. The point is not that $7 billion could be raised by the mechanism of charging $10 per hit for admission to a Web site—it almost certainly can't. The point is that the monies are there, in the supporting general public, to enable a privately funded program of human Mars exploration—and settlement, for that matter. The key is to implement a program of escalating activity that would allow the Mars Society to gain the credibility required to mobilize the vast level of potential funding represented by the broad-based support demonstrated by the 700 million *Pathfinder* hits.

Unless we are unexpectedly lucky with early fundraising, this program needs to start small. The Mars Society conference resolved on two early initiatives. The first, which appears fundable for about $1 million, is to build a simulated Mars exploration base in the Arctic. There is a location, known as the Haughton Impact Crater, on Devon Island at 75 degrees North, that features geology that greatly resembles Mars as well as a cold-dry climate that is not too dissimilar. NASA scientists have been investigating this area for some time in order to use it to explore Mars by analogy. On the suggestion of Pascal Lee, one of the scientists involved in this work, the Mars Society decided to build a base in the area, including prototype Mars habitation modules equipped with suitable field lab instrumentation. This base can be used to support the ongoing field work and as a test bed for Mars life support equipment, field mobility vehicles, permafrost-penetrating drill rigs, astronaut training, and many other invaluable purposes. Current plans are to have the Arctic Mars base up and running by the summer of 2000. The next early initiative, which will probably cost about $10 million, is to build a hitchhiker payload to fly on a NASA or European Mars mission in 2003. One good candidate for this hitchhiker could be a balloon equipped with a camera to enable aerial photography on Mars, although several other options are under consideration at this time as well. If a private organization were to be successful in implementing any mission of this type, the public excitement generated all over the world could well be sufficient to allow much larger funds, on the order of the $100 million required to fund a fully private robotic Mars exploration mission. And if a private organization were to do that, billions could become available, enough to fund human Mars exploration and, eventually, settlement, either by the Society acting alone or perhaps on a cost-sharing basis with NASA or another government space agency.

Thus, the Mars Society could evolve, from exploration advocates, to an exploration and then settlement agency, ultimately becoming the midwife to the birth of a new branch of human civilization on a new world. Such settlement programs initiated by private organizations have happened in the past on Earth and can happen again on Mars, because the real driving force in human history is not the profit motive but the power of ideas. And the opening of Mars to humanity is an idea whose time has come.

But history is not a spectator sport. The talents and energy of every person who supports this ideal are needed if it is to succeed. If you believe in its goals, then the Mars Society needs you. You can join or find out more through our Web site at www.marssociety.org or by writing: Mars Society, Box 273, Indian Hills, CO 80454.

I hope we hear from you.

THE SIGNIFICANCE OF THE MARTIAN FRONTIER

We have come recently to boast of a global economy without thinking of its implications, of how unfortunate we are in finding it. It would be more cheering if news should come that by some freak of the solar system another world had swung gently into our orbit and moved so close that a bridge could be built over which people could pass to new continents untenanted and new seas uncharted. Would those eager immigrants repeat the process they followed when they had that opportunity, or would they redress the grievances of the old earth by a new bill of rights. . . ? The availability of such a new planet, at any rate, would prolong, if it did not save, a civilization based on dynamism, and in the prolongation the individual would again enjoy a spell of freedom . . . The people are going to miss the frontier more than words can express. For four centuries they heard its call, listened to its promises, and bet their lives and fortunes on its outcome. It calls no more.

—WALTER PRESCOTT WEBB
The Great Frontier, 1951

Western humanist civilization as we know and value it today was born in expansion, grew in expansion, and can only exist in a dynamic expanding state. While some form of human society might persist in a non-expanding world, that society will not feature freedom, creativity, individuality, or progress, and, placing no value on those aspects of humanity that differentiate us from animals, it will place no value on human rights or human life as well. Such a dismal future might seem an outrageous prediction, except for the fact that for nearly all of our history most of humanity has been forced to endure precisely such static modes of social organization, and the experience has not been a happy one. Free societies are the exception in human history—they have only existed during the West's four centuries of frontier expansion. That history is now over. The frontier that was opened by the voyage of Christopher Columbus is now closed. If the era of Western humanist society is not to be seen by future historians as some kind of transitory golden age, a brief shining moment in an otherwise endless chronicle of human misery, then a new frontier must be opened.

Humanity needs Mars. An open frontier on Mars will allow for the preservation of cultural diversity, which must vanish within the single global society rapidly being created on Earth. The necessities of life on Mars will create a strong driver for technological progress that will produce a flood of innovations that will upset any tendency toward technological stagnation otherwise inevitable on the unified mother planet. The labor shortage that will exist on Mars will function in much the same way as the labor shortage did in nineteenth-century America—driving not only technological but social innovation, increasing pay and public education, and in every way setting a new standard for a higher form of humanist civilization. Martian settlers, building new cities, defining new laws and customs, and ultimately transforming their planet, will know sensuously and prove to all outside observers that human beings are the makers of their world and not merely its inhabitants. By doing so they will reaffirm in the most powerful way possible the humanist notion of the dignity and value of mankind.

Mars beckons. In taking on the challenge of Mars, humanity will take the essential leap that will transform it from a Type I to a Type II species.

FOCUS ON: THE RIGHTS OF MARS

Why would anyone but a few scientists want to go to Mars, to live in domes and face water rationing and numerous other privations that will undoubtedly accompany the early years of Martian settlement? Well, people will emigrate to Mars, despite any material hardship and personal risk, if by so doing they obtain a higher level of freedom. Such freedom can be created if Martian law is made to embody a deeper and more far-reaching notion of human rights than any currently existing on Earth. Thus, as the young United States did in its time, Mars can serve as another "noble experiment," in which a more progressive version of law is introduced than that prevailing or considered feasible by sophisticated people in previous societies. Mars will succeed, both for itself and for all mankind, if it can retain and innovate further the best forms of law, culture, and society Earth has to offer and leave the worst behind.

Let's therefore start with the best, the rights of man embodied in the U.S. Declaration of Independence, Constitution, and Bill of Rights. The Declaration of Independence makes a commitment to the general notion that "all men are created equal and are endowed by their creator with certain inalienable rights, among them life, liberty and the pursuit of happiness." This commitment is elaborated in the Constitution, Bill of Rights, and subsequent amendments and laws as including:

1. Freedom of Religion, Assembly, Speech, and of the Press
2. The Right to Bear Arms
3. The Right to Due Process and Trial by Jury
4. The Right to Face One's Accusers

5. The Right to Be Free of Arbitrary Arrest or Long Imprisonment without Trial
6. The Right to Vote for Representative Government
7. The Right to Own Property
8. The Right to Be Free of Chattel Slavery
9. The Right to Equal Protection under the Law regardless of Race, Creed, Color, or Country of National Origin

In addition, there is an emerging consensus and body of law tending toward the establishment of an additional right, which can be phrased as follows:

10. The Right to Equal Opportunity Regardless of Race or Sex

This is the best that twentieth-century Earth has to offer. Mars must therefore include it but move beyond it. I therefore propose that Martian law also incorporate the following as fundamental human rights:

11. The Right to Self-Government by Direct Voting
12. The Right to Access to Means of Mass Communication
13. The Right to All Scientific Knowledge
14. The Right to Knowledge of all Government Activities
15. The Right to Be Free of Involuntary Military Service
16. The Right to Immigrate or Emigrate
17. The Right to Free Education
18. The Right to Practice Any Profession
19. The Right to Opportunity for Useful Employment
20. The Right to Initiate Enterprises
21. The Right to Invent and Implement New Technologies
22. The Right to Build, Develop Natural Resources, and Improve Nature
23. The Right to Have Children
24. The Right to a Comprehensible Legal System Based on Justice and Equity
25. The Right to Be Free from Extortionate Lawsuits
26. The Right to Privacy

This list will no doubt be controversial, both by what it includes and by what it omits. While we don't have enough room here to adequately consider all (or any!) of the inclusions or omissions, let's briefly discuss a few.

Consider rights 11 through 14 and 24 through 26, whose purpose is to establish an actual democracy—of the people, by the people, for the people. In America today, people do have individual rights, elaborated by items 1 through 10, which protect them from various kinds of abuse. But we do not, in fact, have a democracy. We have a semi-oligarchy with democratic influences. Ordinary citizens have little control over the government, as their elected representatives mostly do as they please or as their Beltway consultants suggest and respond to the public only when massive pressure is evidenced. In addition, many government operations are secret, and the legal system is unfathomable. Of course, when the United States was founded, such indirect representation was the best approximation to democracy that was feasible. But today, with the availability of the Internet and other forms of instantaneous electronic communication, there is no fundamental technological reason why the general public could not directly engage in voting on legislation, taxation, expenditures, and other issues, up to and including those of war and peace. It might be argued that the general public is not qualified to do so. Personally, as one who has interacted with some of those calling the shots within the present system, I see no evidence for the public's inferiority. Such skepticism of the people's capacity to engage in direct government is reminiscent of similar skepticism offered by sophisticated European observers of the practicality of the Founding Fathers' notions of the viability of representative democracy, freedom of religion, freedom of the press, the right of the people to bear arms, trial by jury, and so on. To the establishment eighteenth-century mind, all of these concepts were prescriptions for chaos. It took a "noble experiment" in a new land to prove their viability. Until that was done, it was impossible to implement most of them in Europe. Similarly, the representative system in the United States will never yield to actual democracy until the latter is proved somewhere. For that, a new "noble experiment" will be necessary.

Rights 15 through 23 have all existed, explicitly or implicitly, to one extent or another, at one time or another, in the United States and many other countries. Many of these, however, have become significantly constricted in many nations in recent years, and there will be increasing pressure to do so further on Earth, especially if the Pax Mundana should result in a stagnant, zero growth world. A Martian civilization that offered these as fundamental rights could become the magnet for the dreams and hopes of millions.

But are they practical? Can the public really govern itself? Judge itself? Can a society really function with people free to practice professions without our present set of caste or guild-like certification systems? Can a society function with people free to invent and implement new technologies without interference by government regulators? Perhaps there are other rights that should be added. Could a society function and progress with people guaranteed food, clothing, shelter, medical care, and other necessities as fundamental rights?

I don't know, and I don't think anybody else does either. To find the answers, a lot of noble experiments will have to be run, with various combinations of these or other rights. The ones that work will lead to societies that succeed, and thus be remembered, and copied.

Like Hegel, who believed that the efficient Prussian bureaucracy was the ultimate development of human government, Professor Fukuyama believes that current representative systems are the final answer. I disagree. I don't know what it is, but I'm sure we can find better. And our best chance of finding it, of taking our next giant step, will occur when once again, in a new and as-yet-unstructured land, a group of serious-minded people equipped with the experience and most advanced thought of their era gather around a blank sheet of paper, and with all the force of their reason, begin to write:

"We hold these truths to be self evident . . ."

CHAPTER 7

Asteroids for Good and Evil

I would find it easier to believe that two Yankee professors would lie than that rocks should fall from the sky.

—THOMAS JEFFERSON, 1807

All civilizations become either spacefaring or extinct.

—CARL SAGAN

IN THE LINE OF FIRE

We're being bombed.

On October 1, 1990, U.S. strategic defense forces detected an explosion in the central Pacific with a power equal to the 10-kiloton atomic bomb that leveled Nagasaki at the end of World War II. The French government, it is known, sometimes tests nuclear weapons in the Pacific, but this was not one of theirs. No Earthly power was behind it. This missile came from outer space. Exploding in the air above the uninhabited central Pacific it caused no known casualties. However, had it arrived some ten hours later, it would have detonated in the Middle East, right in the midst of a powder keg loaded with a vast U.N. coalition and Iraqi armed forces then preparing for all-out war.[1]

The October 1, 1990, projectile came from the enemy's lightest artillery, and it flew wide of any vital target. But there's a lot more where that came from, and we've been hit by far worse. And not so long ago.

For example, on February 12, 1947, a larger extraterrestrial projectile with a 100-kiloton explosive force (ten times the Nagasaki bomb) impacted less than 400 kilometers from the Russian city of Vladivostok. This may have caused casualties among the large labor camp population that adorned the region at that time. Records of the missing, and the reactions of the no doubt bewildered authorities, may perhaps still be found in musty Stalinist archives.

A much larger hit occurred on June 30, 1908, when an extraterrestrial warhead with a force of 20,000 kilotons (20 megatons!) hit Tunguska, Siberia, leveling thousands of square kilometers of forest. Thousands of deer were killed, and probably hundreds of primitive Siberian hunters whose lives were beneath the notice of the Tsarist regime as well. Had the object arrived only three hours later, it would have devastated a large area of European Russia and possibly caused the cancellation of one of the social events at the Tsar's summer palace.

These three well-documented impacts are all from the twentieth century. Similar events doubtless occurred at a similar rate in preceding centuries, but due to the prevalence of mysticism, illiteracy, lack of scientific outlook, and poor global communications in prior times, humanity's capability to reliably observe, communicate, and record earlier impacts was very weak. Perhaps some of the cataclysmic events described in ancient religious and mythological texts, such as the destruction of Sodom and Gomorrah, are folkloric accounts of asteroid impacts. It's hard to say, but since both asteroids and the true origin of meteorites were unknown until the nineteenth century, an earlier impact event could probably only be explained by its surviving observers as a manifestation of the rage of God.

But if pre-modern humanity was an unreliable recorder of asteroid bombardment, there was another scribe on the scene whose accounts, while incomplete, never lie—the Earth itself. The geological record shows without room for doubt that the Earth has been the target for massive asteroid bombardment throughout its history (see Figure 7.1).

Of course, to make it into the Earth's record, an asteroid hit has to be something *significant,* not a mere A-bomb- or H-bomb-sized blast like October 1990 or Tunguska. For example, the kilometer-wide, 200-meter-deep crater in northern Arizona visited by many tourists today is the scar left by the impact of a 150-meter-diameter asteroid that hit the Earth 50,000 years

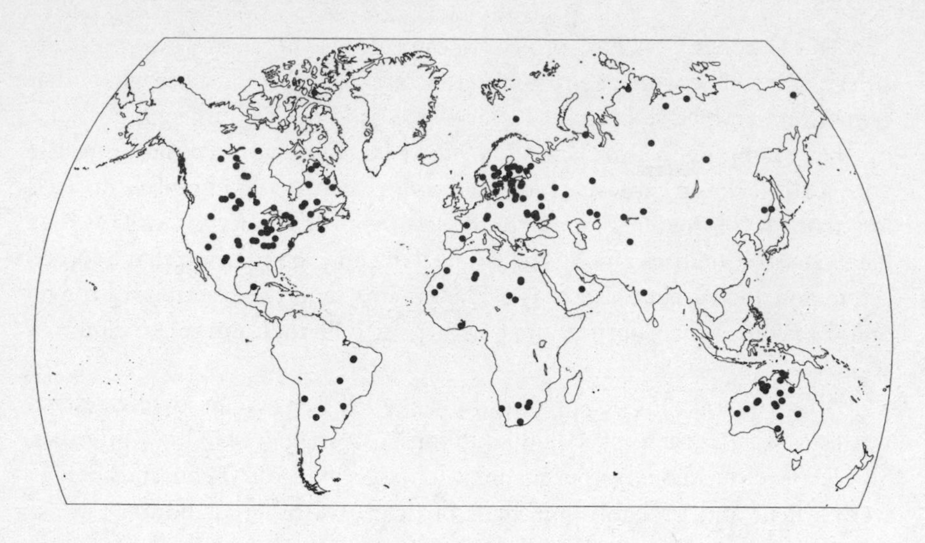

FIGURE 7.1 *Known asteroid impact craters are scattered all over the Earth. (Courtesy Richard Grieve, Geological Survey of Canada)*

ago with a force of about 840,000 kilotons (84,000 Nagasaki bombs) releasing roughly the same explosive power as would have been detonated had an all-out nuclear war occurred between NATO and the Warsaw Pact at the most heavily armed phase of their balance of terror. The impact probably wiped out much of the life in the American Southwest and raised enough dust to throw the entire world into a "nuclear winter" deep freeze for several years.

But the 150-meter rock that hit Arizona was still a *small* asteroid. There are others out there with masses that are thousands and even *millions* of times as great. Sometimes they hit, too. A rough guide to how much force is released when that happens and the estimated frequency of such events is given in Table 7.1.

The asteroid impact of 65 million B.C. is the most famous of the large events, as its discovery has led to a revolution in our understanding of the history of life on Earth. Prior to the 1980s, scientists believed that the evolution of species followed a gradual course, driven only by interaction among species and between life and the Earth's climate. Then, in 1980, the team of Nobel Prize–winning physicist Luis Alvarez and his son, paleontologist Walter Alvarez, discovered a thin layer of iridium in Italian sediments located precisely at the 65-million-year-old boundary between the Cretaceous era, in which dinosaurs dominated the Earth, and the Tertiary era, in which dinosaurs no longer existed.[2] Iridium is rare on Earth but relatively

▪ TABLE 7.1 ▪

Destructive Force of Asteroid Impacts

Asteroid Diameter	Explosive Yield (kilotons)	Frequency	Example
4 m	16	1/year	Central Pacific, October 1, 1990
8 m	128	1/decade	Vladivostok, February 12, 1947
43 m	20,000	1/400 years	Tunguska, June 30, 1908
150 m	844,000	1/6,000 years	Northern Arizona, 50,000 B.C.
1 km	240 million	1/300,000 years	South Pacific, 2.3 million B.C.
10 km	240 billion	1/60 million years	Yucatan, 65 million B.C.

common in meteorites, which represent samples of small asteroids. The sudden disappearance of the long-lasting, widely distributed, and widely varied order of dinosaurs had always been a mystery in paleontology. Discovering asteroid material in Italy deposited precisely at the time of the dinosaur's demise, and then finding it all around the world, the Alvarezes advanced the startling theory that an asteroid impact had wiped the giant reptiles out. Traditional paleontologists wedded to older ideas resisted the hypothesis, but the Alvarezes' case became conclusive in 1991 when Canadian geologist Alan Hildebrand discovered the crater left by the dinosaur's destroyer at Chicxulub in the Yucatan Peninsula of Mexico.

The size of the crater reveals the size of the impact. With an explosive force of over 200 billion kilotons of high explosive, the question is no longer what killed the dinosaurs. The real mystery is how anything else survived. First, the hypersonic shock created when the asteroid tore into the atmosphere turned the air to superhot plasma, instantly baking anything within line of sight of the entry trajectory. Then the asteroid slammed into the Earth itself, shooting vast amounts of ejecta into space, which later reentered at hypersonic speed and heated the entire atmosphere to incandescence. The glowing sky set fires to forests everywhere. Any life that could not find a hiding place underground or underwater was killed. The burning lasted only a few days, but afterward the dust released by the impact and consequent smoky fires caused intense, planet-wide lethal poisonous acid rain and sent the Earth into years or decades of a dust-shrouded deep "nuclear winter." Eventually the dust rained out, allowing the blessed sunlight to warm the Earth again, except that the impact and the fires had put a massive amount of carbon dioxide gas into the atmosphere, which the sparse vegetation of the post-holocaust world could do little to clean up. As a result, a powerful greenhouse effect was created, rapidly driving Earth into an intolerable hothouse that may have lasted for centuries.

Subsequent research has revealed that the dinosaurs' killer was not unique and that many, if not all, of the dozens of mass extinctions that define the successive ages of the Earth's paleontologic history were caused by asteroid impacts. The body count of these events is very high; during *each* of them, between 35 percent and 95 percent of all living species of plants and animals on the planet were exterminated. Over two-thirds of all species that have ever lived on Earth were wiped out by asteroids.

So be proud of your ancestors, who survived all this. They were *very* tough. But it seems unwise to keep pushing our luck. Take a look at Figure 7.2, which shows the orbits of known Earth-crossing asteroids. About 200 are known, but it is estimated that there are at least 2,000 of them with diameters greater than 1 kilometer and 200,000 bigger than 100 meters. About 20 percent of them will hit the Earth sooner or later.[3] In most cases, later means not for hundreds of millions of years. But not all. And somewhere out there, right now, there is an asteroid with your descendants' names written on it. It's not on a direct collision course; it will orbit the Sun

FIGURE 7.2 *The cosmic shooting gallery: Earth is in the line of fire. (Courtesy Richard P. Binzel, Massachusetts Institute of Technology)*

many times before it hits. But as complex as its path might be, if nothing is done the laws of gravity and celestial mechanics dictate that it will unerringly follow its prescribed course to bring death to your progeny and myriads of other species who passively await its coming.

We are targets in a cosmic shooting gallery. Prudence dictates that we take steps to rectify this condition.

THE ASTEROIDS

In 1781, British astronomer Sir William Herschel discovered Uranus, bringing the number of known planets to the mystical figure of seven. Contemplating this in 1800, Professor Hegel, the Fukuyama of his day, pronounced that Herschel's discovery represented the end of astronomy, as the inherent perfection associated with the number seven guaranteed that there could not be any additional planets. Unfortunately, however, the Sicilian astronomer Giuseppe Piazzi neglected to read Hegel and, on January 1, 1801, in blind ignorance of the Prussian philosopher's deep dialectical insight, proceeded to discover an eighth planet, which he named Ceres, orbiting the Sun between Mars and Jupiter.

While displeasing to Hegel, the discovery of Ceres was very gratifying to astronomers, particularly as its orbital distance from the Sun, 2.7 Astronomical Units (AU), was very close to the 2.8 AU predicted for a missing planet by the German astronomers Titius and Bode based on other planetary orbit distance ratios back in the 1770s (1 AU equals 150 million kilometers, the distance at which the Earth orbits the Sun). Ceres proved to be a very small planet, only 900 kilometers in diameter (about half that of the Earth's Moon), but this disappointment was compensated for when, over the next few years, three additional planetoids, Pallas, Juno, and Vesta, were discovered in similar orbits close by. This suggested to the German astronomer Heinrich Olbers that the four objects were fragments of a properly sized planet that had broken up. More fragments could therefore be expected, and as the nineteenth century wore on and telescopes got better, this hypothesis was substantiated as dozens and then hundreds of additional "asteroids" were discovered. By 1890 over 300 were known, all orbiting the Sun in a belt between 2 and 3 AU, nicely bracketed near the geometric mean of 2.8 AU between the orbit of Mars (1.52 AU) and that of Jupiter (5.2 AU).

But then, in 1898, an asteroid was discovered whose maximum distance, or aphelion, from the Sun is 1.78 AU and whose minimum distance, or perihelion, is only 1.14 AU. It thus *crosses the orbit of Mars* and sometimes swings within 20 million kilometers of the Earth. This wandering across orbits was considered truly errant behavior, and so to distinguish it from the well-behaved female deities orbiting the Main Belt so nicely, the new 10-km class asteroid was given a male name, Eros. Soon other male, planet-crossing asteroids were discovered, and some of these crossed not only Mars' orbit but that of Earth as well. In 1932, the asteroid Apollo crossed the Earth's orbit and passed within 10 million kilometers of our world. In 1936, Adonis passed us at a distance of 2 million kilometers. On March 23, 1989, the "small" (800 meters, or 120-million-kiloton impact energy) and therefore unnamed asteroid 1989 FC swept by at a distance of 720,000 km, passing through a point in space that the Earth had occupied less than six hours before.

As mentioned above, about 200 Earth-crossing, "male," or "near-Earth" asteroids larger than 1 km are known today, and it is estimated that there are at least 2,000 of them out there. Over 5,000 "female" asteroids are known to exist in the Main Belt between Mars and Jupiter, including every asteroid larger than 10 km, hundreds larger than 100 km, and one as large as 900 km. Because they orbit farther from both the Sun and the Earth, the smaller female Main Belt asteroids are much harder to see than their Earth-crossing counterparts. It's estimated that there are at least 2 million Main Belt asteroids larger than 1 km. The girls thus outnumber the boys by about a thousand to one, and utterly dwarf them as well. It's good the ladies are well behaved. If they acted like their brothers, all life on Earth would have been exterminated a thousand times over.

Of course there is a reason for the overwhelming dominance of female asteroids. Male asteroids don't live long. They kill themselves by crashing into us.

MOVING ASTEROIDS

So the good news is that all we have to worry about is a couple of thousand little boy asteroids with explosive yields of a few hundred million kilotons of dynamite each, plus maybe a few million more tiny fellows with yields of tens or hundreds of thousands of kilotons. Possibly because political office offers little protection against asteroid impacts, Congress has found this threat sufficiently compelling to spend a few million dollars on a NASA Near-Earth Object Program Office (NEOPO) to gradually identify and begin to track potential doomsday projectiles. Equipped only with a few smallish or obsolescent telescopes, NEOPO will take a couple of decades to chart existing major hazards. Presumably we've got the time. Smaller, 100-meter objects capable of delivering the punch of several hundred H-bombs will, however, not be detected. This seems to me to be a significant shortcoming in the program, because such objects hit us much more frequently than the big planet killers. Also, if we could detect them, it would be much more feasible to take positive action to deflect one of the small guys. Which brings us to the $64,000 question: If we do detect a world-destroying asteroid coming our way, what do we do about it?

There are a number of conceivable alternatives. These include:

1. Sit tight and die. This is the traditional approach, which wants improvement.
2. Evacuate the Earth. This will be technically infeasible for some time to come and always undesirable.
3. Move the Earth out of the way. Amusing, but technically infeasible.
4. Destroy the asteroid prior to impact. This won't work. It's probably infeasible to fragment a 1-km asteroid with weaponry, but even if we could it would do little good. The fragments hitting the Earth would do nearly as much damage as the asteroid would have if left in one piece.
5. Deflect the asteroid so that it misses the Earth. This would be difficult, but is possible in principle. It has thus justly received the most attention by those concerned with planetary defense.

How can we deflect an oncoming asteroid? This was the subject of a workshop held at Los Alamos National Laboratory in January 1992. Most of the technical contributors to this meeting were designers of advanced thermonuclear weapons from Los Alamos and Livermore, so naturally the focus fell to the use of such devices. This has caused some to dismiss the group as a bunch of self-interested bomb makers trying to stir up business now that the Soviet threat has gone, but this is unfair. Atomic explosives are certainly mankind's most potent current physical capability. It's certainly reasonable to examine their practicality as a means of asteroid defense. Let's do that now.

Consider a 1-km-diameter asteroid heading toward the Earth at a typical interception speed of 16 kilometers per second (km/s). There are many different types of asteroids. Some are made of iron-nickel and are as hard as steel. Others are made of stone. Others are made of weaker carbonaceous materials and some even have a significant component of water ice. Let's say ours is a stone, since that is the most common type among the near-Earth group, and is midway in density between the iron and carbonaceous types. If this is the case, our asteroid will have a mass of about 2.5 trillion kg (2.5 billion tonnes). Now let's say we launch a 10-megaton H-bomb and detonate it right beside the asteroid, so as to give it a sideways nudge. The bomb has a mass of about 10 tonnes and releases 4×10^{16} joules (11 terawatt-hours) of energy. If all of this energy goes into the bomb (i.e., none is lost by radiation), the fragments will explode with an average velocity of 2.8 million meters per second. The total impulse generated by the bomb will be 28 billion kg-meter/s, and one-quarter of this will be available to push the asteroid sideways. The asteroid will thus be imparted a velocity change of 7×10^9 kg-m/s $\div 2.5 \times 10^{12}$ kg = 0.0028 m/s. Now the Earth has a diameter of approximately 12,800 km, so we need to deflect the asteroid by about this much to make it miss. Dividing this distance by the 0.0028 m/s velocity increment (a first-order approximation to a calculation of trajectory change that is good enough for our rough estimation purposes here), we find that the bomb would need to be detonated 4.6 billion seconds, or 145 *years,* prior to impact to achieve the desired result. Chances are that we would not have that much time.

A much larger velocity change could be achieved if we inserted the bomb in a ground-penetrating warhead and fired it into the asteroid at high velocity, detonating beneath the surface. This would not work if the asteroid was a solid iron-nickel object and could fail even if the asteroid was mostly stone but had iron lumps in it that could destroy the bomb on impact. But if the asteroid was of the weaker stony or carbonaceous sort, good penetration should be possible. In this case, the 4×10^{16} joules represented by the bomb would be distributed not to 10 tonnes of matter but to a much larger amount, perhaps 1,000 tonnes. In that case, the characteristic velocity of the ejected mass would be reduced by a factor of 10 compared to the previous calculation, but since there would be 100 times as much mass ejected, the net result would be an impulse 10 times as great. So now only a 14.5-year lead time would be needed. If the bomb's energy could somehow be usefully distributed across 100,000 tonnes of asteroid, the required lead time would fall to 1.45 years.

This sounds like it might be feasible, but wait. Where is the asteroid 1.45 years before impact? It is somewhere in deep space, with 730 million kilometers of travel path between it and the Earth. And how do we know how much mass the bomb will eject? To know that we would have to know how deep it will penetrate after impact, which means we have to know the geology of the object, and not just at its surface but underground as well. And how do we know the direction that fragmentation and ejection will occur? Subsurface strength variations could have a strong effect on the ejection vector. Once again, we need to know the subsurface geology in detail. And furthermore, the bomb will not only eject rocks, it will heat up the asteroid as well, possibly causing outgassing of volatile materials. The asteroid is probably rotating, so this outgassing will act to propel the asteroid in unanticipated directions unless the geometry, geology, and kinetic characteristics of the object are thoroughly understood.

There are too many unknowns. The fate of humanity is at stake in the success of the operation. If bombs are to be used as asteroid deflectors, they cannot just be launched willy-nilly. No, before any bombs are detonated, the asteroid will have to be thoroughly explored, its geology assessed, and subsurface bomb placements carefully determined and precisely located on the basis of such knowledge. A human crew, consisting of surveyors, geologists, miners, drillers, and demolition experts, will be needed on the scene to do the job right.

But if a human crew is to be sent, there may be other ways besides bombs to give the asteroid the required push. For example, a spacefaring civilization will almost certainly develop space nuclear reactors with power outputs in the 10 megawatts of electricity (MWe) range for the purpose of driving nuclear electric propulsion (NEP) ion-drive cargo vessels. Let's say we delivered one of these units to an asteroid, and set it up to drive a catapult firing chunks of the asteroid off into space at a velocity of 1 km/s. The catapult would thus act as a kind of rocket engine, using the asteroid's own mass as propellant. The average mass flow of the catapult would thus be about 20 kg/s, and the total thrust generated would be 20,000 Newtons (N) (equivalent to 4,490 pounds of force). This thrust would then be able to accelerate the asteroid in a precisely controlled direction at a rate of 20×10^3 N $\div 2.5 \times 10^{12}$ kg $= 8 \times 10^{-9}$ m/s^2. This might seem imperceptible, but in the course of a year, a velocity increment of 0.25 m/s would develop, sufficient to deflect the asteroid from collision with Earth provided that the push was imparted at least 1.6 years in advance of the impact date.

Alternatively, if the asteroid has ice on it, this can be used as propellant in a nuclear thermal rocket (NTR). NTRs work by employing a solid-core nuclear reactor to heat a working fluid to very high temperatures and then ejecting it from a rocket nozzle as high-temperature gas. In the 1960s, the United States had a program called NERVA, which ground-tested about a dozen NTRs with thrusts ranging from 45,000 N (10,000 pounds) to 1.1 million N (250,000 pounds) and power levels ranging from 200 to 5,000 megawatts of heat (MWt). If hydrogen is used as the propellant, exploiting its low molecular weight to obtain high exhaust velocities, such engines operating at 2,500°C can generate a specific impulse of 900 seconds. The 1960s NERVA engines actually generated about 825 seconds—almost twice that of the best chemical rocket engines possible. Wernher von Braun planned to use these engines for NASA's expeditions to Mars that were supposed to follow Apollo by 1981. Unfortunately, when the Nixon administration gutted the Apollo program and canceled plans for Mars, it derailed the NERVA program too, and the nuclear engines were never flown. However, the technology certainly works, and what's more, has an additional advantage beyond high performance: versatility. In principle, any fluid can be used as a propellant in an NTR. On Mars, carbon dioxide atmosphere is everywhere, so Mars-based NTR-powered rocket hoppers using CO_2 propellant could refuel themselves just by running a pump each time they land. Such "NIMF" vehicles, discussed in *The Case for Mars,* would thus give Martian explorers and settlers complete global mobility. Among the asteroids, ice is frequently available. This could be melted into water and stored in propellant tanks and then turned to steam thrust in the NTR engine. The specific impulse obtained would be about 350 s—nowhere near as good as hydrogen, but, like their Earthly counterparts, asteroid prospectors will need a mule that can live on mountain scrub, not a racehorse that only eats gourmet fodder. NTR steam rockets thus offer a very attractive technology for those wishing to get around among the asteroids. Furthermore, if an asteroid has enough ice, NTRs could be used very effectively to move it.

Let's say we took a 5,000-MWt NTR, no larger than the biggest NERVA engine tested in the 1960s, and placed it on a 1-km asteroid and fed it water propellant. The required mass flow would be 850 kg of water per second, and the thrust would be 2.9 million N (650,000 pounds). This would accelerate the asteroid at a rate of 2.9×10^6 N $\div$ 2.5×10^{12} kg = 1.16×10^{-6} m/s^2, or 36 m/s per year. This is over a hundred times the acceleration possible with the electric catapult or bomb-driven systems discussed above. Using such technology, not only could we nudge the asteroid enough to make it miss the Earth on a particular pass, we could literally tug the asteroid into a substantially different orbit from which it would never threaten the Earth. We could even consider rearranging the orbits of asteroids so as to make groups of them more convenient for mining or other forms of development.

So we need not be helpless in the face of asteroids. Two things are necessary for our defense. We must learn much more about the enemy. And we must become spacefarers.

ASTEROID EXPLORATION

In one sense we already know more about the asteroids than nearly any other extraterrestrial body, except perhaps the Moon, because we have hundreds of thousands of samples. These are the meteorites, fragments of asteroids that have fallen to Earth and are available for collection. The meteorites show a variety of asteroid compositions, ranging from nearly pure metal, to stone, to carbonaceous materials. Because they survive atmospheric entry best, and because they are easiest to distinguish from terrestrial rocks, iron-metal meteorites enjoy preferential representation in terrestrial meteorite collections. However, by comparing the spectral characteristics of meteorites with those reflected from asteroids in space, astronomers have been able to classify the asteroids according to their composition. The principal types are listed in Table 7.2.

No type of asteroid is concentrated in the near-earth objects (NEOs), because the NEOs represent only a tiny minority of the asteroids overall. Also, because they are small, it is hard to get a composition-determining reflection from most of the known NEOs. However, of those that have been

▪ TABLE 7.2 ▪

Principal Types of Asteroids

Type	Composition	Primary Concentration
M	Metal	Inner Main Belt, near Mars
E	Silicate rocks	Inner Main Belt, 1.9 AU
S	Stony-iron	Central Main Belt, 2.4 AU
C	Carbonaceous	Outer Main Belt, 3.3 AU
P	Carbonaceous/volatile	Outermost Main Belt, 4 AU
D	Frozen volatiles	Beyond Jupiter

surveyed by astronomer Lucy Ann McFadden, about 80 percent were found to be S asteroids and 20 percent were C asteroids. Types M and E were probably not observed simply because of the small size of the assessed sample—these are generally rarer than S and C types. Types P and D cannot exist in near-Earth space for long without evaporating. If one came our way we would observe it as a comet.

With the exception of radar imaging of a few close-passing NEOs (such as Toutatis, which was imaged by JPL's Steve Ostro using the Goldstone Deep Space Communications Complex as radar during the asteroid's 3.5-million-kilometer pass by our planet in December 1992), asteroids are too small or too far away to be photographed by Earth-based telescopes. Our first good look at asteroids, therefore, had to wait for images returned by interplanetary spacecraft. The first of these was produced by the Galileo spacecraft when it flew by the Main Belt asteroid Gaspra on its way to Jupiter in October 1991. Galileo's magnetometer also measured a surprisingly strong magnetic field around the asteroid, indicating the presence of a large quantity of metallic iron. The images returned show Gaspra to be a potato-shaped object, 19 km long and 11 km in diameter. Based on counting the number of craters on its surface, mission scientists judged that Gaspra is only about 400 million years old, which is a bit of a mystery since the rest of the solar system is ten times that age. Perhaps Gaspra is a fragment cast off by a catastrophic collision between two larger bodies 400 million years ago.

In August 1993, Galileo, pushing farther out through the Main Belt, was able to get a look at another asteroid, 51-km-long Ida. Surprisingly, Ida was found to have a moon of her own, a 1-km sized object, which mission scientists named Dactyl. Since the rate of Dactyl's orbit about Ida is dependent on Ida's mass, and since Ida's size is known, the mission team was able to use measurements of Dactyl's orbit to calculate that the density of Ida is about two and a half times that of water, consistent with a carbonaceous composition.

But these Galileo fly-bys were just quick snapshots. Much more can be learned by having a dedicated spacecraft rendezvous with an asteroid and hanging around to take close-up photographs and detailed sets of measurements. This will be done by the Near Earth Asteroid Rendezvous (NEAR) mission designed by the Johns Hopkins Applied Physics Laboratory, which is scheduled to reach and orbit Eros, first discovered and largest of the NEOs, in February 2000. NEAR will orbit Eros for over a year, gradually lowering its orbit to within 30 kilometers, which should enable some spectacular photographs.

Other robotic asteroid missions include Deep Space 1, a JPL "New Millennium" program ion-drive micro spacecraft that will fly by an asteroid with a miniaturized imaging spectrometer in 2000. A NASA Discovery mission called STARDUST will also venture out into the asteroid belt to intercept and capture, via impact on aerogel, particles of the comet Wild 2 in 2004. Most ambitious of the currently announced asteroid exploration missions is MUSES C, developed by the Japanese space agency ISAS, which will launch from Earth in 2002, land on the NEO Nereus in 2003, gather surface samples, and return to Earth with its scientific booty in 2006. To assist in the sample collection, MUSES C may employ a small insectoid robotic rover, which will exploit the asteroid's very low gravity environment to hop from place to place.

GAIASHIELD: A HUMAN ASTEROID MISSION

However, as impressive as NEAR and MUSES C are compared to previous asteroid exploration efforts, they will only scratch the surface. Given the importance of gaining a detailed knowledge of NEOs to the future security of humanity, the costs associated with human exploration are more than justified. If we were to launch the Mars Direct program described in chapter 6, the Ares launch vehicles and habitation modules developed for Red Planet exploration could also be used to perform rendezvous and return missions to near-Earth asteroids. Indeed, since the rocket propulsion requirements required to leave low Earth orbit (LEO) for a one-way trajectory and landing on Mars (a ΔV of about 4.2 km/s) are nearly identical to those for a round-trip from LEO to many NEOs, hardware designed for Mars Direct–type missions enjoys natural commonality with that needed for NEO exploration. But because asteroids have no atmosphere and little gravity, eliminating the need for reentry and landing systems, and because asteroids are small, eliminating the need for ground vehicles and split field exploration and base crews, a minimal piloted asteroid exploration mission can be launched with a significantly smaller and more limited set of hardware than that needed for Mars exploration. Such a mission could be flown within four years, using launch vehicles and technology available today. If there is no manned Mars exploration program, it may help produce one, because in the course of flying to an asteroid and back the astronauts will destroy the bugaboos of putative cosmic-ray, subnormal gravity, and human factors barriers to long-distance spaceflight that are used as excuses for lack of initiative by apologists for current go-nowhere space policies.

I call my asteroid mission plan "Gaiashield" because it will be an important first step in giving humanity the knowledge and spacefaring capability that it needs to protect the Earth's biosphere from another mass extinction.

The Gaiashield mission will employ a simple cylindrical habitation module, 5 meters in diameter and 20 meters long, somewhat similar to those used on the International Space Station. This module can be launched in one piece by the Space Shuttle, or alternatively by a Titan IV, Proton, or Ariane V, all of which have the necessary 20-tonnes-to-LEO lift capability. The module will be equipped with a set of photovoltaic panels, which will be deployed outward from it like wings, somewhat in the manner of the United States' 1970s-era space station, Skylab. After it is launched, a series of four Protons (or Shuttles, Ariane V's, or Titan IV's, or any combination thereof) will deliver four 20-tonne storable chemical propulsion stages, which will be stacked onto the rear of the crew module like a train. Finally, the two-person crew will be brought up in a Shuttle launch that will deliver them together with an Apollo-like reentry capsule, which will mate and dock with the front of the cylindrical crew module.

When all is ready, the chemical propulsion stages will be fired in succession, delivering the 3.5 km/s ΔV required to send the crew module onto transasteroid injection (TAI). Three of the four chemical stages will be completely exhausted and be discarded. The fourth will be only partially used and will remain with the vehicle for further maneuvers. Once on the transasteroid trajectory, small reaction control thrusters will fire, causing the spacecraft to spin in the same plane in which the solar panels are located, with the spin axis and the solar panels pointing at the Sun. The length of the spin arm between the crew module center of gravity and the decks at the far end of the module is about 10 meters. As a result, by spinning at 4 rpm, lunar-equivalent gravity could be generated at the "lowest" decks. By spinning at 6 rpm, Mars-level gravity can be created. While NASA officials wishing to justify Space Station research programs on the human health effects of long-duration zero gravity exposure frequently affect deep concern over the possible disorientation caused by Coriolis forces and other concomitants of artificial gravity systems, experiments done in the 1960s show that humans can adapt and operate well in vehicles rotating as rapidly as 6 rpm. Many current artificial gravity researchers, such as Professor Larry Young of M.I.T., believe that rotation rates as high as 10 rpm are viable. By employing artificial gravity, the Gaiashield crew will be protected against the severe negative health impacts that have afflicted cosmonauts and others who failed to implement strenuous exercise programs when flying for long periods in zero gravity. Figure 7.3 shows the layout of the Gaiashield spacecraft.

The ship will take about six months to reach the asteroid, as despite its close distance to Earth a Hohmann transfer elliptical trajectory from one side of Earth's orbit to the other will probably be needed to get there. Upon interception, the crew will despin the ship and use most of the propellant in the remaining propulsion stage to effect a ΔV of perhaps 0.4 km/s to establish the spacecraft in orbit a few kilometers from the asteroid, where it will remain for one year. The crew will then proceed to explore the asteroid in detail, using backpack gas thrusters similar to the Space Shuttle MMU unit to fly to it from the ship and hop about the body at will. A small portable drilling rig will be used to take repeated deep samples from all over the body.

At the end of a year of intensive exploration, the propulsion stage will fire its last allotment of propellant to give the ship the required 0.4 km/s ΔV needed to send it on trans-Earth injection. After a voyage of another six months, the ship will approach Earth, and the crew will bail out in the reentry capsule and be picked up by a boat, much as the Apollo astronauts did a generation before. Empty of crew, the ship itself will remain in a cycling or-

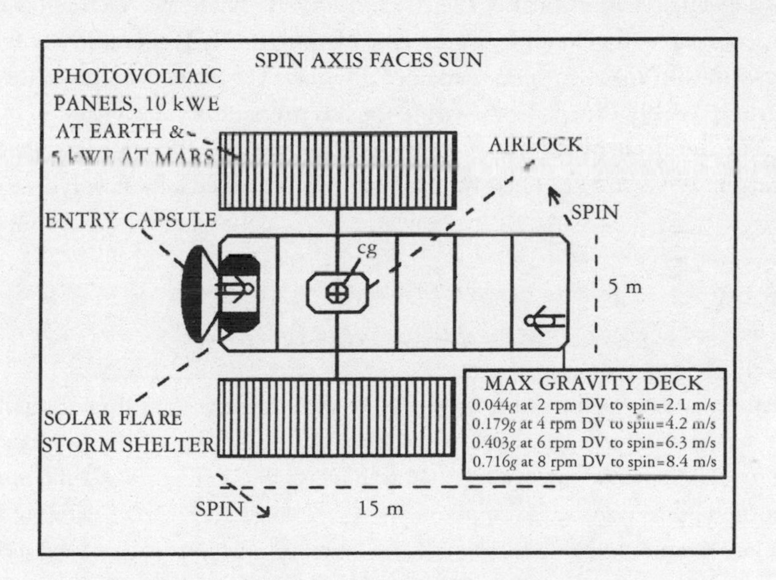

FIGURE 7.3 *Layout of the Gaiashield spacecraft (cg = center of gravity).*

bit between the Earth and the asteroid, possibly to be refitted for further use when appropriate capabilities for doing so are developed.

The crew will thus spend two years in interplanetary space, about twice that required for a round-trip to Mars (which spends six months traveling each way and 1.5 years on the surface). In the course of this trip, they will take about 100 rem of cosmic radiation, which represents about a 2 percent statistical risk of fatal cancer later in life for each member of the crew. (In contrast, an average smoker incurs a 20 percent risk.) This is small compared to other risks associated with piloted space missions, and there is no doubt that many astronauts would be more than willing to take it on.

The estimated cost of the Gaiashield mission is about $2.4 billion, on the same order as such upper-end robotic interplanetary probes as Galileo and Cassini. The basis for that estimate and the required mission mass is given in Tables 7.3, 7.4, and 7.5.

Gaiashield would be a terrific asteroid science mission, but it would be more: It would be an *icebreaker* mission. Since the end of Apollo, Mars has been staring NASA in the face as the next challenge for human exploration. The public knows this, and no amount of dancing around the issue by NASA officials can hide it.

Two things, however, have kept NASA from sending human explorers to Mars. The first is the notion that such missions must be incredibly expensive. The second is fear of the risks involved. These two factors have fed off each other—for example, in the case of the 90 Day Report, where fear of long-duration space voyages made NASA place the Mars mission at the end of an impossibly expensive 30-year series of preparatory activities.

The debilitating effects of long-duration spaceflight are not caused by radiation. No astronaut or cosmonaut has ever received a radiation dose dur-

▪ TABLE 7.3 ▪

Cost of Gaiashield Mission

Shuttle launches: two needed at $600 million each	$1,200 million
Proton launches: four needed at $70 million each	$280 million
Habitat development: (use space station technology)	$400 million
Reentry capsule: (use Apollo capsule or space station ACRV)	$100 million
Mission operations	$100 million
Reserves and contingency (20 percent)	$300 million
Total cost	**$2,390 million**

▪ TABLE 7.4 ▪

Mass Allocations for Gaiashield Mission Plan

Habitat structure	4.0 tonnes
Life support system	2.0
Consumables	7.7
Electric power (5 kWe solar)	1.0
Reaction control system	0.5
Comm and information management	0.2
Science equipment	0.2
Crew	0.2
4 EVA suits	0.4
Furniture and interior	0.5
Reentry capsule	4.0
Spares and margin (25 percent)	5.2
Habitat total	**25.9 tonnes**

▪ TABLE 7.5 ▪

Consumable Requirements for Gaiashield Mission with Crew of Two

Item	Need/ Man-Day	Fraction Recycled	Wasted Man-Day	Crew Requirements for 900-Day Mission[a]
Oxygen	1.0 kg	0.8	0.2	360 kg
Dry food	0.5	0.0	0.5	900
Whole food	1.0	0.0	1.0	1,800
Potable water	4.0	0.8	0.0[b]	0
Wash water	26.0	0.9	2.6	4,680
Total	**32.5 kg**	**0.87**	**4.3 kg**	**7,740 kg**

[a] Sufficient consumables are available for a 900-day mission, providing significant margin against the 730 days actually required.

[b] In Table 7.5 the reason why there is no wastage of potable water is because water lost by the potable water loop is replaced by water added to the system from the use of (water-rich) whole food.

ing flight large enough and prompt enough to create any visible effects. Rather, all the well-known ill effects are due to long-duration zero gravity exposure and ensuing complications.

The Gaiashield mission demonstration of such an artificial-gravity–

piloted interplanetary spacecraft would kill forever the dragons of cosmic-ray threat and of zero gravity space sickness that are barring us from the solar system. Furthermore, it would destroy the myth that interplanetary manned exploration need be impossibly costly. It would also directly accomplish a significant fraction of the nonrecurring development that needs to be performed for a human Mars mission.

Before Copernicus, Ptolemaic astronomers believed that humanity was walled off from the heavens by a set of crystal spheres. In a way those spheres are still there, made not of glass but of fear. The Gaiashield mission would smash them.

MINING THE ASTEROIDS

So far in this chapter we have discussed the asteroids primarily as a potential threat to humanity and the rest of the terrestrial biosphere. But just as fire, a deadly menace to animals and children who do not understand it, becomes in the hands of competent adults one of humanity's greatest boons, so the asteroids, which offer nothing but mass death for the pre-sentient biosphere or Earthbound Type I humanity, hold the promise of vast riches for a Type II spacefaring civilization.

The asteroid belt is known to contain vast supplies of very high grade metal ore in a low gravity environment that makes it comparatively easy to export to Earth. For example, in his book *Space Resources,* Professor John Lewis of the University of Arizona considers a single small type S asteroid just 1 km in diameter—a run-of-the-mill asteroid.[4] This body would have a mass of around 2 billion tonnes, of which 200 million tonnes would be iron, 30 million tonnes high-quality nickel, 1.5 million tonnes the strategic metal cobalt, and 7,500 tonnes a mixture of platinum group metals whose average value, at current prices, would be in the neighborhood of $20,000 per kilogram. That adds up to $150 billion just for the platinum group stuff! There is little doubt about this—we have lots of samples of asteroids in the form of meteorites. As a rule, meteoritic iron contains between 6 percent and 30 percent nickel, between 0.5 percent and 1 percent cobalt, and platinum group metal concentrations at least ten times the best terrestrial ore. Furthermore, since the asteroids also contain a good deal of carbon and oxygen, all of these materials can be separated from the asteroid and from each other using variations of the carbon monoxide–based chemistry needed for refining metals on Mars.

As mentioned earlier, there are over 5,000 asteroids known today, of which about 96 percent are in the Main Belt lying between Mars and Jupiter, with an average distance from the Sun of about 2.7 AU. The Main Belt group includes all the known asteroids residing within the orbit of Jupiter with diameters greater than 10 km. The remaining 4 percent, all small, are the NEOs. The 4 percent figure, however, greatly overstates the proportion of NEOs to Main Belters, because their relative closeness to the Earth and Sun makes them much easier to see. Of the Near Earth asteroids, about 90 percent orbit closer to Mars than to the Earth.

As should be clear from Lewis's example, these asteroids collectively represent enormous economic potential. The Near Earth group is the one we need to address with respect to planetary defense. However, the relative numbers of the two classes make it clear that for mining purposes, the real action is going to be in the Main Belt, where *millions* of 1-km class ($150 billion worth of platinum class!) objects undoubtedly reside.

Because the asteroids have a clearly identifiable potential material cash export, some people, such as Professor Lewis, have pointed to them as the best targets for human colonization. However, while water and carbonaceous material can also readily be found among the asteroids (making them as a group far richer than the Moon), it is not necessarily the case that such volatiles can be found on those asteroids that are most rich in exportable metals. Quite the contrary, the metal-rich type M asteroids are nearly volatile free. Moreover, while many of the Main Belt asteroids contain all the carbon, hydrogen, and oxygen needed to support agriculture, nitrogen is generally rare. Furthermore, sunlight in the Main Belt is too dim to support agriculture, which means that plants would have to be grown by artificially generated light. This is a massive disadvantage for asteroid colonization, because plants are enormous consumers of light energy, and it is doubtful whether growing plants with electric lights to support any significant population is practical with near-term space power sources. In addition, while collectively the asteroids may someday possess a significant mining workforce, until very advanced robotic technology becomes available, it is unlikely that any one asteroid will have the sufficient manpower required to develop the division of labor necessary for true multifaceted industrial development.

Mining bases, yes. Farms and industries, no—at least not until the widespread use of controlled fusion makes very large scale employment of artificial power possible in the Main Belt. For the twenty-first century and some time beyond, most of the supplies needed to support the asteroid prospectors and miners will have to come from somewhere else. As shown in Table 7.6, Mars has an overwhelming positional advantage as a shipping location.

In Table 7.6, Ceres is chosen as a typical Main Belt asteroid destination, as it is the largest asteroid and positioned right in the heart of the belt. You'll notice, however, that I also give Earth's Moon as a potential port of call. Despite the fact that it is much closer to Earth physically, it can be seen that from the point of view of rocket propulsion, it is much easier to reach the Moon from Mars than it is from the Earth! (i.e., the required mass ratio is only 12.5 going from Mars to the Moon, whereas it is 57.6 from Earth). This would be even more forcefully the case for travel from either Earth or Mars to nearly any Near Earth asteroid as well.

In Table 7.6 all entries except the last two are based upon a transportation system using CH_4/O_2 engines with an Isp of 380 s and ΔV's appropriate for trajectories employing high-thrust chemical propulsion systems. These were chosen because CH_4/O_2 is the highest-performing space-storable chemical propellant and can be manufactured easily on either Earth, Mars, or a carbonaceous asteroid. H_2/O_2, while offering a higher Isp (450 s), is not storable for long durations in space. Moreover, it is an unsuitable propellant for a cheap reusable space transportation system, since it costs over an order of magnitude more than CH_4/O_2 (thus ruling it out for true cheap surface-

■ TABLE 7.6 ■

Transportation in the Inner Solar System

	EARTH		MARS	
	ΔV (KM/S)	MASS RATIO	ΔV (KM/S)	MASS RATIO
Surface to low orbit	9.0	11.4	4.0	2.9
Surface to escape	12.0	25.6	5.5	4.4
Low orbit to lunar surface	6.0	5.1	5.4	4.3
Surface to lunar surface	15.0	57.6	9.4	12.5
Low orbit to Ceres	9.6	13.4	4.9	3.8
Surface to Ceres	18.6	152.5	8.9	11.1
Ceres to planet	4.8	3.7	2.7	2.1
NEP round-trip LO to Ceres	40.0	2.3	15.0	1.35
Chemical to LO, NEP round-trip to Ceres	9/40	26.2	4/15	3.9

to-orbit systems) and its bulk makes it very difficult to transport to orbit in any quantity using reusable single-stage-to-orbit (SSTO) vehicles. The last two entries in the table are based on nuclear electric propulsion (NEP) using argon propellant, available at either the Earth or Mars, with an Isp of 5,000 s for in-space propulsion, with CH_4/O_2 used to reach low orbit (LO) from the planet's surface. Such SSTO and NEP systems, while somewhat futuristic today, represent a conservative baseline for interplanetary transportation technology in the mid twenty-first century.

It can be seen that if chemical systems are used exclusively, then the mass ratio required to deliver dry mass to the asteroid belt from Earth is fourteen times greater than from Mars. This implies a still (much) greater ratio of payload to takeoff mass ratio from Mars to Ceres than from Earth, because all the extra propellant requires massive tankage and larger-caliber engines, all of which require still more propellant and therefore more tankage, and so on. In fact, looking at Table 7.6, it can safely be said that useful trade between Earth and Ceres (or any other body in the Main Belt) using chemical propulsion is probably impossible, whereas from Mars it is easy.

If nuclear electric propulsion is introduced the story changes, but not much. Mars still has a sevenfold advantage in mass ratio over Earth as a port of departure for the Main Belt.

But those are just mass ratios, which understates Mars' advantage, because rocket propellant is needed not just to accelerate payloads but also to accelerate the tanks necessary to hold the propellant and the engines and other spacecraft systems that use it. If you work the problem in detail, as I showed in *The Case for Mars,* what you find is that regardless of whether chemical or NEP propulsion is employed, the total gross liftoff mass required to send a given cargo to Ceres is *50 times greater* if launched from the Earth than if launched from Mars.

The result that follows is simply this: Anything that needs to be sent to the asteroid belt that can be produced on Mars will be produced on Mars.

The outline of mid-future interplanetary commerce in the inner solar system thus becomes clear. There will be a "triangle trade," with Earth supplying high-technology manufactured goods to Mars, Mars supplying low-technology manufactured goods and food staples to the asteroid belt and possibly the Moon as well, and the asteroids sending metals and perhaps the Moon sending helium-3 to Earth. This triangle trade is directly analogous to the triangle trade of Britain, her North American colonies, and the West Indies during the colonial period. Britain would send manufactured goods to North America, the American colonies would send food staples and needed craft products to the West Indies, and the West Indies would send cash crops such as sugar to Britain. A similar triangle trade involving Britain, Australia, and the Spice Islands also supported British trade in the East Indies during the nineteenth century.

However, while the asteroids' multiplicity represents a disadvantage for societal development in the near and medium term, in the far term it will be a great advantage. Mars, while huge, is after all one world. A multiple of social experiments will start there, but eventually these are likely to be resolved and fuse into a single, or at most a few, new branches of human civilization. But the resource utilization, labor-saving, space transportation, and energy production technologies developed for the colonization of Mars will open the way to the settlement of the asteroids, which will force both the technologies and the aptitudes that created them even further. This will make available thousands of potential new worlds, whose cultures and systems of law need never fuse. Perhaps some will be republican, others anarchist. Some communalist, others capitalist. Some patriarchal, others matriarchal. Some aristocratic, others egalitarian. Some religious, others rationalist. Some Epicurean, others puritanical. Some traditional, others novel. For a long time to come, groups of human beings who think they have found a better way will have places to go where they can give it a try. As among the city-states of the ancient Greek islands, a bewildering myriad of societies may flower and bloom.

The rest of humanity will watch and learn from their experiences. That which works will be repeated. So shall we continue to progress.

ASTEROIDS AND LIFE

While averting doomsday asteroids is an important subject, ultimately it is merely a subset of a much more important endeavor, that of conquering the space frontier. Indeed, it is the absence of this insight that reduces much of the talk about the asteroid menace to mere alarmism.

Life on Earth has survived and prospered because at an early date it was able to take control, dictating the physical and chemical conditions of the planet in defiance of both solar and geological cycles. If it had not done this, terrestrial Life would long since have gone extinct, as the Sun today is more than 40 percent brighter than it was at the time of Life's origin. Without

Life's ability to control terrestrial temperatures by regulating the CO_2 and other greenhouse gas content of the atmosphere, our ancestors would have all been cooked billions of years ago. Moreover, by replacing the CO_2 content of the atmosphere with oxygen, Life transformed the terrestrial chemical environment to favor the development of species with the capability for increased activity and intelligence and ever more rapid evolution of still higher and more complex forms.

Within the history of the biosphere itself, the same phenomenon repeats. Those groups of species—whether natural ecosystems or human civilizations—whose activities effectively control their surrounding environments to favor their own growth are those that survive. Those which do not risk extinction. In the game of life, the only way to win is to have a part in making the rules.

On the short timescale, the relevant environment for most species is the Earth, and most ecosystems can get by for a fair while if they can deal with developments below the stratosphere. But over the long haul this is not true. Since the success of the Alvarez hypothesis in explaining the mass extinctions that occurred at the end of the age of dinosaurs as having been caused by asteroid bombardment, it is now apparent that the relevant environment for life on Earth is not merely the planet of residence, but the whole solar system.

Few people today understand this, yet it is true that subtle events in the asteroid belt determined the fates of their ancestors, and may well in the future determine the fates of their descendants. It seems unbelievable that invisible happenings so far away could matter so much here. Similarly, throughout history, few inhabitants of rustic villages going about their daily lives were aware of the machinations of politicians and diplomats in their nations' capitals, which periodically would sweep the villagers off to die in catacylsmic wars on distant battlefields.

What you can't see can kill you. What you can't control probably will.

Humanity's home, humanity's environment, is not the Earth—it is the solar system. We've done well for ourselves so far, by taking over the Earth and changing it in our own interests. Most people today, at least in the world's more advanced sectors, can walk about without fear of being dismembered by giant cats, are assured of sufficient food and fuel to survive the next winter, and can even drink water without risk of death. Saber-toothed tigers, locusts, and bacteria—for now at least, we've beaten them all. But in a larger sense we're still helpless. We may feel safe, having thrown our village's hoodlums in jail, but in the capital, behind the scenes, diplomats are meeting with generals, and plans are being made. . . .

Our environment is the solar system, and we won't control our fate until we control it. The geological record is clear. Asteroids do hit. Mass extinctions, of sets of species every bit as dominant on Earth in their day as humanity is in ours, do occur.

You can't shoot down an incoming asteroid with an anti-aircraft gun, air-to-air missile, or Star Wars defense system. If you want to prevent asteroid impacts, you have to be able to direct the course of these massive objects while they are still hundreds of millions of kilometers away from the Earth. A Type I civilization, however prosperous it might become, is intrinsically a helpless target in the asteroid shooting gallery. If it remains stunted at that level, its long-term prospects for survival are limited. If we are to be in charge of our fate, we must be able to control our *true* environment in the way that only a Type II civilization can: We must take charge of the asteroids.

In short, the lesson of the asteroids is this: If humanity wants to either progress *or* survive, we have to become a spacefaring species. In the end, it is creativity, not austerity, that will be the key to our survival.

FOCUS ON: CHEMISTRY FOR SPACE SETTLERS

Just as the pioneers of old needed to know how to find the edible plants and methods of hunting the game available in their environments, so space settlers will need to know how to extract useful resources from their new worlds. The following is a brief compendium of some of the key techniques.

On the Moon

On the Moon, oxygen can be produced from the mineral ilmenite, which is found in up to 10 percent concentrations in some lunar soils. The reaction is

$$FeTiO_3 + H_2 \rightarrow Fe + TiO_2 + H_2O \qquad (7.1)$$

The water produced is then electrolyzed to produce hydrogen, which is recycled back into the reactor, and oxygen, which is the net useful product of the system. The feasibility of this system has been demonstrated by researchers working at Carbotek in Houston, Texas. If you don't want to go prospecting for ilmenite, you can try carbothermal reduction, a system pioneered by Sanders Rosenberg at Aerojet, which will work with a larger variety of lunar rocks, including the very common silicates:

$$MgSiO_4 + CH_4 \rightarrow MgO + Si + CO + 2H_2O \qquad (7.2)$$

The water is then electrolyzed to produce oxygen, while the carbon monoxide and hydrogen from the electrolysis are combined to remake the methane:

$$CO + 3H_2 \rightarrow CH_4 + H_2O \qquad (7.3)$$

Reactions 7.1 and 7.2 are very endothermic (i.e., they need energy input) and must be done at high temperatures (above 1,000°C). Reaction 7.3 is exothermic (i.e., it produces energy) and occurs rapidly at 400°C. The carbon and hydrogen reagents are extremely rare on the Moon, so the systems must be designed for very efficient recycling.

On Mars

On Mars, the most accessible resource is the atmosphere, which can be used to make fuel, oxygen, and water in a variety of ways. The simplest technique is to bring some hydrogen from Earth and react it with the CO_2 that comprises 95 percent of the Martian air:

$$CO_2 + 4H_2 \rightarrow CH_4 + 2H_2O \qquad (7.4)$$

Reaction 7.4 is known as the Sabatier reaction and has been widely performed by the chemical industry on Earth in large-scale one-pass units since the 1890s. It is exothermic, occurs rapidly, and goes to completion when catalyzed by ruthenium on alumina pellets at 400°C. I first demonstrated a compact system appropriate for Mars application uniting this reaction with a water electrolysis and recycle loop while working at Martin Marietta in Denver in 1993. The methane produced is great rocket fuel. The water can be either consumed as such or electrolyzed to make oxygen for propellant or consumable purposes, and hydrogen, which is recycled.

Another system that has been demonstrated for Mars resource utilization is direct dissociation of CO_2 using zirconia electrolysis cells. The reaction is

$$CO_2 \rightarrow CO + \frac{1}{2}O_2 \qquad (7.5)$$

Reaction 7.5 is very endothermic and requires the use of a ceramic system with high-temperature seals operating above 1,000°C. Its feasibility was first demonstrated by Robert Ash at the Jet Propulsion Laboratory in the late 1970s, and the performance of such systems has since been significantly improved by Kumar Ramohali and K. R. Sridhar at the University of Arizona. Its great advantage is that no cycling reagents are needed. Its disadvantage is that it requires a lot of power—about five times that of the Sabatier/Electrolysis process to produce the same amount of propellant.

Still another method of Mars propellant production is the reverse water gas shift (RWGS):

$$CO_2 + H_2 \rightarrow CO + H_2O \qquad (7.6)$$

This reaction is very mildly endothermic and has been known to chemistry since the nineteenth century. Its advantage over the Sabatier reaction is that all the hydrogen reacted goes into the water, from where it can be electrolyzed and used again, allowing a nearly infinite amount of oxygen to be produced from a small recycling hydrogen supply. It occurs rapidly at 400ºC. However, its equilibrium constant is low, which means that it does not ordinarily go to completion, and it is in competition with the Sabatier reaction (7.4), which does. Working at Pioneer Astronautics in 1997, Brian Frankie, Tomiko Kito, and I demonstrated that copper on alumina catalyst was 100 percent specific for this reaction, however, and that by using a water condenser and air separation membrane in a recycle loop with an RWGS reactor, conversions approaching 100 percent could be readily achieved.

Running the RWGS with extra hydrogen, a waste gas stream consisting of CO and H_2 can be produced. This is known as "synthesis gas" and can be reacted exothermically in a second catalytic bed to produce methanol (reaction 7.7), propylene (reaction 7.8), or other fuels. Such use of RWGS "waste" gas to make methanol was also demonstrated during the 1997 Pioneer Astronautics project, while the propylene production reaction was demonstrated by the same team during a follow-on program in 1998.

$$CO + 2H_2 \rightarrow CH_3OH \qquad (7.7)$$

$$6CO + 3H_2 \rightarrow C_3H_6 + 3CO_2 \qquad (7.8)$$

On Mars, buffer gas for breathing systems, consisting of nitrogen and argon, can be extracted directly from the atmosphere using pumps, as these gases comprise 2.7 percent and 1.6 percent of the air there, respectively. Water can also be extracted from the atmosphere using zeolite sorption beds, as shown by Adam Bruckner, Steve Coons, and John Williams at the University of Washington. Alternatively, it can be baked out of the soil, which the Viking landers revealed to consist of at least 1 percent water by weight (and probably several percent in fact), or mined from the permafrost. Subsurface liquid water may also be accessible using drilling rigs.

Iron can also be produced on Mars very readily using either reaction 7.9 or 7.10. I say "very readily" because the solid feedstock, Fe_2O_3, is so omnipresent on Mars that it gives the planet its red color and thus indirectly its name.

$$Fe_2O_3 + 3H_2 \rightarrow 2Fe + 3H_2O \qquad (7.9)$$

$$Fe_2O_3 + 3CO \rightarrow 2Fe + 3CO_2 \qquad (7.10)$$

Reaction 7.9 is mildly endothermic (energy consuming) and can be used with a water electrolysis recycling system to produce oxygen as well. Reaction 7.10 is mildly exothermic (energy producing) and can be used in tandem with an electrolyzer and an RWGS unit to also produce oxygen. The iron can be used as such or turned into steel, as carbon, manganese, phosphorus, silicon, nickel, chromium, and vanadium, the key elements used in producing the principal carbon and stainless steel alloys, are all relatively common on Mars.

The carbon monoxide produced by the RWGS can be used to produce carbon via

$$2CO \rightarrow CO_2 + C \qquad (7.11)$$

This reaction is exothermic and occurs spontaneously at high pressure and temperatures of about 600°C. The carbon so produced can be used to produce silicon or aluminum via reactions 7.12 and 7.13:

$$SiO_2 + 2C \rightarrow 2CO + Si \qquad (7.12)$$

$$Al_2O_3 + 3C \rightarrow 2Al + 3CO \qquad (7.13)$$

Both SiO_2 and Al_2O_3 are common on Mars, so finding feedstock will be no problem. Reactions 7.12 and 7.13 are both highly endothermic, however. So except for specialty applications where aluminum is really required, steel will be the metal of choice for Martian construction.

On the Asteroids

The asteroids are rich in metals and also possess carbon; so much of the carbon-based resource utilization reactions developed for Mars could also be used there. Of special interest to asteroid miners will be a means of acquiring pure samples of various metals for purposes of commercial export. One way to do this is to produce carbonyls, as pointed out by the University of Arizona's Professor Lewis.

For example, carbon monoxide can be combined with iron at 110°C to produce iron carbonyl [$Fe(CO)_5$], which is liquid at room temperature. Then iron carbonyl can be poured into a mold and then heated to about 200°C, at which time it will decompose. Pure iron, very strong, will be left in the mold, while the carbon monoxide will be released, allowing it to be used again. Similar carbonyls can be formed between carbon monoxide and nickel, chromium, osmium, iridium, ruthenium, rhenium, cobalt, and tungsten. Each of these carbonyls decomposes under slightly different conditions, allowing a mixture of metal carbonyls to be separated into its pure components by successive decomposition, one metal at a time.

An additional advantage of this technique is the opportunities it offers to enable precision low-temperature metal casting. You can take the iron carbonyl, for example, and deposit the iron in layers by decomposing carbonyl vapor, allowing hollow objects of any complex shape desired to be made. For this reason, carbonyl manufacturing and casting will no doubt also find extensive use on Mars.

According to the ballad, Davy Crockett "knew every tree." His successors on the space frontier will need to know every rock.

CHAPTER 8

Settling the Outer Solar System

But let that man with better sense advise
That of the world least part to us is red;
And daily how through hardy enterprise
Many great regions are discovered,
Which to late ages were never mentioned.
Who ever heard of th' Indian Peru?
Or who in venturous vessel measured
The Amazon huge river now found true?
Or fruitfullest Virginia who did ever view?
Yet all these were when no man did them know,
Yet have from wisest ages hidden been;
And later times more things unknown shall show.
Why then should witless man so much misweene
That nothing is, but that which he hath seen?
What if within the Moon's fair shining sphere?
What if in every other star unseen
Of other worlds he happily should hear?
He wonder would much more: yet such to some appear.

—EDMUND SPENSER,
The Faerie Queene, 1590

THE OUTER SOLAR system is a vast arena including within its domain four spectacular giant planets, a minor planet, six moons of planetary size and scores of smaller ones, and several known and probably a myriad of unknown asteroidal and cometary objects of every description imaginable or unimaginable.[1] We have only begun to explore this realm, and what we know or will know from our telescopes, the Pioneer and Voyager spacecraft of the 1970s and 1980s, the Galileo probe now orbiting Jupiter, and the Cassini mission on its way to Saturn will barely scratch the surface of the worlds of secrets nature has placed there. A generation of more capable probes will be needed, with nuclear power to allow for active sensing through thick atmospheres and data communication rates thousands of times Voyager's, and still we will not know. The human mind will have to follow, and the challenge of the distances will demand the development of propulsion technologies far more capable than those needed for human missions to the Moon or Mars. A measure of time thus will pass before the outer solar system becomes the domain of human activity, but it surely will. For though the future can be but dimly seen, we already know that these outer worlds contain the keys to continued human survival, and progress, and our posterity's hopes for the stars.

THE SOURCES OF POWER

To glimpse the probable nature of the human condition a century hence, it is first necessary for us to look at the trends of the past. The history of humanity's technological advance can be written as a history of ever-increasing energy utilization. If we consider the energy consumed not only in daily life but in transportation and the production of industrial and agricultural goods, then Americans in the electrified 1990s use approximately three times as much energy per capita as their predecessors of the steam and gaslight 1890s, who in turn had nearly triple the per-capita energy consumption of those of the preindustrial 1790s. Some have decried this trend as a direct threat to the world's resources, but the fact of the matter is that such rising levels of energy consumption have historically correlated rather directly with rising living standards and, if we compare living standards and per-capita energy consumption of the advanced sector nations with those of the impoverished Third World, continue to do so today. This relationship between energy consumption and the wealth of nations will place extreme demands on our current set of available resources. In the first place, simply to raise the entire present world population to current American living standards (and in a world of global communications it is doubtful that any other arrangement will be acceptable in the long run) would require increasing global energy consumption at least ten times. However, world population is increasing, and while global industrialization is slowing this trend, it is likely that terrestrial population levels will at least double before they stabilize. Finally, current American living standards and technology utilization are hardly likely to be the ultimate (after all, even in late-twentieth-century America, there is still plenty of poverty) and will be no more acceptable to our descendants a century hence than those of a century ago are to us. All in all, it is clear that the exponential rise in humanity's energy utilization will continue. In 1998, humanity mustered about 14 terawatts of power (1 terawatt, TW, equals 1 million megawatts, MW, of power). At the current 2.6 percent rate of growth we will be using nearly 200 TW by the year 2100. The total anticipated power utilization and the cumulative energy used (starting in 1998) is given in Table 8.1

By way of comparison, the total known or estimated energy resources are given in Table 8.2.

In Table 8.2, the amount of He3 given for each of the giant planets is

▪ TABLE 8.1 ▪

Projected Human Use of Energy Resources

Year	Power (TW)	Energy Used after 1998 (TW-Years)
2000	15	29
2025	28	545
2050	53	1,520
2075	101	3,380
2100	192	7,000
2125	365	13,700
2150	693	26,400
2175	1,320	50,600
2200	2,500	96,500

■ TABLE 8.2 ■

Solar System Energy Resources

Resource	Amount (TW-Years)
Known terrestrial fossil fuels	3,000
Estimated unknown terrestrial fossil fuels	7,000
Nuclear fission without breeder reactors	300
Nuclear fission with breeder reactors	22,000
Fusion using lunar He3	10,000
Fusion using Jupiter He3	5,600,000,000
Fusion using Saturn He3	3,040,000,000
Fusion using Uranus He3	3,160,000,000
Fusion using Neptune He3	2,100,000,000

that present in their atmospheres down to a depth where the pressure is ten times that of the Earth's at sea level. If one extracted at a depth where the pressure was greater, the total available He3 would increase in proportion. If we compare the energy needs for a growing human civilization with the availability of resources, it is clear that, even if the environmental problems associated with burning fossil fuels and nuclear fission are completely ignored, within a couple of centuries the energy stockpiles of the Earth and its Moon will be effectively exhausted. Large-scale use of solar power can alter this picture somewhat, but sooner or later the enormous reserves of energy available in the atmospheres of the giant planets must be brought into play.

Thermonuclear fusion reactors work by using magnetic fields to confine a plasma consisting of ultra-hot charged particles within a vacuum chamber where they can collide and react. Since high-energy particles have the ability to gradually fight their way out of the magnetic trap, the reactor chamber must be of a certain minimum size so as to stall the particles' escape long enough for a reaction to occur. This minimum size requirement tends to make fusion power plants unattractive for low-power applications, but in the world of the future where human energy needs will be on a scale tens or hundreds of times greater than today, fusion will be far and away the cheapest game in town.

A century or so from now, nuclear fusion using the clean-burning (no radioactive waste) D-He3 reaction will be one of humanity's primary sources of energy, and the outer planets will be the Persian Gulf of the solar system.

THE PERSIAN GULF OF THE SOLAR SYSTEM

Today the Earth's economy thirsts for oil, which is transported over oceans from the Persian Gulf and Alaska's North Slope by fleets of oil-powered tankers. In the future, the inhabitants of the inner solar system will have the fuel for their fusion reactors delivered from the outer worlds by fleets of spacecraft driven by the same thermonuclear power source. For while the ballistic interplanetary trajectories made possible by chemical or nuclear thermal propulsion are adequate for human exploration of the inner solar system and unmanned probes beyond, something a lot faster is going to be needed to sustain interplanetary commerce encompassing the gas giants.

Fusion reactors powered by D-He3 are a good candidate for a very advanced spacecraft propulsion. The fuel has the highest energy-to-mass ratio of any substance found in nature, and, further, in space the vacuum the reaction needs to run can be had for free in any size desired. A rocket engine based upon controlled fusion could work simply by allowing the plasma to leak out of one end of the magnetic trap, adding ordinary hydrogen to the leaked plasma, and then directing the exhaust mixture away from the ship with a magnetic nozzle. The more hydrogen added, the higher the thrust (since you're adding mass to the flow), but the lower the exhaust velocity (because the added hydrogen tends to cool the flow a bit). For travel to the outer solar system, the exhaust would be over 95 percent ordinary hydrogen, and the exhaust velocity would be over 250 km/s (a specific impulse of 25,000 s, which compares quite well with the specific impulses of chemical or nuclear thermal rockets of 450 s or 900 s, respectively). Large nuclear electric propulsion (NEP) systems using fission reactors and ion engines, a more near-term possibility than fusion, could also achieve 25,000 s specific impulse. However, because of the complex electric conversion systems such NEP engines require, the engines would probably weigh about eight times as much as fusion systems and, as a result, the trips would take twice as long. If no hydrogen is added, a fusion configuration could theoretically yield exhaust velocities as high as 15,000 km/s, or 5 percent the speed of light! Although the thrust level of such a pure D-He3 rocket would be too low for in-system travel, the terrific exhaust velocity would make possible voyages to nearby stars with trip times of less than a century.

Extracting the He3 from the atmospheres of the giant planets will be difficult, but not impossible. What is required is a winged transatmospheric vehicle that can use a planet's atmosphere for propellant, heating it in a nuclear reactor to produce thrust. I call such a craft a NIFT (for Nuclear Indigenous Fueled Transatmospheric vehicle). After sortieing from its base on one of the planet's moons, a NIFT would either cruise the atmosphere of a gas giant, separating out the He3, or rendezvous in the atmosphere with an aerostat station that had already produced a shipment. In either case, after acquiring its cargo, the NIFT would fuel itself with liquid hydrogen extracted from the planet's air and rocket out of the atmosphere to deliver the He3 shipment to an orbiting fusion-powered tanker bound for the inner solar system.

In Table 8.3 we show the basic facts that will govern commerce in He3 from the outer solar system. Flight times given are one-way from Earth to the planet, with the ballistic flight times shown being those for minimum-energy orbit transfers. These can be shortened somewhat at the expense of propellant (gravity assists can help, too, but are available too infrequently to support regular commerce) but in any case are too long for commercial traffic to Saturn and beyond (even if the vessels are fully automated, time is money). The NEP and fusion trip times shown assume that 40 percent of the ship's initial mass in Earth orbit is payload, 36 percent is propellant (for one-way travel; the ships refuel with local hydrogen at the outer planet), and 24 percent is engine. Jupiter is much closer than the other giants, but its gravity is so large that even with the help of its very high equatorial rotational velocity, the velocity required to achieve orbit is an enormous 29.5 km/s. A NIFT is basically a nuclear thermal rocket with an exhaust velocity of about 9 km/s, and so even assuming a "running start" air speed of 1 km/s,

• TABLE 8.3 •

Getting Around the Outer Solar System

Planet	Distance from Sun (AU)	One-Way Flight Time (years)			Velocity to Orbit (km/s)	NIFT Mass Ratio
		Ballistic	NEP	Fusion		
Jupiter	5.2	2.7	2.2	1.1	29.5	23.7
Saturn	9.5	6.0	3.0	1.5	14.8	4.6
Uranus	19.2	16.0	5.0	2.5	12.6	3.6
Neptune	30.1	30.7	6.6	3.3	14.2	4.3

the mass ratio it would need to achieve such an ascent is over 20. This essentially means that Jupiter is off limits for He3 mining, because it's probably not possible to build a hydrogen-fueled rocket with a mass ratio greater than 6 or 7. On the other hand, with the help of lower gravity and still large equatorial rotational velocities, NIFTs with buildable mass ratios of about 4 would be able to achieve orbit around Saturn, Uranus, or Neptune.

TITAN

As Saturn is the closest of the outer planets whose He3 supplies are accessible to extraction, it will most likely be the first of the outer planets to be developed. The case for Saturn is further enhanced by the fact that the ringed planet possesses an excellent system of satellites, including Titan, a moon that, with a radius of nearly 2,600 km, is actually larger than the planet Mercury.

It's not just size that makes Titan interesting. Saturn's largest moon possesses an abundance of all the elements necessary to support life. It is believed by many scientists that Titan's chemistry may resemble that of the Earth during the period of the origin of life, frozen in time by the slow rate of chemical reactions in a low-temperature environment. These abundant pre-biotic organic compounds comprising Titan's surface, atmosphere, and oceans can provide the basis for extensive human settlement to support the Saturnian He3 acquisition operations.

Because of Titan's thick cloudy atmosphere, its surface is not visible from space, and many basic facts about this world remain a mystery. Here's what we know.

Titan's atmosphere is composed of 90 percent nitrogen, 6 percent methane, and 4 percent argon. The atmospheric pressure is 1.5 that of Earth sea level, but because of the surface temperature of 100 K (–173°C), the density is 4.5 Earth sea level. The surface gravity is one-seventh that of the Earth, and the wind conditions are believed to be light. The latest evidence from radiotelescopes using Earth-based radar indicates that the surface consists of a mixed terrain, including at least one solid continent in a methane ocean. These results are crude, but the radar on the Cassini probe should allow us to map out Titan's continents and oceans fairly well. In addition, a Titan probe carried by Cassini will parachute into the moon's murky atmosphere and return reams of atmospheric data and more than 1,000 images of Titan's environs. The presence of higher hydrocarbons and other organic compounds within the bodies of liquid methane is highly probable, but the precise chemical nature of the mixture of unknown. Hydrocarbon and ammonia ice may also exist.

The same nuclear thermal rocket (NTR) engines that power NIFT vehicles mining Saturn's atmosphere could use the methane abundant in Titan's atmosphere as propellant to enable travel not only all over Titan, but throughout most of the Saturnian system. For example, because of Titan's thick atmosphere and low gravity, an 8-tonne nuclear thermal flight vehicle operating in an air-breathing mode in Titan's atmosphere at a flight speed of 160 km/hr would require a wing area of only 4 square meters to stay aloft—in other words, no wings at all. Employing the methane as rocket propellant in an NTR engine, a specific impulse of about 560 seconds (5.5 km/s exhaust velocity) could be achieved. The ΔV required to take off from Titan and go onto an elliptical orbit with a minimum altitude just above Saturn is only 3.2 km/s. Because the specific impulse of the rocket is high and the required mission ΔV is low, the mass ratio of the Titan-Saturn NTR ferry would only have to be about 1.8, which means that it could deliver a great deal of cargo. The downward shipped cargo would be released in pods equipped with aeroshields that would allow them to brake from the elliptical transfer orbit down to the low circular orbit of a Saturn helium-3 processing station, which supports the operation of the Saturn-diving NIFTs. After releasing the cargo pods, the ferry would continue on its elliptical orbit until it reached its apogee at Titan's distance from Saturn, just six days after its initial departure. Because Titan's orbital period is sixteen days, it would not be there to meet the ferry. So a small rocket burn would be effected that would raise the orbit's periapsis (low point) a bit, thereby adjusting the orbital period of the ferry to ten days, allowing it to rendezvous with Titan and aerobrake and land on the next go round. Most of the cargo delivered to low Saturn would be supplies or crew for the orbiting NIFT base. However, some could be pods filled with methane propellant. These could be stockpiled at the orbiting station. When enough are accumulated to enable the 9 km/s ΔV needed to travel from low Saturn orbit onto a trans-Titan trajectory, a ferry could aerobrake itself and go to the station, and then be used to ship crew or cargo back to Titan.

Alternatively, it might also be found desirable to use some of Saturn's lower moons (several of which are quite sizable and may represent developable worlds in their own right) as intermediate bases. This could make ferry operations a lot easier.

The propulsion requirements to travel from Titan to Saturn's other moons are shown in Table 8.4. Each excursion involves landing on the destination moon twice, engaging in activity at two locations separated by up to 40 degrees of latitude or longitude, and then returning to aerobrake and refuel at Titan.

Since methane is more than six times as dense as hydrogen, NTR vehicles using methane propellant should be able to achieve mass ratios greater than 8. It can be seen that with such capability, Titan-based NTR vehicles will be able to travel to and from all of Saturn's moons, except Mimas, virtually at will.

In certain ways, Titan is the most hospitable extraterrestrial world within our solar system for human colonization. In the almost Earth-normal atmospheric pressure of Titan, you would not need a pressure suit, just a dry suit to keep out the cold. On your back you could carry a tank of liquid oxygen, which would need no refrigeration in Titan's environment, would weigh almost nothing, and could supply your breathing needs for a week-long trip outside of the settlement. A small bleed valve off the tank would allow a trickle of oxygen to burn against the methane atmosphere, heating your breathing air and suit to desirable temperatures. With one-seventh Earth gravity and 4.5 times terrestrial sea-level atmospheric density, humans on Titan would be able to strap on wings and fly like birds. (Just as in

▪ TABLE 8.4 ▪

Titan-Based Methane-Propelled NTR Excursions to Saturn's Other Satellites

Destination	Distance from Saturn (km)	Radius (km)	ΔV (km/s)	Mass Ratio
Mimas	185,600	195	13.17	11.0
Enceladus	238,100	255	11.25	7.77
Tethys	294,700	525	10.05	6.24
Dione	377,500	560	8.60	4.79
Rhea	527,200	765	6.91	3.52
Titan	1,221,600	2,575	0.00	1.00
Hyperion	1,483,000	143	3.84	2.01
Iapetus	3,560,100	720	6.90	3.52
Phoebe	12,950,000	100	8.33	4.56

the story of Daedalus and Icarus—though being more than nine times distant from the Sun than Earth, such fliers wouldn't have the worry of their wings melting.) Electricity could be produced in great abundance, as the 100 K heat sink available in Titan's atmosphere would allow for easy conversion of thermal energy from nuclear fission or fusion reactors to electricity at efficiencies of better than 80 percent. Most important, Titan contains billions of tonnes of easily accessible carbon, hydrogen, nitrogen, and oxygen. By utilizing these elements together with heat and light from large-scale nuclear fusion reactors, seeds, and some breeding pairs of livestock from Earth, a sizable agricultural base could be created within a protected biosphere on Titan.

COLONIZING THE JOVIAN SYSTEM

We have discussed colonization of Saturn and the major planets beyond. Why not Jupiter, which is much closer to Earth and has four giant moons to Saturn's one? The answer is that as interesting as Jupiter's system is scientifically, its development will probably follow that of Saturn, primarily because the giant planet's enormous gravitational field makes extracting its atmospheric helium-3 supplies extremely difficult. Another problem facing the development of Jupiter is its extremely powerful radiation belts, within which many of its moons orbit.

Table 8.5 shows Jupiter's satellite system and gives the radiation dose that would be experienced by an unshielded human on the surface of each. The radiation doses are my own calculation based on data produced by James Van Allen during the Pioneer 10 and 11 missions.

In Table 8.5, radiation doses of "0" mean negligible doses from Jupiter's radiation belts as such. There still would be the normal cosmic-ray doses of about 0.14 rem/day. Also, while doses from Jupiter's belts would be negligible under normal circumstances, they would be much higher when the satellites occasionally pass through Jupiter's enormous magnetotail, which extends in the anti-sunward direction from Jupiter for hundreds of millions of kilometers. Presumably, however, people could take cover underground during these occurrences.

A radiation dose of 75 rem or more, if delivered during a short time compared to the cell repair and replacement cycles of the human body, say

▪ TABLE 8.5 ▪

The Jupiter System

Moon	Distance from Jupiter (km)	Radius (km)	Radiation Dose (rem/day)
Metis	127,960	20	18,000
Adrastea	128,980	10	18,000
Amalthea	181,300	105	18,000
Thebe	221,900	50	18,000
Io	421,600	1,815	3,600
Europa	670,900	1,569	540
Ganymede	1,070,000	2,631	8
Callisto	1,883,000	2,400	0.01
Leda	11,094,000	8	0
Himalia	11,480,000	90	0
Lysithea	11,720,000	20	0
Elara	11,737,000	40	0
Ananke	21,200,000	15	0
Carme	22,600,000	22	0
Pasiphae	23,500,000	35	0
Sinope	23,700,000	20	0

30 days, will generally cause radiation sickness, while doses over 500 rem will result in death. It can be seen that on Europa and all moons farther in, such fatal doses would be administered to unshielded humans within a single day. On Ganymede the dose rate is not too bad, provided that people generally stayed in shielded quarters and only came out on the surface for a few hours now and then to perform essential tasks. On Callisto and those moons farther out, Jupiter's radiation belts are not an issue, except during the time of magneto tail pass-through, as discussed above.

So, of Jupiter's planetary-sized satellites, only Callisto and perhaps Ganymede can be considered reasonable targets for human settlement. They are big places and possess such necessary elements as water, carbonaceous material, metals, and silicates. Sunlight is too dim for solar to be a viable power supply, but there is a reasonable chance that geothermal power generated by tidal interaction with Jupiter may be available on Ganymede (it certainly is on Io, the innermost major moon, whose active volcanoes are so numerous that some were photographed during eruption by the *Voyager* probe) and possibly on Callisto as well. The moons beyond Callisto are probably captured asteroids. Their main attraction compared to those in the Main Belt would be that they are permanently stationed in the Jupiter system—if a significant branch of human society should develop on Callisto, their exploitation could readily be supported from that location.

Jupiter's curse is its gravity field. Paradoxically, that may also prove to be its greatest resource. Jupiter is unmatched in its ability to slingshot a spacecraft on an exceptionally fast trajectory with no propellant costs. You simply "drop" your spacecraft so it falls toward Jupiter but misses to perform a fast swing-by instead. If your spacecraft is not bound to Jupiter—say it is heading into the outer solar system from Earth—this is easy to do. By whipping past Jupiter, you can use such a "gravity assist" to add a great deal of velocity to the spacecraft at no cost in propellant. This was the trick that enabled the Voyager missions.

But even if you are on an orbit that is bound within Jupiter's system, you can still use its gravity to generate fast departure velocities. I know this statement sounds bizarre, especially to people who know their basic physics, but it's true.

Let's say you are living on the manufacturing colony of Callisto and you want to ship out some supplies fast to the helium-3 mining operation based on Saturn's moon Titan. There's ice on Callisto, so you can use this to make hydrogen/oxygen rocket propellant to get to orbit and perform high-thrust maneuvers in the Jupiter system. It will require a ΔV of about 2.4 km/s to take off from Callisto and reach a highly elliptical parking orbit about the moon. There you refuel, or transfer yourself and some propellant to a dedicated interplanetary spacecraft, and then execute a ΔV of 1.4 km/s to depart Callisto onto an elliptical orbit with its closest approach to Jupiter at 489,000 km from the planet's center. This orbit will have a period exactly half that of Callisto's, so after two of your orbits you will meet Callisto again (16.7 days later). Along the way, you make it your business to pass close by either Europa or Ganymede and use their gravity to distort your orbit a bit, so as to give you an increased encounter velocity when you return to Callisto. At that point you perform still another gravity assist to lower your closest approach to Jupiter to 78,640 km from the giant's center, which means you will pass above its surface at an altitude of 7,150 km. This will take you through the thick of Jupiter's radiation belts, and any crew or sensitive electronics aboard will have to be well shielded. Because you have dived so low, your velocity at minimum altitude will be an enormous 55.7

km/s (nearly 125,000 mph). Jupiter's escape velocity at that altitude is 56.8 km/s, so a little extra push of 1.1 km/s would allow you to depart into interplanetary space. But instead of giving a little push, you give a big one, firing your chemical rocket to deliver a ΔV of, say, 6 km/s. Rocket propulsion systems don't know or care how fast you are flying; they only know how much velocity they add. But the energy of a spacecraft is a function of the square of its velocity. So the faster you are already going, the more energy you add to the trajectory with a given velocity addition. The relevant equation is:

$$V_d^2 = V_{max}^2 - V_e^2 \qquad (8.1)$$

where V_d is the velocity that the spacecraft departs the planet, V_{max} is the maximum velocity achieved right after the spacecraft fires its engine during its fast dive through low orbit, and V_e is the planet's escape velocity at the lowest point of the orbit. The ramifications of this equation are discussed in *The Case for Mars.* Without going into the math here, let's show the effect of applying it to the case at hand, firing our rocket engine during the above-described low pass over Jupiter. The results are shown in Table 8.6.

▪ TABLE 8.6 ▪

Departing Jupiter at High Velocity Using High-Thrust Rockets (initial orbit is 78,640 km × 1,883,000 km around Jupiter's center)

Rocket ΔV (km/s)	Maximum Velocity (km/s)	Departure Velocity (km/s)
1.1	56.8	0
1.5	57.2	6.8
2	57.7	10.2
3	58.7	14.8
4	59.7	18.4
5	60.7	21.4
6	61.7	24.1
7	62.7	26.6
8	63.7	28.8
9	64.7	31.0
10	65.7	33.0

So, in exchange for your rocket's own velocity increment of 6 km/s imparted during your orbital dive, you can go screaming out of the Jupiter system at the phenomenal clip of 24 km/s! That's an initial speed for your spacecraft of nearly 5 AU per year, with nothing but chemical propulsion. Advanced propulsion systems on board, such as nuclear electric propulsion or fusion, could then be used to accelerate the system even more after departure.

Thus, once there is helium-3 commerce to be supported in the outer solar system, Jupiter, using the resources of its outer moons and its gravity well, could develop as an important solar system transportation node.

Nineteenth-century New Englanders thought they had an unmatchable racket selling ice. Imagine the envy of those sharp-minded old-time Yankees if they could awake from their graves and look into the future to see Callisto colonists selling . . . gravity!

MOVING ICETEROIDS

As we move out through the asteroid belt toward Jupiter, we find bodies increasingly composed of volatile material. This is to be expected. An asteroid made of ice would vaporize if it orbited for long near the Earth, and other volatiles, such as ammonia or hydrocarbons, would evaporate from asteroids orbiting near Mars or even in the central Main Belt. It is therefore reasonable to assume that this trend continues beyond Jupiter and that the outer solar system should be rich in asteroid-sized objects consisting almost entirely of frozen volatiles, such as water, ammonia, and methane ice. As of this writing, only one major ice asteroid or "iceteroid" is known, but that one, Chiron, orbiting between Saturn and Uranus, is rather large (180 km in diameter) and it's a rule of thumb in astronomy that a lot of small objects can be found for every big one. In all probability, the outer solar system contains thousands of asteroids that we have yet to discover because they shine so dimly compared to those in the Main Belt (the brightness of an asteroid as seen from Earth is inversely proportional to the fourth power of its distance from the Sun). Furthermore, at this point, it is known conclusively that starting not far beyond Neptune (which orbits at 31 AU) there is an enormous zone known as the Kuiper Belt that contains millions of ice objects. Beyond that lies the astronomically still vaster domain of the Oort

Cloud, stretching out more than a light-year (64,000 AU) and home to trillions of frozen objects. It is from these regions that comets originate.

Once nuclear thermal rockets become available, an object made of volatiles is basically an object made of rocket propellant. Such objects, including very large ones, can therefore be moved about the solar system in accord with human designs.

The late twenty-first century will see widespread human activity throughout the inner solar system ranging from mining asteroids and the Moon to terraforming Mars. Many of these activities may require the importation of large quantities of volatiles for their support. Now, the easiest way to move a lot of stuff around the solar system is in the form of an asteroid. But why go to the outer solar system for it? The reason, strange as it may seem, is that it is easier to move an asteroid from the outer solar system to Mars, for example, than it is to do so from the Main Belt or any other inner solar system orbit. This odd result follows from the laws of orbital mechanics, which cause an object farther away from the Sun to orbit it slower than one that is closer in. Because an object in the outer solar system moves slower, it takes a smaller velocity change (or ΔV) to alter its orbit from a circular to an elliptical shape. Furthermore, the orbit does not have to be so elliptical that it stretches from Mars to the outer solar system—it is sufficient to distort the object's orbit so that it intersects the path of a major planet, after which a gravity assist can do the rest. The results are shown in Figure 8.1. It can

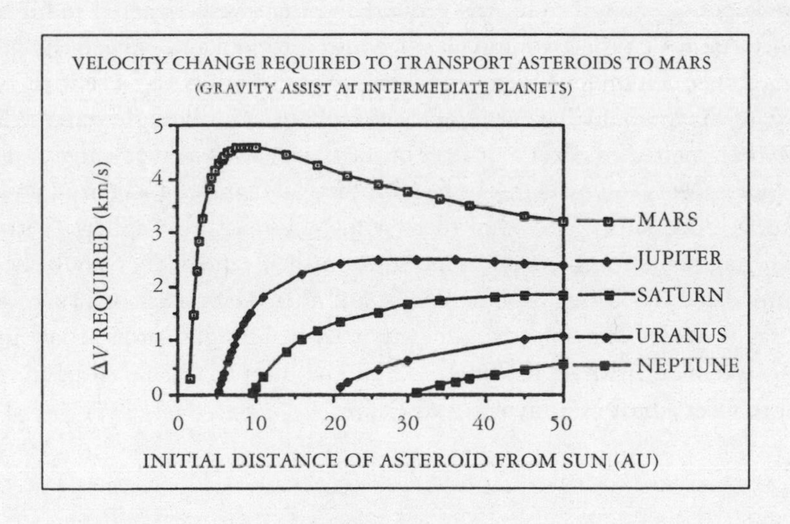

FIGURE 8.1 *Moving outer solar system asteroids.*

be seen that moving an asteroid positioned in a circular orbit at 25 AU, by way of a Uranus gravity assist to Mars, requires a ΔV of only 0.3 km/s, compared to a ΔV of 3.0 km/s to move an asteroid directly to Mars from a 2.7-AU position in the Main Belt.

Consider an asteroid made of frozen ammonia with a mass of 10 billion tonnes orbiting the Sun at a distance of 12 AU. Such an object, if spherical, would have a diameter of about 2.6 km, and changing its orbit to intersect Saturn's (where it could get a trans-Mars gravity assist) would require a ΔV of 0.3 km/s. If a quartet of 5,000-MW NTR engines powered by either fission or fusion were used to heat some of its ammonia to 1,900°C (5,000-MW fission NTRs operating at 2,200°C were tested in the 1960s), they would produce an exhaust velocity of 4 km/s, which would allow them to move the asteroid onto its required course using only 8 percent of its material as propellant. Ten years of steady thrusting would be required, followed by about a 20-year coast to arrival. If the object were suitably fragmented in advance, it could be allowed to enter and vaporize in Mars' atmosphere. In the course of doing so, it would release about 10 TW-years of energy, enough to melt 1 trillion tonnes of water (a lake 140 km on a side and 50 meters deep). In addition, the ammonia released by a single such object would raise the planet's temperature by about 3°C and form a shield that would effectively mask the planet's surface from ultraviolet radiation. Forty such missions would double the nitrogen content of Mars' atmosphere by direct importation, and could produce much more if parts of the iceteroid were kept big enough to hit the ground and then were targeted to hit beds of nitrates, which they would volatilize into nitrogen and oxygen upon impact. If one such mission were launched per year, within half a century or so most of Mars would have a temperate climate, and enough water would have been melted to cover a quarter of the planet with a layer 1 meter deep. Of course, Mars colonists might be a bit leery of using big iceteroid chunks as surface impactors, and even mere wholesale mass atmospheric entry of small fragments might make some squeamish. Even so, they could be accommodated. By using Jupiter, Venus, and then Mars itself in a succession of gravity assists, the object could eventually be brought into a sedate inner solar system orbit where its contents could be chipped off and shipped off at will to meet a host of human purposes.

EXPLORATION OF THE OUTER SOLAR SYSTEM

Human exploration of the outer solar system began in a serious way with Galileo, who was the first to turn the telescope in that direction. The government of Venice rewarded Galileo handsomely for perfecting the spyglass, as it had great utility for naval warfare, but the Church authorities were much less than pleased by what he saw in the heavens. Pointing his instrument into the night of January 7, 1610, Galileo identified the four major satellites (the "Galilean satellites") of Jupiter: Io, Europa, Ganymede, and Callisto, each comparable in size to the planet Mercury. This discovery of astronomical bodies orbiting a planet other than the Earth was a major blow to the Ptolemaic-Aristotelian worldview, upon which the Church had pinned its authority. It also had extensive practical significance for the coming age of maritime discovery. By providing navigators with a completely reliable clock in the sky, the system of Galilean satellites allowed explorers to establish their longitude anywhere in the world (since if you can set your watch in agreement with some absolute standard, the time of sunrise will give you your longitude). Thus, by engaging in the apparently completely impractical activity of studying Jupiter, Galileo finally made it possible for humans to map and reliably navigate the Earth. The result was an age of long-range maritime commerce that generated fortunes that made the hoards of the Venetian merchant city-state seem quite petty in comparison.

As telescopes grew and improved, other important finds followed with the discovery of Saturn's rings and its moon Titan by Christiaan Huygens in 1655; Jupiter's giant red spot by Giovanni Domenico Cassini in 1665; the planet Uranus by William Herschel in 1781; the planet Neptune by John Couch Adams and Urbain-Jean-Joseph LeVerrier in 1846; Neptune's giant moon Triton by William Lassell in 1846; Pluto by Clyde Tombaugh in 1930; Titan's atmosphere by Gerard Kuiper in 1944; and the giant iceteroid Chiron by Charles Kowal in 1977. By the 1960s, Jupiter was known to have twelve satellites, Saturn nine, Uranus five, and Neptune two.

Outer solar system exploration was revolutionized in the 1970s with the advent of robotic exploration spacecraft, starting with the Pioneer 10 mission to Jupiter, which flew by the giant planet in 1972. This was followed by Pioneer 11 in 1973 and Voyager 1 in 1977, both of which made a close pass by Jupiter and then flew on to visit Saturn, reaching the ringed planet in 1979 and 1980, respectively. This was the first demonstration of multiple planet flyby missions using gravity assists, a technique that Voyager 2 took to a brilliant conclusion by visiting Jupiter (1979), Saturn (1981), Uranus (1986), and Neptune (1989) in succession. The political maneuvers that made Voyager 2's mission possible were almost as tricky as the celestial mechanics. To save money, the Carter- and Reagan-era NASA headquarters brass as well as bureaucrats in the Office of Management and Budget wanted to limit Voyager 2's mission to Jupiter and Saturn only. To get the mission launched, JPL's management had to assure them this would be the case, only later gaining agreement to return data from Uranus ("since the spacecraft was on its way there anyway") and then, pulling the same trick again, to survey Neptune as well.[2]

The Voyager program was a tour de force, and stands with Viking and Apollo as one of NASA's three greatest accomplishments to date. In addition to returning volumes of spectacular color images of the giant planets, the Voyagers imaged all the known moons and discovered literally dozens of additional ones. Ring systems around Jupiter, Uranus, and Neptune were discovered, as were many new features in Saturn's rings. Magnetic fields around the giant planets were measured to reveal an enormously powerful magnetic field and associated radiation belts circling Jupiter. Measurements were taken that showed the interior of the giant planets to be much warmer than anticipated. Both Voyagers actually imaged volcanoes in the process of erupting on Io, which was very unexpected, and which made abundantly clear the heating power of geothermal energy generated by tidal forces on bodies orbiting Jupiter. The importance of this was emphasized by another of Voyager's finds—namely images of Europa showing the entire moon to be covered with ice. As revealed by Voyager, Europa's ice had fractures in it, suggesting to some a thick layer of sea ice lying above an ocean of liquid water. Prior to Voyager's observations of volcanoes on Io, no one would have thought that liquid water could possibly exist in the Jovian system. But if tidal forces could heat the interior of Io to the melting point of rock, was it not possible that the interior of Europa, the next major satellite outward from the giant planet, could be warmed tidally to above the melting point of ice? And if there were liquid water and heat beneath the ice of Europa, could that not potentially represent a home for life?

This speculation was amplified in 1996 and 1997, when Galileo imaged Europa again, at much finer resolution than Voyager, showing conclusively that Europa's ice covering is in fact sea ice, perhaps 50 km thick, floating over an ocean of liquid water that is probably 100 km deep. Not only is there liquid water on Europa, there is more liquid water on Europa than there is on Earth! During the 1980s, oceanographers had discovered deep sea life subsisting on a food chain based not on photosynthesis but on chemosynthesis linked to hot vents on the sea floor. There seems to be no fundamental reason why similar ecosystems could not exist in the depths of Europa's ocean.

Exploring Europa's ocean is now a major target for NASA's exobiological research program. The main problem is how does one penetrate 50 km of ice? At least at the surface, the ice is supercold—about –160°C. Ice at that temperature is as hard as rock. Even if human crews could be sent to Europa to set up drill rigs (which could be tough, as Europa is right in the middle of Jupiter's very dangerous radiation belts), drilling through 50 km of such material would be an incredible task. An alternative idea that I have suggested to NASA is that a radioisotope-heated sphere, built as strong as a cannonball, be released from a spacecraft and allowed to impact Europa's surface at high velocity. The sphere would thus bury itself beneath the ice and very slowly begin to melt its way down. The surface layer of meltwater around the probe would contain the chemicals, and perhaps frozen microbes, of the Europan ocean of the past. As the probe penetrated deeper, more recently created ice would be encountered and its captured contents would be made available for analysis. Thus, as it went deeper and deeper, the probe would produce a scan of Europa's ocean over a long period of geologic time. As long as the probe stayed in the ice, it could transmit data back to an orbiter using low-frequency radio. If it reached the ocean, it would lose contact (since radio can penetrate ice much better than water) and sink rapidly, but it could still take measurements. Then, when it hit the bottom, ballast could be released, allowing the probe to float back up through the ocean, taking more measurements, until it hit the ice, when radio contact with the orbiter could be reestablished.

Since the probe would penetrate the ice quite slowly, perhaps only a few meters per day, it would be very advantageous to aim it for locations where the ice is thinnest, if we hope to reach the liquid ocean in a reasonable amount of time. This could be done by equipping the carrier spacecraft with long-wavelength ice-penetrating radar, which could map the thickness of the ice sheet covering Europa from orbit. If a thin region were identified, it could then be targeted when the probe is released.

NASA is planning a Europa orbiter mission for 2003. In my view, a probe like this would make an excellent hitchhiker.

NUCLEAR POWER AND OUTER SOLAR SYSTEM EXPLORATION

In discussing current-day and near-term robotic exploration of the outer solar system, one must inevitably address the issue of nuclear power. Solar energy diminishes as the square of the distance from the Sun. At Jupiter it is only 3.7 percent as strong as it is on Earth, at Saturn 1.1 percent, at Uranus 0.28 percent, and 0.1 percent at Neptune. To make matters worse, not only does solar energy die beyond Jupiter, but spacecraft power requirements rise, since more power is needed to heat the spacecraft and to transmit data over longer distances. Indeed, without nuclear power sources, specifically the standard 300-W radioisotope thermoelectric generators (RTGs) and numerous small 1-W radioisotope heating units (RHUs), none of our outer solar system probes, including Pioneer, Voyager, Galileo, and Cassini, would have been possible.

This fact is disputed by anti-nuclear activists, such as Professor Michio Kaku, a string theorist from City College of New York, who have demonstrated and filed lawsuits to attempt to block the launch of every recent radioisotope-equipped probe. According to them, the launching of RTGs represents an intolerable risk to the Earth's environment, because in the event of a launch failure the plutonium contained by such devices could break up on reentry and pollute the world. Furthermore, they maintain, such devices are unnecessary. In a debate with NASA's former nuclear program director Dr. Gary Bennett prior to the Galileo launch, for example, Professor Kaku claimed that the mission could be just as well performed powered by batteries instead.

In fact, the anti-nuclear activists are wrong on both counts. An RTG contains about 100,000 Curies (Ci) of plutonium-238. On a personal level this is a nontrivial amount—you certainly wouldn't want it around your house—but on a global level it is utterly insignificant. To put it in perspective, if a launch were to fail and an RTG were to break up and be dispersed into the world's biosphere (instead of staying intact and sinking as a solid brick into the subseabed in the Atlantic downrange from Cape Canaveral, which is actually what would happen), it would release a radiological inventory approximately 1/100th of 1 percent as great as that released by a typical nuclear bomb test. It would represent an even smaller fraction of the radiological release emitted by each and every one of the half dozen or so sunken U.S. and Soviet nuclear submarines (such as the *Thresher*) currently rusting away on the ocean floor. Furthermore, the plutonium-238 used in RTGs is not the right kind to use in atomic bombs and has a half-life of 88 years, so it does not last as a long-lived feature of the Earth's environment. In RTGs, it is present not as a metal, but as plutonium oxide, in which form it is chemically inert. The statement that a reentering RTG could represent a significant threat to the Earth's environment is simply untrue.

Equally wrong is Professor Kaku's assertion that outer solar system probes could be powered by batteries. To see how silly this idea is, consider the Galileo spacecraft, which is powered by two 300-W RTGs and warmed by several hundred 1-W RHUs, for about 800 W in all. For the sake of discussion, let's grant that this is overkill and assume that the mission really could get by with just 200 W of power. Good primary batteries can store about 300 W-h/kg. Galileo left Earth in October 1989, and as of August 1998, or 70,000 hours later, was still functioning. At 300 W-h/kg, that would be about 47,000 kg of batteries! (The two RTGs currently on board weigh about 60 kg each; the RHU mass is negligible.) Of course, with this much battery mass, the power requirement would be much greater than 200 W, since the spacecraft would require additional power to keep the batteries from freezing. To keep 47,000 kg of batteries (about 5,000 gallons' worth) warm, we would probably need to expend at least 2,000 W. But to supply that power, we would need 470,000 kg of batteries, which would need 20,000 W to keep warm, which would require 4,700,000 kg of batteries, and so on. The mission is clearly impossible on battery power.

In fact, outer solar system exploration needs to move in the direction of significantly higher power levels if it is to be executed efficiently. Not only do we need RTGs, we need to move beyond them to actual space nuclear power reactors that use nuclear fission, rather than mere radioisotope decay, to generate tens or hundreds of kilowatts. The reason for this is very simple. On Earth, it has been said, knowledge is power. In the outer solar system, power is knowledge.

Data transmission rates are linear in proportion to transmitter power. An outer solar system probe equipped with a 30-kWe nuclear reactor can return 100 times as much data as a conventional mission equipped with a standard 300-W RTG, and returning data is what a science mission is all about. Equipped with such a power supply, an outer solar system probe would be able to use multi-kilowatt communication systems, similar to those now employed by the U.S. military, instead of the 40 W of radiated power (W rf) traveling wave tube antennas that are the current NASA interplanetary mission standard. Because it is generating 30 kWe of power, the spacecraft can also employ nuclear electric propulsion, which could probably double its payload as well, but that's just an extra. The fact that space nuclear power can increase the data return of a planetary mission by a factor of 100 is much, much more important than the fact that it can increase the payload by a factor of 2. Such multiple order of magnitude increases in mission science return easily justify the added expense entailed by a nuclear system, regardless of whether mission science payload is increased at all.

The higher data rates produced by space nuclear power would make possible very high resolution multi-spectral images (with high resolutions both spatially and spectrally), a form of science hitherto impossible in the outer solar system. Increase the data rate by a factor of 100 and you can increase the spatial resolution by a factor of 10. Instead of seeing things the size of cars, you'll be able to see things the size of cats. The number of pictures you can return also grows in direct proportion to data rate. That means that instead of returning stills, you can return movies, real motion pictures of atmospheric phenomena globally and on a small scale, a meteorologist's feast. Furthermore, having a large number of pictures greatly increases the probability of capturing transient phenomena such as lightning, avalanches, floods, waterspouts, and volcanic activity. There's no telling what we'd find, because with the tools we've had available to date, we've hardly been able to look.

Table 8.7 shows a comparison of the data transmission rates for a probe in orbit around other planets, assuming either a 300-W RTG (60 W rf) for transmitter power with a fully functional 5-meter-diameter (the size of the stuck dish on Galileo) X-band dish or a 30-kWe nuclear spacecraft transmitting with two smaller 3-meter-diameter X-band dishes to one of the big 70-meter Deep Space Network antennas. If the cheaper-to-use 34-meter dishes are used as receivers instead, data rates for both options would be one-quarter those shown.

Assuming standard data compression techniques are used, 1 kb/s translates into about three good photographs transmitted per hour. Even if the nuclear mission were to cost twice as much as the conventional one, it's pretty clear which mission offers the higher payoff.

But there's more. The high-power system would also augment the mission science return by enabling active sensing, probing the planet with electromagnetic waves. This is not science fiction—two state-of-the-art techniques are readily available now in the form of radar and radio occulta-

Funded by the Strategic Defense Initiative, the DC-X changed people's thinking about the feasibility of reusable single-stage-to-orbit launch vehicles. *(Photo courtesy of NASA)*

NASA responded to the success of DC-X by initiating the X-33 reusable launch vehicle program. Shown here are the concepts bid by Lockheed Martin, Rockwell, and McDonnell Douglas. Lockheed Martin won with a design combining an aerodynamic lifting body with an aerospike engine. *(Artwork courtesy of NASA)*

The Kistler company has developed a concept for a reusable two-stage-to-orbit vehicle and has succeeded in raising considerable sums for its development. *(Artwork courtesy of Kistler Aerospace)*

Rocketplanes, like Pioneer Rocketplane's *Pathfinder*, could be used to launch satellites or deliver packages quickly across transoceanic distances. *(Artwork by Michael Carroll, courtesy of Pioneer Rocketplane Corp.)*

Space asset servicing may be a viable near-term opportunity for orbital enterprise. *Shown:* Shuttle astronauts fix the Hubble Space Telescope. *(Photo courtesy of NASA)*

The International Space Station will test the utility of orbital research labs. *(Artwork courtesy of NASA)*

If launch prices can be brought down sufficiently, space business parks providing facilities for siting orbital labs, factories, and hotels may become commercially viable. *(Artwork courtesy of Orbital Properties LLC)*

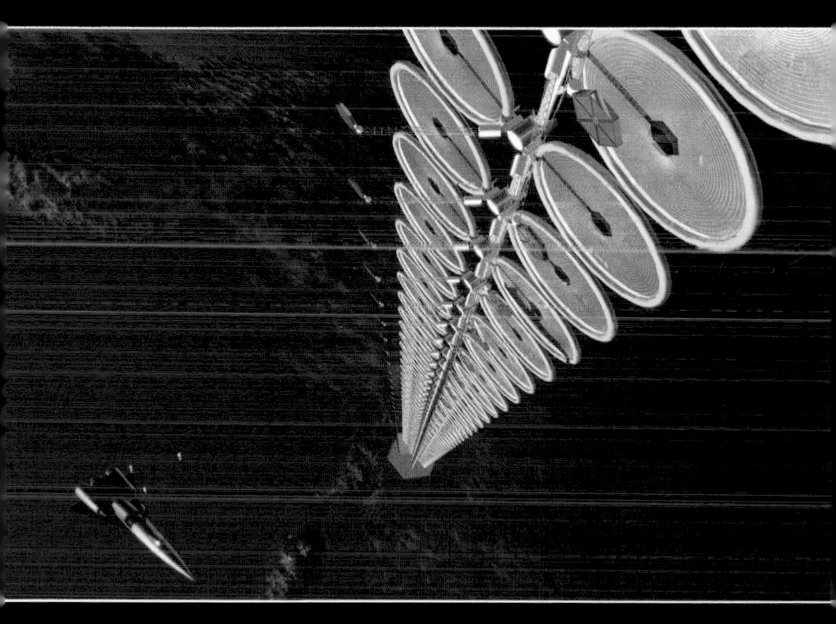

Solar-powered satellites could produce electricity in space and beam it down to Earth via microwaves. However, launch prices must drop by a factor of 2,000 to make such systems competitive with terrestrial power. Their commercial feasibility is thus doubtful. *(Artwork by Pat Rawlings, courtesy of NASA)*

Believers in solar-powered satellite systems have envisioned huge orbital cities, or "L-5 colonies," that could be constructed in space from their profits. *(Artwork courtesy of NASA)*

Helium-3, mined on the Moon, could provide a source of fuel for fusion reactors on Earth. *(Artwork by Pat Rawlings, courtesy of NASA)*

An array of telescopes on the Moon, or optical interferometer, could potentially resolve features of Earth-like planets circling other stars at distances of up to 100 light-years. *(Artwork by Pat Rawlings, courtesy of NASA)*

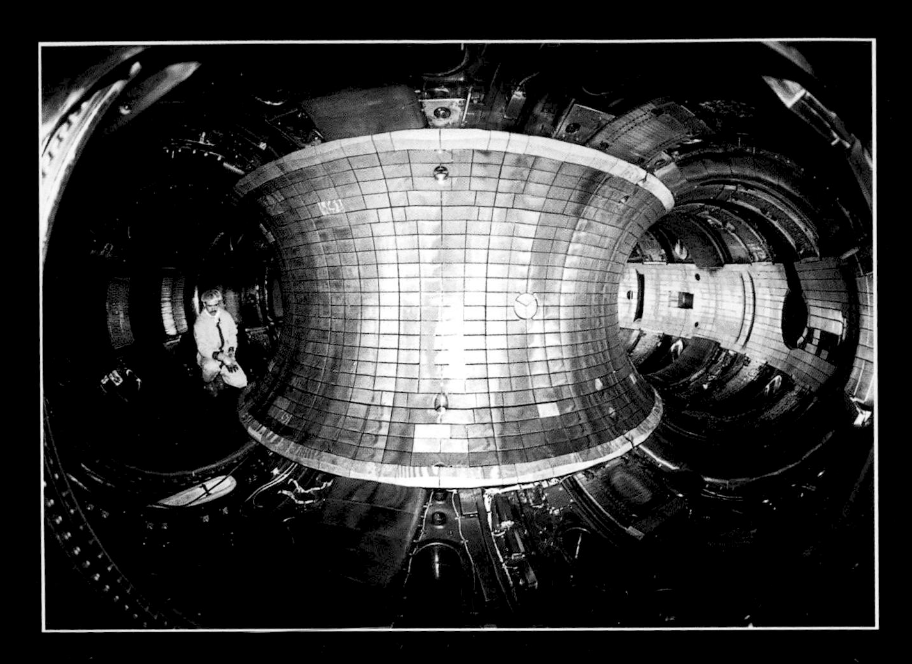

Tokamak fusion experimental reactor. In the past 30 years, experimental fusion yields have multiplied thousands of times and now approach breakeven. Further progress, however, has been blocked by shortsighted budget cuts. *(Photo courtesy of Princeton Plasma Physics Lab)*

The Alan Hills meteorite, ALH84001, originated on Mars and shows controversial evidence of life in that planet's distant past. Inset: possible microfossils. *(Photo courtesy of NASA)*

Sojourner, the first mobile system delivered to Mars, wandered successfully within a Martian flood channel in 1997 until its parent lander failure left the little robot an orphan. *(Photo courtesy of NASA/JPL)*

The Mars Direct plan. First an unfueled Earth Return Vehicle (ERV, *right*) is delivered to Mars where it manufactures its propellant from the Martian atmosphere. The crew then flies to Mars in the tuna-can-shaped hab module, which also provides living quarters, lab, and workshop facilities for a 1.5-year Mars stay. *(Artwork courtesy of Robert Murray, Pioneer Astronautics)*

Technology to manufacture rocket propellant on Mars was demonstrated by the author *(right)* and Larry Clark *(left)* at Martin Marietta in 1993. Using local resources to "live off the land" greatly reduces mission costs. *(Photo courtesy of Lockheed Martin)*

As more local resource-utilization technologies are developed, an increasingly self-sufficient initial Mars exploration base could grow into a true settlement. *(Artwork courtesy of Michael Carroll)*

The Mars Society was founded in Boulder, Colorado, in August 1998 to further the exploration and settlement of Mars by both public and private means. Over 700 people from 40 countries attended the founding convention. *(Photo by Gilbert Chew, courtesy of Mars Society)*

For its first project, the Mars Society proposes to deploy a prototype human Mars exploration base in the high arctic polar desert of

It is now believed that the dinosaurs were wiped out by the catastrophic results of an asteroid collision with Earth. Such impacts have caused numerous mass extinctions in our planet's past. *(Artwork by Don Davis, courtesy of NASA)*

Today's threat, tomorrow's bonanza. Asteroids contain trillions of dollars' worth of precious metals and could potentially be mined for profit by a spacefaring civilization. *(Artwork by William Hartman)*

Nuclear thermal rocket (NTR) engines could use asteroidal or cometary volatiles to propel them off collision course with Earth. *Shown:* NTR test firing in Nevada in 1968. *(Photo courtesy of LANL.)*

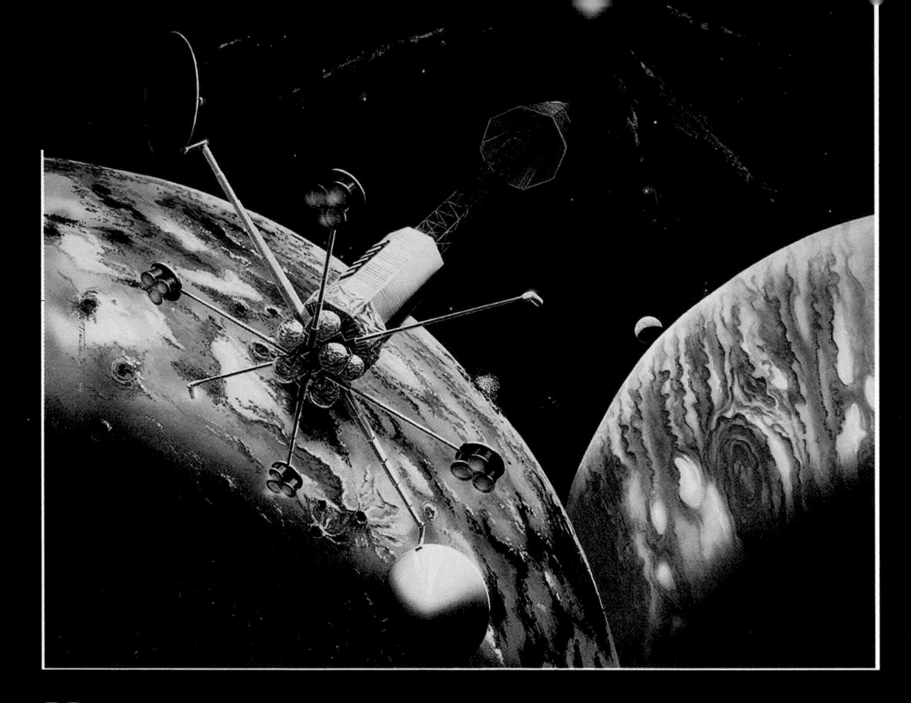

Nuclear electric propulsion can enable outer solar system robotic probes that return thousands of times the data of existing systems. *(Artwork by John Tieleman, courtesy of Lockheed Martin)*

The nuclear indigenous-fueled Titan explorer could engage in long-duration jet-propelled flight in Titan's thick atmosphere, probing the surface continents and hydrocarbon oceans with small robotic tilt-rotor aircraft. *(Artwork by John Tieleman, courtesy of Lockheed Martin)*

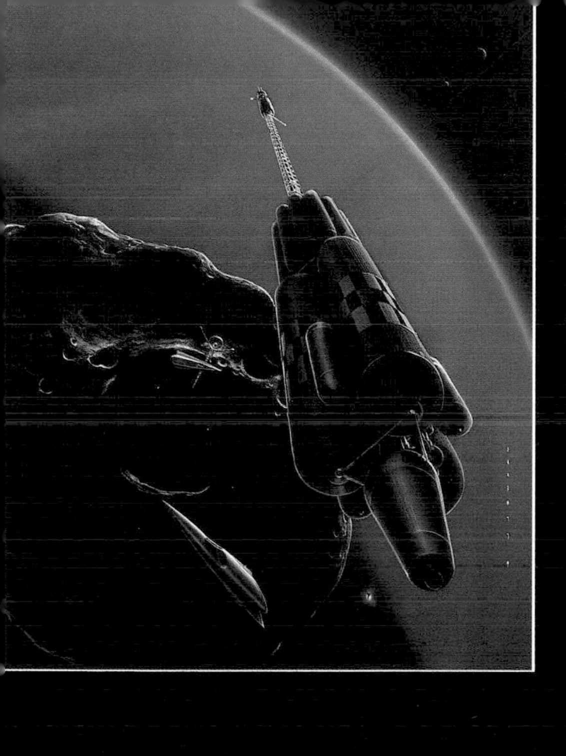

The enormous helium-3 reserves of the atmosphere of the giant planets may one day make the outer solar system humanity's primary source of energy. *Shown:* Nuclear thermal transatmospheric vehicles transport precious helium-3 cargo to a station orbiting Neptune. *(Artwork courtesy of Mark Maxwell)*

Helium-3 fusion could enable starships with velocities of about 10 percent lightspeed. *Shown:* the Daedalus spacecraft designed by the British Interplanetary Society. *(Artwork courtesy of Adrian Mann)*

Solar sails use a thin foil to reflect light, thereby generating thrust without using propellant. Ultrathin solar sails could attain 1 percent of the speed of light pushed by sunlight alone or 30 percent light speed if pushed from behind by batteries of high-power lasers. *(Artwork courtesy of Adrian Mann)*

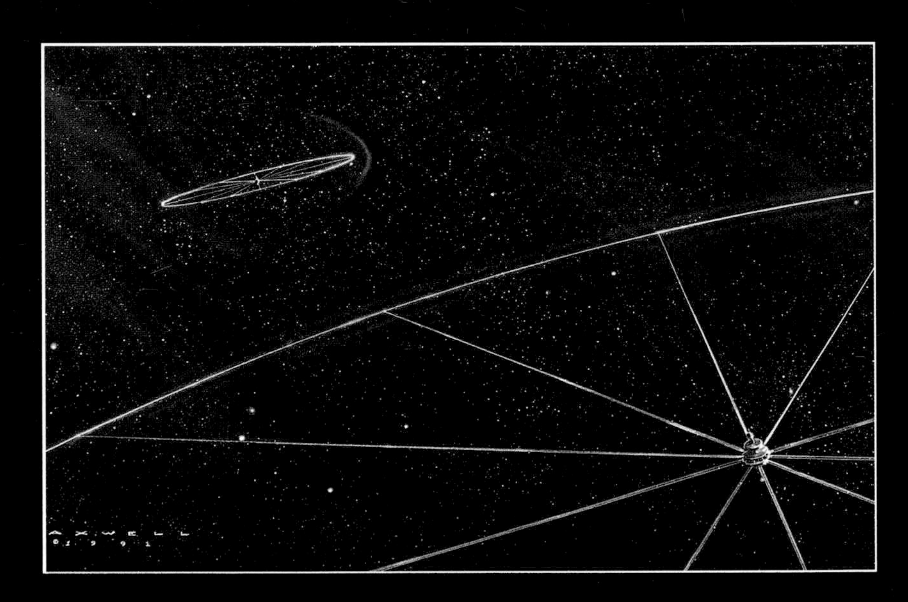

Magnetic sails use a magnetic field to deflect plasmas found in space. Using the solar wind, such devices offer promise for low-cost interplanetary travel. Dragging against the interstellar medium, they can be used to stop an ultrafast spacecraft without the need for propellant. *(Artwork courtesy of Mark Maxwell)*

▪ TABLE 8.7 ▪

Data Transmission Rates from the Outer Solar System

Planet	Distance (AU)	300-W RTG Data Rate (kb/s)	30-kWe Nuclear Data Rate (kb/s)
Mars	2	750	56,250
Asteroids	3	330	24,750
Jupiter	5	120	9,000
Saturn	9.5	33	2,475
Uranus	19.2	8.1	608
Neptune	30	3.3	248
Pluto	35	2.4	180

tion science. But they need more power if they are going to produce their best results. Consider mapping rates: The linear resolution of radar imaging improves linearly with the power available, as does the areal rate at which mapping can be done for a given spatial resolution. Such mapping rates are very important for mission designs involving single or repeated flybys. For example, the 60-We Cassini radar mapper, which will fly repeatedly past Titan over several years, will only be able to map less than 10 percent of Titan because of the slow pulse rate available. Furthermore, increasing the power also allows mapping to occur at longer ranges, a fact which is very important for flybys and for operation within systems like that of Jupiter, where strong radiation belts may preclude long-duration spacecraft operation too close to the planet. Higher power also means that ground-penetrating radars can penetrate much deeper into the subsurface of bodies such as the Galilean satellites, and topographic studies can be done at longer range with higher resolution. Radio occultation science is similarly enhanced by high transmitter power. The low power of the Voyager transmitter made its radio occultation investigation at Neptune essentially ineffective—a multikilowatt transmitter would have solved the problem completely. The spatial resolution achieved in ring occultations is also directly proportional to transmitter power. If a meaningful investigation of the dynamic structure of Saturn's rings is to be done via radio occultation, an increase in transmitter power to the 10-kWe range is essential.

High power could also enable other active sensing techniques that are not yet used at all. Tunable lasers could undertake sounding the atmospheres of the major planets for chromophores. Surface chemistry of solid bodies could also be assessed by exciting them with lasers or ion or electron particle beams.

But put most generally, higher power would allow active sensing investigations everywhere to penetrate much deeper into a planet's atmosphere, or much farther underground. It's important to understand how key this is. The vast majority of every planet is to be found beneath its surface. A human limited to staring at the surface of the ocean knows almost nothing about what is going on in the ocean, and the world of life to be found in a coral reef will astonish him the first time he puts on a diving mask. Similarly, observing a planet passively from orbit gives us a very limited view. There could be underground rivers and oceans on Mars or Io, for all we know. We don't know. Without the kind of power required for deep active sensing we're almost blind.

The future human development of the outer solar system will require nuclear power systems generating tens to hundreds of megawatts. Near-term outer solar system robotic exploration needs nuclear power units offering tens to hundreds of kilowatts, which is about the same size as that required to support human activities on the Moon and Mars. Prior to the Clinton administration, both NASA and the U.S. Air Force had programs to develop such power sources. However, as part of its program (curiously dubbed "Building America's Bridges to the 21st Century") the administration took swift measures to wreck both during its first year in office. This decision needs to be reversed.

THE ROAD TO THE STARS

The two main obstacles to settling the outer solar system are power and transportation. As mentioned earlier, solar energy in the realm of the gas giants and beyond is negligible. However, in the era we are discussing, we can expect that fusion powered by helium-3 will be the dominant energy source. Indeed, the need to acquire helium-3 to fuel such systems will be one of the prime motivations for the colonization of the far worlds of the outer solar system.

As for the issue of transportation, I call current space transportation systems first generation. These are sufficient for launch into Earth orbit, for manned missions to the Moon, Mars, and near-Earth asteroids, and for limited-capability unmanned probes to other planets. For colonization of the inner solar system, out to the Main Belt, we need to move on to second-generation systems, typified by nuclear thermal rocket propulsion, nuclear electric propulsion, and advanced aerobraking technology. Such second-generation systems also open up capabilities for vastly expanded unmanned exploration of the outer solar system. They are, however, marginal for manned colonization of Titan, as the three- to four-year one-way flight times they impose on this mission are excessive. However, as the fusion economy initiated by the Moon's supply of He3 grows, demands will be developed that can be satisfied only by the vastly larger stocks of this substance available in the outer solar system. By improving the in-space life support systems associated with second-generation technologies (and by moving from the second generation's simple air and water recycling in the direction of closed cycle ecology as the basis for very long term life support), a few pioneers will make their way to Titan using second-generation transportation technologies. Once even a small base is established on Titan, there will be a tremendous incentive to develop third-generation systems, such as fusion propulsion (especially since we will then have the abundant He3 supplies needed to fuel them). This will allow for quick trips and rapid development of Titan and the rest of the outer solar system. Such third-generation propulsion systems, however, together with fully third-generation closed cycle ecological life support, will enable travel beyond the nine known planets to the Oort Cloud and, when advanced to their limits, will create a basis for interstellar missions, with flight times to nearby stars on the order of 50 to 100 years.

Humans will go to the outer solar system not merely to work, but to live, to love, to build, and to stay. But the irony of the life of pioneers is that if they are successful, they conquer the frontier that is their only true home, and a frontier conquered is a frontier destroyed. For the best of humanity, then, the move must be ever outward. The farther we go, the farther we will become able to go, and the farther we shall need to go. Ultimately the outer solar system will simply be a way station toward the vaster universe beyond. Just as Columbus's discovery of the New World called into being the full rigged sailing ships, steamers, and Boeing 707s that allowed the rest of humanity to follow in his wake, so those brave souls who dare the great void to our neighbor stars with ships of the third generation will draw after them a set of fourth-generation space transportation systems, whose capabilities will open up the galaxy for humankind.

For while the stars may be distant, human creativity is infinite.

FOCUS ON: THE KUIPER BELT AND OORT CLOUD

It is generally considered that beyond the Sun's family of planets there is absolute emptiness extending for light-years until you come to another star. In fact it is likely that the space around the Solar System is populated by huge numbers of comets, small worlds a few miles in diameter, rich in water and other chemicals essential to life . . . comets, not planets, are the major potential habitat of life in space.

—FREEMAN DYSON, 1972

As mentioned earlier, beyond Neptune lie two zones of asteroid-sized objects rich in volatiles. The innermost region is the Kuiper Belt. Consisting of millions of iceteroids orbiting more or less in the same plane as the planets ("the ecliptic"), it begins at about 40 AU and extends to perhaps 10,000 AU. Beyond 10,000 AU, the orbits become randomly oriented, and the "belt" diffuses into a spherical cloud, the Oort Cloud, whose trillions of iceteroids populate the space surrounding our solar system all the way out to 100,000 AU—roughly halfway to the nearest star. Because they are so far from the Sun, such objects orbit very slowly. At 10,000 AU, for example, the speed required to orbit the Sun is just 300 m/s (compared to the Earth's 30,000 m/s), so it takes only a very mild velocity change to radically perturb such objects' orbits. It is believed that such perturbations occasionally occur naturally when passing stars, black dwarfs, brown dwarfs, or unbound objects of planetary size pass through the Oort Cloud and disturb its orbits with their gravitational fields. When that happens, one or more of the iceteroids can be displaced from their peaceful existence in the outer darkness. As they fall toward the Sun, they speed up enormously and, with volatiles boiling away, they come blazing into the inner solar system as gigantic young comets.

The recent Hale-Bopp comet was one of these. Because it was so huge, it was detected by amateur astronomers while still beyond Saturn, farther out than any comet had ever been spotted before. But by that time—the spring of 1995—it was already moving at incredible speed and crossed Earth's orbit just two years later. Hale-Bopp didn't come near the Earth, but if it had been heading our way, by the time it was identified it would have been coming on much too strong to deflect.

Other comets have hit, with results similar to those resulting from asteroidal impacts. However, unlike near-Earth asteroids, which spend their lives in the inner solar system and which can, in principle, be spotted and have their trajectories mapped many orbits before a potential Earth-smashing collision, comets can emerge from the dark and come in fast and hard with the advantage of surprise. The only way to control them is to detect and deflect them while they are still very far out. This means that someday, for security purposes if no other, there will be a need for a substantial human presence and technical capability in both the Kuiper Belt and the Oort Cloud.

But there may be other reasons that drive humans to populate this vast archipelago of cosmic islands. Based on analysis of comets, it's fairly clear that the volatile iceteroids of the Oort Cloud are rich not just in water, but carbon and nitrogen, much of it in the form of the usual compounds of organic chemistry and life. In addition, some of the most essential elements of industry, including iron, silicon, magnesium, sulfur, nickel, and chromium, are present in modest but possibly sufficient concentrations. This has caused some,[3,4] notably the visionary Princeton professor Freeman Dyson, to identify these bodies as a major arena for the human future.

It's rather futuristic, but not impossible. The inhabitants of such places wouldn't really need much steel for their constructions. For most purposes, ice—lightweight, cheap, and superstrong at 20 K (–253°C)—would serve quite well. Incredible degrees of both robotic automation and human versatility will be required to compensate for the limited division of labor possible in such small, widely scattered colonies, but perhaps earlier human experience in coping with a lesser degree of this same problem settling the asteroid belt will pave the way. The main missing ingredient is energy. While some have suggested concentrating starlight, it doesn't really make sense. To get a single megawatt of power, the mirror would have to be the size of the continental United States. The only viable alternative based on currently known physics is fusion. In the Kuiper Belt, it might be possible to get helium-3 shipped out from mining operations around Neptune. Oort Cloud settlements would be too far out to obtain much from the solar system, though deuterium should be available in all iceteroids, so perhaps the colonists might choose to build reactors based on that fuel alone. However, helium can exist in the liquid phase below 5 K (–268°C), which is the environmental temperature at about 3,000 AU. It is therefore not impossible that liquid helium could exist within Oort Cloud objects beyond that distance. Helium is the second most abundant element in the cosmos, and at very low temperatures it could accrete within hydrogen ice objects into a helium-rich iceteroid. For Oort Cloud colonists, such an object would be quite a find!

Perhaps the same wanderlust and reach for diversity that drove the old folks to settle the asteroid Main Belt in the twenty-first century will move their descendants a century or more later to try their luck among the million untamed worlds of the Kuiper frontier. Why go? Why stay? Why live on a planet whose social laws and possibilities were defined by generations long dead, when you can be a pioneer and help to shape a new world according to reason as you see it? The need to create is fundamental. Once started, the outward movement will not stop.

TYPE III

Entering Galactic Civilization

As the geometer intently seeks
to square the circle, but he cannot reach,
through thought on thought, the principle he needs,
so I searched that strange sight; I wished to see
the way in which our human effigy
suited the circle and found place in it—
and my own wings were far too weak for that.
But then my mind was struck by light that flashed
and, with this light, received what it had asked.
Here force failed my high fantasy; but my
desire and will were moved already—like
a wheel revolving uniformly—by
the Love that moves the sun, and the other stars.

—DANTE ALIGHIERI
Paradiso
Canto XXXIV, Lines 133–145;
translated by Allen Mandelbaum

CHAPTER 9

The Challenge of Interstellar Travel

Too low they build, who build beneath the stars.

—EDWARD YOUNG,
Night Thoughts on Life, Death, and Immortality, 1745
(inscribed on a wall of the Library of Congress)

INTERSTELLAR TRAVEL is the holy grail of astronautical engineering. The challenge is daunting, but the rewards are potentially infinite.

The most obvious challenge is that of distance. Distances to the nearest known stars are tens of thousands of times greater than those to the farthest planets in our solar system. The Earth travels at a distance of 150 million kilometers, or 1 astronomical unit (AU), from the Sun. Mars orbits at 1.52 AU, Jupiter at 5.2, Saturn at 9.5, Uranus at 19 AU, Neptune at 30, and Pluto at 39.5. In contrast, the nearest known stellar system, Alpha Centauri (consisting of the Sun-like type-G star Alpha Centauri A and the dwarf stars Alpha Centauri B and Proxima Centauri), is 4.3 light-years, or 270,000 AU, distant. Our fastest spacecraft to date, Voyager, took thirteen years to reach Neptune. That's an average of about 2.5 AU per year. However, due to the fact that it employed successive gravity assists to speed up at Jupiter, Saturn, Uranus, and Neptune, Voyager managed to depart the solar system with a final velocity of about 3.4 AU per year (17 km/s). At that rate, it would take more than 790 centuries to reach Alpha Centauri. If such a probe had been launched from Earth the day *Homo sapiens* first set foot in Europe, it would still have another 30,000 years to go.

Furthermore, not only is it far between stars, but it is hard to find much in the way of supporting resources along the way. Sunlight in deep space is nil, so power for an interstellar spacecraft must be nuclear. Even there the answer is not easy—a typical space nuclear reactor can power itself at 100 percent capacity for only seven years or so on a full load of fuel. That's great compared to the few hours or days possible from any system burning chemical fuels, but insignificant should the output requirement be for tens of millennia.

Even communication is difficult. For example, a typical robotic spacecraft today might use 100 W of power and a 2-meter-diameter dish to transmit data via X-band radio at a rate of 40 kb/s from Mars to one of the 70-meter-diameter Deep Space Network receiving stations on Earth. If the same gear were to be used to transmit data at the same rate from Alpha Centauri, the power needed would be a trillion watts (a terawatt), or roughly 8 percent of all the power currently used by human civilization.

In the face of such imposing challenges, a literature has been created showing interstellar travel as dependent on the exploitation of exotic or fantastical physical phenomena such as wormholes, space warps, cosmic strings, and so forth. While some of these concepts are mathematically consistent with the currently known laws of the universe, there is no evidence that they actually exist or, if they do, that there is any method by which they could be manipulated by humans to produce a practical technology for space propulsion.

Therefore, many people believe that interstellar travel is impossible.

I disagree. Interstellar travel is incredibly difficult, perhaps as difficult to us today as a flight to Mars would have appeared to Christopher Columbus or other would-be transoceanic navigators 500 years ago. Indeed, the ratio of the distance from Earth to Mars compared to Columbus's voyage from Spain to the Caribbean—80,000:1—is roughly the same as the ratio of the distance to Alpha Centauri compared to a trip to Mars. Thus, the key missions required to establish humanity successively as a Type I, Type II, and Type III civilization all stand in similar relation to each other, and if the 500 years since Columbus have sufficed to multiply human capabilities to the point where we now can reach for Mars, so a similar span into the future might be expected to prepare us for the leap to the stars. Actually, it should not take so long, because with its much larger population of inventive minds and better means of communication, the Type II civilization that will spread throughout our solar system over the next several centuries should be able to generate technological progress at a considerably faster rate than was possible by the emerging Type I civilization of our recent past.

I'm all for breakthroughs in physics that will give us capabilities as yet unknown. We may well get them someday. But even without such, methods can already be seen in outline by which currently known physics and greatly developed and refined versions of currently understood engineering can get us to the stars. That development and refinement will occur as part and parcel of the process of maturation of humanity as a Type II species.

When mature, Type II civilizations give birth to Type III civilizations.

Here's how we'll do it.

CHEMICAL PROPULSION

Since we currently have efficient chemical rocket systems, it is worth asking if these can be used to accomplish interstellar missions. On the surface, the idea seems absurd—the maximum possible exhaust velocity for a chemical rocket is about 5 km/s (our current hydrogen/oxygen rocket engines already achieve 4.5 km/s, or 90 percent of what is theoretically feasible), and as we have already discussed, the maximum practical velocity increment that can be delivered by a rocket engine to a spacecraft is about twice the exhaust velocity. So an advanced chemical rocket system might be able to give us a 10 km/s push, which is 2 AU per year, or 135,000 years to reach Alpha Centauri.

This is a bit on the slow side, but, as we have seen, by using planetary gravity assists Voyager was able to leave the solar system with double this speed, reducing the required flight time to a mere 79,000 years. Obviously, this is still unacceptable, but Voyager wasn't trying for a high solar system escape speed, and all of its gravity assists were done without active thrusting. If we really pushed the technique of powered gravity assist and went all out for speed, how fast could we go?

Well, the most massive object in our solar system is the Sun, with an escape velocity at its surface of 617 km/s. We could send a spacecraft to Jupiter and use a gravity assist there to send a well-insulated, thermally protected spacecraft on a screaming dive into the inner solar system on a path that would take it just 40,000 km above the surface of the Sun. At that altitude the escape velocity would be 600 km/s, and since the spacecraft would have fallen there from Jupiter, it would be traveling at least that fast. We then fire our rocket engine and impart a 10 km/s ΔV to the spacecraft, raising the velocity to 610 km/s. As discussed in chapter 8, the relevant equation is:

$$V_d^2 = V_{max}^2 - V_e^2 \qquad (9.1)$$

where V_d is the departure velocity, V_{max} is the maximum velocity (610 km/s), and V_e is the escape velocity (600 km/s). The result: After firing our engine to create a ΔV of 10 km/s at lowest altitude and then slowing down during the climb away from the Sun, we will still have a departure velocity of 110 km/s, about six times that of Voyager, allowing us to reach Alpha Centauri in 12,300 years.

If we really forced the engineering to wild extremes and piled on numerous stages, we might, in principle, be able to generate a ΔV of 25 km/s. This would result in a departure velocity of 175 km/s, or 7,700 years to Alpha Centauri. With chemical rocket technology, that's as good as it gets.

There are two methods that have been proposed to enable space voyages of this length. One is to put the crew in suspended animation, perhaps cryogenically frozen so they do not age. There are massive problems using this latter technique, because water expands when it freezes, thereby causing cell walls to rupture when a body is frozen. Using drugs to induce hibernation is probably possible, as illustrated by woodchucks, but aging and metabolism would still proceed, albeit at a reduced rate, making such expedients of marginal value for millennia-long voyages.

The other method is to build a spaceship large enough to house a sizable number of people for their entire lives—perhaps a nuclear-powered O'Neill colony—and send it on its way. The initial crew would raise a generation of children to carry on, who would raise another and so on for 7,700 years until the destination star is reached. While the engineering of such a vessel would be formidable, there is nothing in the laws of physics or biology that would preclude such a mission. However, the idea that the sense of purpose of the initial crew could be preserved generation after generation for a span greater than that of all recorded human history seems rather fantastical. We therefore turn our attention to more advanced propulsion concepts that can reduce the travel time to the stars to no more than one or two human lifetimes.

FISSION PROPULSION

The fundamental physical reason why chemical rocket engines cannot produce exhaust velocities greater than 5 km/s is because the energy per unit mass, or enthalpy, of chemical fuels is limited to about 13 megajoules per kilogram (MJ/kg) by the laws of chemistry. Nuclear fission, on the other hand, offers fuels with an enthalpy of 82 million MJ/kg, more than 6 million times as great as the best possible chemical propellants. Now the maximum theoretical exhaust velocity of a rocket propellant is equal to the square root of twice the enthalpy; thus, 5.1 km/s for chemicals, 12,800 km/s for nuclear fission. That's a lot better. The speed of light is 300,000 km/s. A fission rocket could thus, in principle, generate an exhaust velocity of 4 percent the speed of light. Since a spacecraft can generally be designed to obtain a ΔV equal to twice its exhaust velocity, a theoretically perfect fission drive could get us to 8 percent lightspeed. Since Alpha Centauri is 4.3 light-years away, that would mean a one-way transit in 54 years. If half the ΔV is used to slow down at the destination, maximum speed would be 4 percent of light, and the transit time would be increased to 108 years.

There are a number of problems, however. One of them is being able to take advantage of all the energy available. Primitive nuclear propulsion systems, such as nuclear thermal rockets, do a very poor job of this. By using a solid nuclear reactor to heat a flowing gas, the maximum exhaust velocities attained are only in the 9 km/s range—good by comparison with chemical rockets, but nowhere near the performance needed for interstellar missions. If the nuclear fuel is allowed to become gaseous (a "gas-core"[1–3] nuclear thermal rocket—NASA did a fair amount of work on such systems in the 1960s), exhaust velocities of 50 km/s could be achieved. This would be excellent for interplanetary travel but is still not in the interstellar class. If a nuclear reactor is used to generate electric power to drive an ion engine (NEP propulsion), exhaust velocities of up to several hundred kilometers per second could be obtained, if hydrogen is employed as propellant. But the systems required are very massive, the thrust (and thus rate of acceleration) they can produce is low, and the exhaust velocity is still not good enough for interstellar missions. What is needed is a way to turn the nuclear energy directly into thrust. One answer is so straightforward it has been known since 1945: Use atomic bombs.

It's pretty clear that if one detonates a series of atomic explosives right behind a spaceship you can push it along rather well. Of course, if you don't go about it correctly, you might also vaporize the spaceship, blow it to pieces, turn the crew to jelly with 100,000*g* of acceleration, or kill everyone on board with a lethal dose of gamma rays. As we say in the engineering business, "These concerns need to be addressed." So you must do it correctly. But if you can, you've got yourself one hell of a propulsion system.

This was the idea behind Project Orion,[4] a top secret program funded by the U.S. Atomic Energy Commission that ran between 1957 and 1963. The original idea came from Los Alamos bomb designer Stanislaw Ulam, and the program drew the talents of such visionary weapon makers as Ted Taylor and Freeman Dyson. A diagram of one of the Orion designs is shown in Figure 9.1. In it, a magazine filled with nuclear bombs is amidship. A series of bombs is fired aft down a long tube to emerge behind the "pusher plate," a very sturdy object backed up by some heavy-duty shock absorbers.

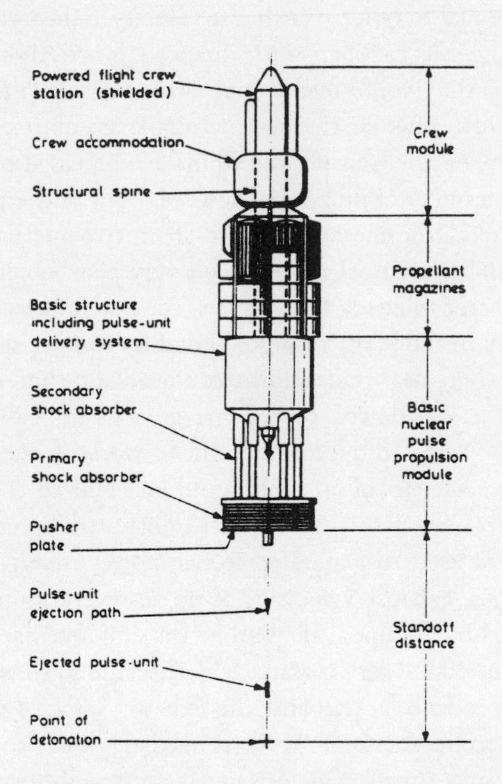

FIGURE 9.1 *Project Orion nuclear-bomb-driven spacecraft. (Courtesy British Interplanetary Society)*

When the bomb goes off, the pusher plate shields the ship from the radiation and heat and takes the impact of the blast, which is then cushioned by the shock absorbers. Since the bombs are detonated one after another in rapid succession, the net effect would be that a fairly even force is felt by the ship and its payload and crew, who are positioned forward of the bomb magazine. The pusher plate scheme is much less efficient at converting explosive force to thrust than a conventional bell-shaped rocket nozzle (perhaps only 25 percent compared to the 94 percent that is state of the art), but it has much more force to play with. So maybe the real effective exhaust velocity would only be about 1 percent the speed of light. That puts a bit of a crimp on our plans for fission-driven interstellar flight, but still, an exhaust velocity of 3,000 km/s in a high-thrust rocket has got to be considered pretty good.

However, for better or worse, Project Orion came to a screeching halt in 1963 when the Test Ban Treaty between the United States and the Soviet Union banned the stationing or detonation of nuclear weapons in outer space.

The Test Ban Treaty will expire someday. But still, it seems like a good idea to avoid stationing ships in space filled with thousands of atomic bombs (and an even better idea to avoid having factories mass-producing such bombs for sale to space travelers). I proposed a way around this problem in the early 1990s with a concept called a nuclear salt water rocket[5] (NSWR), shown in Figure 9.2.

In the NSWR, the fissionable material is dissolved in water as a salt, such as uranium bromide. This is stored in a bundle of tubes, separated from each other by solid material loaded with boron, which is a very strong neutron absorber and therefore cuts off any neutron traffic from one tube to another. Since each tube contains a subcritical mass of uranium, and the

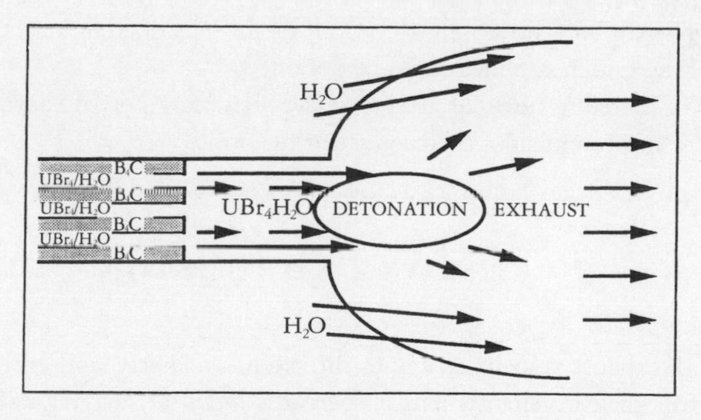

FIGURE 9.2 *Nuclear salt water rocket.*

boron cuts off any neutron communication from one to another, the entire assembly is subcritical. However, when thrust is desired, valves are opened simultaneously on all the tubes and the salt water, which is under pressure, shoots out of all of them into a common plenum. When the moving column of uranium salt water reaches a certain length in the common plenum, a "prompt critical" chain reaction develops and the water explodes into nuclear-heated plasma. This then expands out a rocket nozzle that is shielded from the heat of the plasma flow by a magnetic field. In effect, a standing detonation similar to chemical combustion in a rocket chamber is set up, except that the enthalpy available is millions of times greater. The nozzle would be much more efficient than the Orion pusher plate, but because the uranium content of the propellant is "watered down," the exhaust velocity would also be decreased significantly below nuclear fission's theoretical maximum of 4 percent lightspeed, perhaps to about the same 1 percent achievable by a nuclear-fission-bomb–driven Orion. But at least the need for mass-produced bombs would be eliminated

With exhaust velocities of about 1 percent the speed of light, starships driven by such systems might be able to attain 2 percent lightspeed, allowing Alpha Centauri to be reached in about 215 years. Voyages with trip times on this order might be able to use rotating hibernations to allow a crew to reach the destination. Alternatively, there at least would be some chance that a multi-generation starship could reach its goal with its purpose remaining intact.

However, in addition to offering only marginal performance for interstellar travel, such fission drives have another problem—fuel availability. The amount of fissionable uranium-235 or plutonium-238 needed to fuel such systems would be enormous, perhaps 10,000 tonnes to send a 1,000-tonne (small for a slow, long-duration starship) payload on its way. It is unclear where such supplies could be obtained.

We therefore turn our attention to a still more potent source of energy for starship propulsion: thermonuclear fusion.

FUSION PROPULSION

High exhaust velocity is key to interstellar rocketry, and enthalpy is the key to exhaust velocity. Nuclear fission looks attractive at 82 million MJ/kg, but nuclear fusion is better. For example, if pure deuterium is used as fuel and burned together with all intermediate fusion products (a series of reactions know as "catalyzed D-D fusion"), 208 million MJ/kg of useful enthalpy is available for propulsion, plus 139 MJ/kg of energetic neutrons, which, while useless for propulsion, can be used to produce on-board power. If a mixture of deuterium and helium-3 is used as fuel, the useful propellant enthalpy is a whopping 347 million MJ/kg. As a result, thermonuclear fusion using catalyzed D-D reactions has a maximum theoretical exhaust velocity of 20,400 km/s (6.8 percent of lightspeed, or 0.068*c*), while a rocket using the D-He3 reaction could theoretically produce an exhaust velocity of 26,400 km/s, or 0.088*c*.

Now we're talking starflight! With quadruple the enthalpy of nuclear fission and much more plentiful fuel, nuclear fusion holds the potential for a real starship propulsion system.[6] As in the case of nuclear fission, fusion offers both pulsed explosions and steady-burn options for rocket propulsion, but in the case of fusion both are more practical to implement.

Fission bombs must be of a certain minimum size, because for a fission chain reaction to occur a "critical mass" of fissile material must be assembled. Unless one chooses to simply waste energy by designing an inefficient explosive (not a viable option for interstellar propulsion), this critical mass implies a minimum yield for a fission bomb of about 1,000 tonnes of dynamite.

Fusion is different. There is no critical mass for nuclear fusion, so in principle fusion explosives could be made as small as desirable. Current military fusion explosives—H-bombs—have very high yields because they use a fission atomic bomb to suddenly compress and heat a large amount of fusion fuel to thermonuclear detonation conditions. If one wished to be crude, one could use such hydrogen bombs in an Orion-type propulsion system, with considerably higher performance and much cheaper fuel than the A-bomb–driven version. However, with fusion there are other ways to achieve the required detonation effect on a much smaller scale.

For example, one can use a set of high-power lasers to focus on a very small pellet of fusion fuel, thereby heating, compressing, and detonating it. Preliminary experiments have proved the feasibility of such "laser fusion" systems, and one, the National Ignition Facility (NIF), is currently under construction at the Lawrence Livermore Laboratory in California. A starship utilizing such a system for propulsion would eject a series of pellets with machine-gun rapidity into an aft region of diverging magnetic field. As each pellet entered the target zone, it would be zapped from all sides by an array of lasers. It would then detonate with the force of a few tonnes of dynamite, and the ultra-hot plasma produced would be directed away from the ship by a magnetic nozzle to produce rocket thrust.

Alternatively, it may be possible to implode and detonate fusion pellets using an appropriately shaped set of chemical explosives. I say "may" because a great deal of top secret work has been done to achieve this goal in both the United States and the former Soviet Union, but the results are unpublished. If feasible, such chemically ignited fusion micro-bomblets would eliminate the need for a heavy laser system aboard ship.

As a third alternative, one could implement fusion propulsion without bombs, lasers, or micro-bomblets by using a large magnetic confinement chamber to contain a large volume of reacting thermonuclear fusion plasma (Fig. 9.3). This is presumably the type of system that would be used to produce fusion power in the future, except in such a fusion drive most of the ultra-hot (tens of billions of degrees, or several megavolts) fusion products would be allowed to leak out of one end of the reactor to produce thrust, whereas the rest would be used to heat the plasma to 500 million degrees C (50 kilovolts) or so, which is the proper temperature for fusion reactors. Some of the lower-temperature plasma would also leak out, but because of

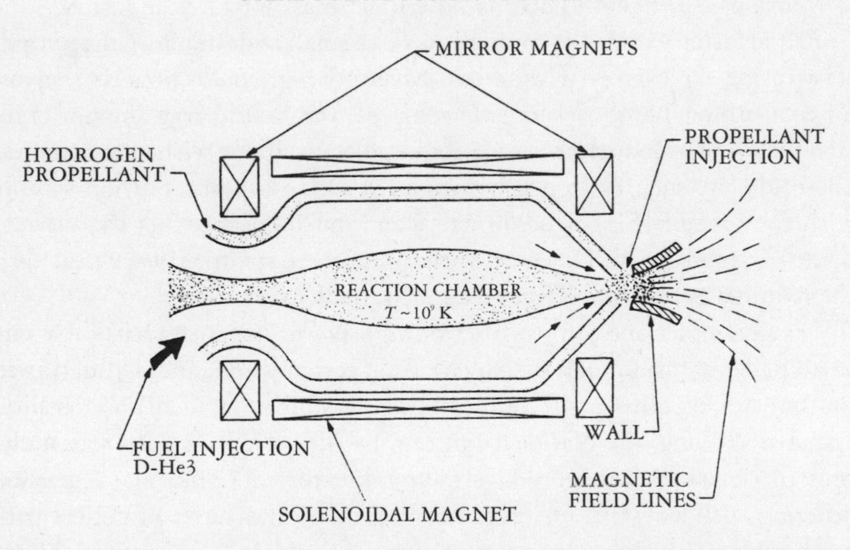

FIGURE 9.3 *Magnetic confinement fusion propulsion system. (Courtesy of John Wiley and Sons, Inc.)*

its lower energy it could be decelerated by an electrostatic grid and used to produce electric power for the ship.

The magnetic nozzles used by fusion propulsion systems would not be as good as the 94 percent efficient bell nozzles used in chemical rocket engines but would be much more effective at channeling thrust than the 25 percent efficient pusher plates of the old Orion. Probably an efficiency of about 60 percent could be achieved. Assuming that to be the case, then a D-He3 fusion rocket should be able to attain an exhaust velocity of about 5 percent the speed of light. Since practical spacecraft can be designed to reach a speed about twice their engines' exhaust velocity, this implies that such fusion propulsion systems could make 10 percent lightspeed. Ignoring the small amount of extra time needed to accelerate, that means one-way to Alpha Centauri in 43 years, or 86 years if we need to use the propulsion system to slow down.

ANTIMATTER

While D-He3 fuel has the highest enthalpy of any substance that can be found in nature, there is an artificial material that has a much higher enthalpy still—antimatter.

Antimatter is mass with the charges of the subatomic particles reversed. In ordinary matter electrons are negative; in antimatter they are positive. Ordinary protons are positive; antiprotons are negative. Because oppositely charged particles attract, antiparticles attract their ordinary-matter mates. The attraction is fatal, though, as the two annihilate each other, transforming their combined mass into energy in accord with Einstein's famous formula $E = mc^2$ (energy equals mass times the speed of light squared).

Antimatter is such a staple of science fiction that many people believe that it *is* science fiction, but antimatter is real. We don't ordinarily encounter it in daily life because the universe, or at least our region of it, was created with an excess of ordinary matter over antimatter. Thus all the antimatter (or all the antimatter in our galaxy) has been annihilated, leaving nothing but the common stuff. But, as a result of the fact that energy can also be turned into matter in accord with the Einstein formula, evanescent antiparticles are created by cosmic-ray impacts with the Earth's atmosphere. We have also been able to create antiparticles in high-energy accelerators and have succeeded in combining antiprotons with antielectrons (or positrons) to produce antihydrogen atoms. These antihydrogen atoms have been further combined together to form antihydrogen molecules. Antiprotons can be stored in special jars called "Penning traps" in which magnetic fields are used to keep the ions from hitting the wall (where they would annihilate). In this way, up to several million antiprotons at a time can be stored for extended periods. Using the big collection rings at leading high-energy physics accelerator facilities such as Fermilab and CERN, up to a trillion antiprotons at a time have been collected. This represents about 1.7 picograms (a picogram is a trillionth of a gram) of antimatter. If this much antimatter were allowed to annihilate, it would release about 300 joules (J) of energy, enough to light a 60-W light bulb for five seconds. Tiny amounts of antihydrogen atoms and molecules have also been confined, using the pushing power of lasers to herd them away from chamber walls.[7]

Now let's say we could do much better than this, and freeze antihydrogen gas into solid crystals. We could then give these crystals a static electric charge, allowing us to store them without touching by levitation inside a magnetic or electrostatic trap. Then we could use this material as fuel on a starship, annihilating it with ordinary hydrogen to produce energy. How much energy? Lots. Because the speed of light, *c,* is such a large number—300,000 km/s—Einstein's formula is generous. If we were to annihilate a single half kilogram of antimatter with a half kilogram of ordinary matter, we would release 90 billion MJ of energy. That's an enthalpy of 90 billion MJ/kg, 259 times greater than D-He3 fusion, over 1,000 times greater than nuclear fission, and nearly 7 billion times as great as an equivalent amount of hydrogen-oxygen rocket propellant. Put another way, a single kilogram of antimatter annihilating with a kilogram of ordinary matter will release as much energy as 40 million tonnes of TNT. The maximum theoretical exhaust velocity of an antimatter rocket would be the speed of light.

That's theory; in practice things are not quite that good. In the first place, about 40 percent of the energy from antimatter annihilation is released in the form of gamma rays with energies of over 200 million volts. This is hundreds of times greater than the typical gamma rays released by nuclear fission reactors and will put a very heavy shielding burden on the spacecraft. Then there is the issue of how thrust will be created. One idea would be to use antimatter to generate an extremely high energy, magnetically confined plasma with average energies of hundreds of millions of volts. In this case, only the portion of the antimatter annihilation energy that comes off as charged particles would be usable, since the gamma rays and uncharged particles would escape from the system before they could heat the plasma. In addition, such a high-temperature plasma would waste massive amounts of energy through both cyclotron and bremsstrahlung radiation. These losses, taken together with the roughly 60 percent efficiency possible with magnetic nozzles, would reduce the attainable effective exhaust velocity of an antimatter plasma drive down to perhaps 30 percent the speed of light.

An alternative method of antimatter propulsion would be to use the energy of annihilation to heat the surface of a stern-mounted solid cylinder composed of a high-temperature material such as graphite or tungsten to incandescence and then direct the light radiated by the glowing object rearward with mirrors (Fig. 9.4). Particles of light, called photons, have momentum, and if they are all directed rearward a net forward force would be created. Such a system is termed a photon rocket.

The exhaust of a photon rocket has the speed of light (because it is light), but not all the energy of the antimatter annihilation will go into it. Most of the energy of the highly penetrating gamma rays will be lost before it can be used to heat the solid cylinder, and the neutrinos and other highly penetrating uncharged particles will carry their energy out of the system be-

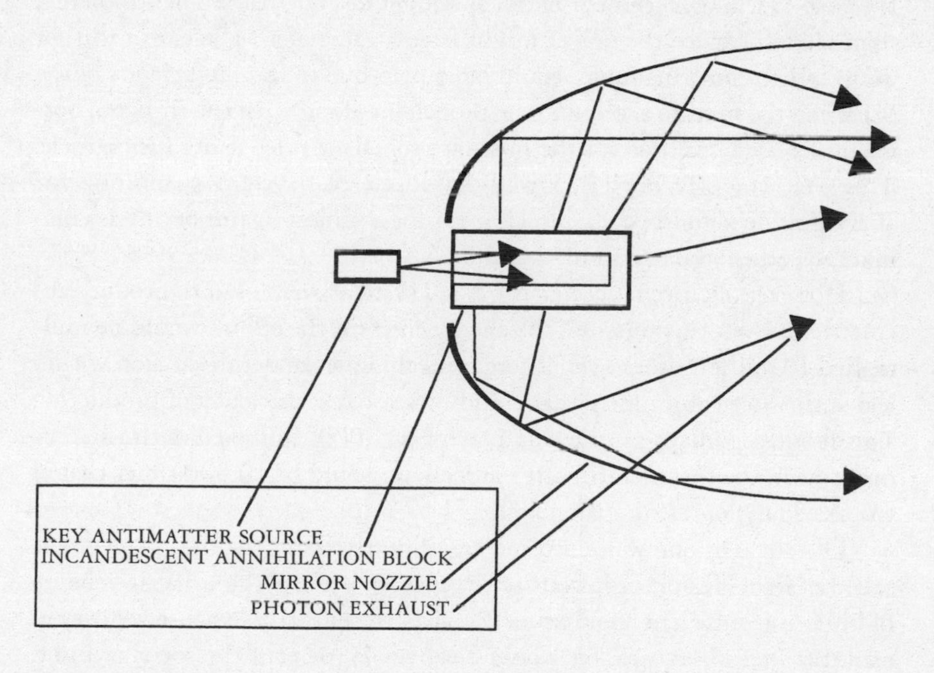

FIGURE 9.4 *Photon rocket.*

fore it can do any good. This reduces the effective exhaust velocity (specific impulse) of the system considerably. But antimatter has energy to spare. Even when all the losses are taken into account, photon rocket effective exhaust velocities on the order of 50 percent lightspeed appear to be attainable.

The photon rocket is simpler than an antimatter plasma drive and offers higher performance. Therefore, if antimatter does become available in sufficient quantities to power interstellar voyages, photon rockets will probably be the engines of choice.

But availability is an issue. Using our current accelerator-based techniques for manufacturing antimatter, it requires over 10 million times as much electric power to create a unit of antimatter as the antimatter energy is worth.

Consider what this means. Let's say we want to get a 1,000-tonne starship up to 10 percent the speed of light. The kinetic energy of the ship at flight speed would be 450 trillion MJ (4.5 × 1,020J), or 125 trillion kWh. Due to the law of conservation of energy, this is the minimum amount of energy needed to accelerate the ship. At current electric power prices of $0.05/kWh, this much energy would be worth about $6 trillion, roughly the current U.S. government national budget for four years. The fusion fuel required to produce the power might have a value of 10 percent of this, or $600 billion, but this figure could be tripled due to the inefficiency of the drive and the need to accelerate the propellant along with the ship. So, bottom line, say $2 trillion for the mission propellant price using fusion fuels. This is rather costly, but a rich, well-developed solar-system–spanning Type II civilization should be able to afford it for a project as important as colonizing another stellar system.

However, if current accelerator-based systems were used to produce antimatter fuel for this mission, the energy costs for the efforts would be multiplied 10 million times over. There are techniques under discussion within the antimatter community that could increase the efficiency of production significantly, perhaps as much as a factor of 1,000. But even with this improvement, the cost of antimatter propellant would be 10,000 times that of the same mission using fusion fuel.

Of course, if one wanted to go significantly faster than 10 percent the speed of light, fusion drops out of the picture because the exhaust velocity of $0.05c$ is insufficient. As discussed above, the effective exhaust velocity of an antimatter photon rocket would be about 50 percent the speed of light, making flights at up to 90 percent lightspeed theoretically possible. This would get a ship to Alpha Centauri in about five years, which would seem like three to the crew due to the effects of relativistic time dilation. But the society that launched such a mission would have to be one that was so rich that cost was simply not an issue.

LIGHT SAILS

Nearly 400 years ago the famous German astronomer Johannes Kepler observed that regardless of whether a comet is moving toward or away from the Sun its tail always points away from the Sun. This caused him to guess that light emanating from the Sun exerts a force that pushes the comet's tail away. He was right, although the fact that light exerts force had to wait until 1901 to be proved by Russian physicist Peter N. Lebedev, who made mirrors suspended on thin fibers in vacuum jars turn by shining light upon them. A few years later, Albert Einstein provided the theoretical basis for this phenomenon, explaining why light exerts force in his classic paper on the photoelectric effect, for which he later received the Nobel Prize.

Well, if light can push comet tails around, why can't we use it to move spaceships around? Why can't we just deploy big mirrors on our spacecraft, solar sails if you will, and have sunlight push on them to create propulsive force? The answer is that we can, but it takes an awful lot of sunlight to exert any significant amount of push. For example, at 1 AU, the Earth's distance from the Sun, a solar sail the size of a square kilometer on a side would receive a total force of 9 N, about 2.0 pounds, pushing on it from the Sun. For such a large object, that's not a lot of force. Consider, if the 1 km^2 sail were made of plastic as thin as writing paper (about 0.1 mm) it would weigh 100 tonnes, and it would take a full year of sunlight at 1 AU to accelerate it through a ΔV of 3 km/s.

This is not an especially impressive performance, but writing paper is hardly the thinnest thing we can manufacture. Let's say we made the sail 0.01 mm thick (10 microns—depending on the brand, kitchen trash bags are 20–40 microns thick). This is about the thickness of the films used on many high-altitude balloons. In that case, the sail would weigh only 10 tonnes, and it could accelerate itself 30 km/s—roughly the round-trip ΔV needed to go from low Earth orbit to Mars and back on a low-thrust trajectory—in just about a year. Of course, if the sail were hauling a payload equal to its own weight, that would slow down by a factor of 2. Still, a 10 micron thick solar sail would be in the ballpark for an effective propulsion device supporting Earth-Mars transportation.

The advantage of the solar sail is clear—it needs no propellant or onboard power supply. Fundamentally the technology is simple, cheap, scaleable, elegant, and, in a word, beautiful. The idea of ships equipped with huge ultralightweight shiny sails coursing effortlessly through space on the power of reflected sunlight alone is romantic in the extreme, recalling as it does the age when sailing ships opened the oceans of Earth to explorers, merchants, and adventurers of every type. Moreover, solar sails may hold enormous potential to similarly open the lanes of interplanetary commerce. For this reason many people, including noted science fiction author and space visionary Arthur C. Clarke and Planetary Society Executive Director Louis Friedman, have long been staunch advocates of this technology.[8] Its development for interplanetary propulsion purposes does not appear to be especially formidable, being mostly a matter of mastering some packaging and mechanical deployment issues, and the fact that solar sails are not already in use is abundant testimony to the stagnation in the space program over the past several decades. Certainly a mature Type II civilization will not only possess solar sails, but employ them widely for interplanetary commerce and numerous other applications.

But we are talking about interstellar propulsion here. How can a system that derives its motive force from the Sun be used to drive a ship through the darkness of interstellar space?

One answer, the simplest and most elegant, is simply to make the solar sail so thin that it can use sunlight to accelerate the spacecraft to interstellar speeds while it is still within the solar system. Such ultrathin solar sails would have to be manufactured in space using techniques that are currently unavailable. To save weight, we would discard the plastic backing and just use a thin layer of aluminum created by molecular deposition in vacuum on a lightweight webbing for the sail. Table 9.1 shows the maximum speed that such a system can achieve driving a 1,000-tonne spacecraft, assuming that the payload spacecraft has a mass equal to that of the sail and that the mission begins 0.1 AU from the Sun.

It's hard to get the solar sail much thinner than 0.001 microns, because this thickness represents a layer of material just four atoms across. (In fact, to avoid being transparent, the aluminum probably has to be at least 0.01 microns thick, but an average density equivalent to a 0.001 micron thick sail can nevertheless be achieved by perforating the sail. Provided the holes are a lot smaller than the 0.5 micron wavelength of visible light, the sail

▪ TABLE 9.1 ▪

Thin Solar Sails for Interstellar Travel

Sail Thickness (microns)	Acceleration at 1 AU (m/s^2)	Sail Radius (km)	Final Velocity (km/s)
0.3	0.006	220	95 (0.03%*c*)
0.1	0.018	234	212 (0.07%*c*)
0.01	0.18	2,108	728 (0.26%*c*)
0.001	1.8	2,343	2,322 (0.77%*c*)

will still reflect light in the same way that a chicken-wire radio antenna reflects radio waves.) Starting the mission closer to the Sun than 0.1 AU is conceivable, but the final speed will increase only in proportion to the inverse square root of the distance (i.e., if we get nine times closer, we will only end up going three times faster), and the acceleration for our 0.001 micron spacecraft when it starts at 0.1 AU is already a stiff 18*g*. So the bottom line is that a light sail driven only by sunlight is unlikely to be able to get a starship much above 1 percent the speed of light. An advantage of such a system would be that it would be cheap (energy cost is zero), simple, and reliable, and, provided the target star has comparable luminosity to the Sun, the same solar sail used to accelerate the mission could also be used to decelerate at the destination star, navigate within its solar system, or even to return. But the flight time to Alpha Centauri would be on the order of five centuries. Perhaps such a system might be acceptable for a multi-generation ship or one employing suspended animation techniques. Perhaps. It might also be found acceptable for use by a species whose natural life span is considerably longer than current humans.

But if they are to serve as a practical means of interstellar propulsion for people as we know them, light sails will need an additional shove to get to speed. One way to do that would be to push them with high-energy lasers,[9] an idea first proposed by physicist Robert Forward in 1962.

Let's take the 1,000-tonne starship together with the 343-km-radius, 0.001 micron thick light sail discussed above and illuminate the sail with laser light five times as bright as sunlight is on Earth (i.e., about as bright as sunlight is at 0.45 AU). The ship will then be accelerated at the comfortable clip of 9 m/s^2 (0.92*g*), reaching 15 percent the speed of light inside of two months. At the end of that time, the ship will have traveled 121 billion kilometers or 806 AU. To keep focused on the light sail at this distance, the laser projector would have to have a lens about 100 meters in radius. This is only about twelve times larger than the largest telescope yet built or under construction (the Keck 16-meter diameter) and so may be considered a modest challenge compared to the rest of the project. However, the amount of power the laser would need is formidable: 240 terawatts (TW). This is about twenty times the total power humanity currently generates each year. However, since it would be needed for only two months (i.e., one-sixth of a year), the total energy would be about what humanity currently consumes in three years. Obviously, even if we had the technology, such an expenditure of power would be out of the question today. But humanity's power production is growing at a rate of 2.6 percent per year. If this trend continues, in the year 2200 we will be producing and consuming energy at a rate of 2,500 TW, and using 240 TW, or 9.6 percent, of this for two months to get to the stars might well be considered affordable.

The laser projector would be kept pointed at the target star. The crew or computer aboard the spacecraft would know in advance the position of the projector as it orbits the Sun and use this knowledge to keep their vessel squarely in the center of the light beam. At 15 percent the speed of light, they would reach Alpha Centauri in about twenty-nine years. If the laser lens was four times as big (i.e., 400 meters instead of 100 meters), we could keep the light on the sail for twice as long, go twice as fast (30 percent *c!*), and reach the destination in half this time.

There's just one problem: No way to stop.

MAGNETIC SAILS

In 1960, the visionary physicist Robert Bussard published one of the classic papers on interstellar travel. In it he proposed a kind of fusion ramjet that would gather interstellar hydrogen as it flew, and then burn it to produce thrust using the same proton-proton fusion reaction that powers the Sun.[10] Bussard's concept was elegant. It was based on more or less known physics, yet because the propellant was gathered in flight, there was no mass ratio limit and the spacecraft could accelerate continuously to asymptotically approach the speed of light.

There are a number of problems, however. One is that the proton-proton reaction is very hard to drive and occurs slowly, so that igniting a fusion reactor using this fuel is 20 orders of magnitude more difficult than the deuterium-tritium, deuterium-deuterium, or deuterium-He3 systems that humanity is currently struggling to make work. Astronomer Daniel Whitmire in 1975 proposed an improvement by recommending that carbon be added to catalyze the reaction using the same carbon-nitrogen-oxygen (CNO) catalytic fusion cycle that drives the process of proton fusion in certain hot stars. This raises the reactivity of the system to the point where it is only a million times more difficult to ignite than deuterium, which certainly helps, but even with CNO catalysis artificial proton fusion reactors remain a difficult and distant prospect.

The other problem is how to gather the material. Because of the diffuse nature of the interstellar medium, the scoop has to be huge, so using a physical inlet is out of the question. The only viable options seem to be some sort of scooping device based on magnetic or electrostatic fields.

In 1988, Boeing engineer Dana Andrews decided to try to take a small step toward a Bussard ramjet by proposing a concept in which a magnetic scoop would be used to gather hydrogen ions in interplanetary space for use as propellant in an ordinary ion engine, which would be powered by an onboard nuclear reactor. This concept thus eliminated the need for the proton fusion required by Bussard's ramjet. The performance of the state-of-the-art nuclear electric propulsion system was too low to be relevant for interstellar missions, but for interplanetary travel the self-fueling ion drive would be terrific. There was a problem, though. As far as Andrews could calculate, the magnetic scoop employed by the system generated more drag against the interplanetary medium than the ion engines produced thrust. The drive was apparently useless.

I was living in Seattle at the time, and Andrews and I were well acquainted. Because I have a good background in plasma physics, Andrews told me about his concept and the problem he was running into. At first I thought there might be a solution, because Andrews was using certain approximations to calculate the plasma drag that were very rough for the situation he was dealing with. So we worked together to write a computer program to calculate the drag more precisely, only to discover that the actual drag was much greater than Andrews had first estimated. At that point, I suggested that we abandon the ion thrusters entirely, and rather than seek to minimize the drag, try to maximize it—to use the magnetic field not as a scoop but as a sail. In this manner, we would derive the spacecraft's motive force from the dynamic pressure of the solar wind, the plasma that flows outward from the Sun. Andrews agreed, and we went back to the drawing board with a new approach. Thus was born the magnetic sail, or magsail.[11,12]

The idea was timely. In 1987, Professor Ching Wu (Paul) Chu of the University of Houston had just invented the first high-temperature superconductors, materials that can conduct electricity without any resistive power dissipation and that operate at reasonable temperatures (previously, the only known superconductors operated at temperatures approaching absolute zero). The magsail could use these to create a powerful magnetic field that could deflect the solar wind, thereby imparting its force to propel a spacecraft. If practical high-temperature superconducting wire could be developed that could conduct currents with the same density as state-of-the-art low-temperature superconductors (about a million amps per square centimeter), then magsails could be developed that could produce fifty times the thrust-to-weight ratio of near-term (10 micron) solar light sails. (As of this writing, high-temperature superconducting wire with 20 percent of the current capacity of low-temperature superconductors is available.) Despite the fact that magsail thrust is always nearly outward from the Sun (as opposed to light sails, which can use the mirror effect to aim their thrust through a wide angle), I was able to derive equations showing how the system could be navigated almost at will throughout the solar system. The maximum possible speed of a solar-wind–pushed magsail is the speed of the solar wind—500 km/s—which is too slow for interstellar flight, and a practical magsail could probably only do half of this. But Andrews has investigated propulsion options including pushing magsails with plasma bombs ("MagOrion"[13]) and charged particle beams that offer significant promise.

However, the most interesting and important thing about the magsail is not what it can do to speed up a spacecraft—what's important is its capability for slowing one down. The magsail is the ideal interstellar mission brake! No matter how fast a spaceship is going, all it has to do to stop is deploy and turn on a magsail, and the drag generated against the interstellar plasma will do the rest. Just as in the case of a parachute deployed by a drag racer, the faster the ship is going, the more "wind" is felt, and the better it works.

The magsail thus provides the missing component needed for interstellar missions using laser-pushed light sails. Alternatively, if fusion rockets (or any type of rocket) are used to accelerate, having a magsail on board means that no fuel will be needed to decelerate. All of the available ΔV can be used to speed up; none is needed to slow down. As a result, the ship can perform its mission twice as fast.

In starflight, deceleration is half the battle. Half the battle is already won.

USING OORT CLOUD OBJECTS

The elegance of using the magsail's interaction with the interstellar medium as a starship deceleration system is a healthy reminder of a general principle that holds in all space endeavors: Wherever possible, exploit local resources. Starflight is so difficult that it behooves us to search for other resources that may help us along the way.

Besides the interstellar medium, what else is available? The answer that most readily comes to mind is Oort Cloud objects, the multitudes of asteroid-sized blocks of frozen volatiles that orbit our Sun all the way out to halfway to the next solar system, which presumably have Oort Clouds of their own from there on in. How could these be used?

One way that has been proposed is that human beings, by settling the Oort Cloud, will simply diffuse across interstellar space from one object to another, eventually reaching our neighbor stars without any particular fuss. That might be possible, but it would take millions of years, and is not what I am talking about here, which is using indigenous resources to assist in high-speed flight. Having a starship stop repeatedly at Oort Cloud objects along the way to refuel is out of the question, as the repeated decelerations required would defeat the purpose. If human stations equipped with high-powered lasers were positioned in Kuiper Belt and Oort Cloud objects along the path to destination stars, they could assist in pushing light-sail craft along, but this seems like a rather infrastructure-intensive approach (although it is perhaps one service that Kuiper Belt or Oort Cloud settlers could sell to outsiders for a high markup).

A more basic approach to utilizing Oort Cloud objects is to blast them into propellant on the fly. Consider a MagOrion, a magsail pushing itself along by exploding thermonuclear plasma bombs to its rear. The plasma reflected off the sternside magnetic field might have an exhaust velocity of 5 percent the speed of light. To get the ship to 1 percent lightspeed, a supply of bombs with a total mass equal to 22 percent the dry weight of the spaceship will be needed. But now let's say that instead of exploding the bombs in the middle of nowhere, we decide to detonate them close by a series of small (car-sized) Oort Cloud objects, which we fly by in rapid succession. Assuming that each bomb vaporizes and ionizes 25 times its mass in Oort stuff, the total propellant available will be multiplied 25-fold. Since we've watered down the exhaust 25-fold, the exhaust velocity will drop by a factor of 5 (because exhaust velocity is proportional to the square root of the propellant's energy/mass ratio), but since we have 25 times the propellant, the net result is that the total impulse imparted to the system is increased 5-fold. Therefore, instead of 220 tonnes of explosives being needed to get the 1,000-tonne starship up to 1 percent lightspeed, only about 44 tonnes will be necessary.

In the above example, I chose a modest maximum flight speed for the mission, because in order for the iceteroid-augmented MagOrion concept to work, it's important that the watered-down exhaust velocity (the expansion speed of the plasma cloud from the ionized Oort object) be at least comparable to the ship's speed. If it is much less, the ship will have moved away before the expanding plasma cloud has a chance to reach and push on the spacecraft's magnetic field. If we want to use this trick at higher speeds, a more energetic explosive than fusion is needed. The obvious answer is antimatter. A small pellet of antimatter fired into an Oort ice block with 400 times its mass would produce an exhaust velocity of 5 percent the speed of light, which would allow efficient acceleration of the ship to this speed (at a 20-fold saving of antimatter), after which we could choose to dilute the antimatter pellets 100-fold, get a 10 percent lightspeed exhaust velocity and accelerate the ship to 0.1 *c* (at a 10-fold saving of antimatter for this part of the acceleration), then dilute 25-fold to get to 20 percent *c*, and so on.

It is believed that 100-km Oort Cloud iceteroids may occur with a frequency of one every 10 AU or so. If that is the case, a reasonable conjecture would be that 10-km objects occur every AU, 1-km objects every 0.1 AU, 100-m objects every 0.01 AU, 10-m objects every 0.001 AU (150,000 km), 1-m objects every 15,000 km, and 0.1-m objects every 1,500 km. If the ship is moving at 1,000 km/s (0.3 percent *c*), it can thus be expected to come reasonably close to a 1-m (~1-tonne) object every 15 seconds and a 0.1-m (~1-kg) object every 1.5 seconds. Active guidance potentially could increase the odds. Targeting the antimatter pellets to hit the ice blocks with high frequency would be a trick requiring great technological sophistication, but there's nothing within the laws of physics that prevents it.

UNKNOWN STARS

So much for interstellar plasma and Oort Cloud iceteroids, the mass we currently know to lie between the stars. But the question is worth asking: Is that all there is? Could there not be much more impressive, interesting, and useful currently unknown objects between us and the Alpha Centauri system? Perhaps in using our lack of knowledge to assume a void we are taking too much on faith. Maybe we should even be asking ourselves, *Are the closest stars really those in Centaurus?* Might we not have stellar neighbors that are much closer?

At first glance the question appears to be outrageous. Certainly one might think if there were stars closer than the Centauri system they would have been identified long ago. As obvious as such an objection might seem, it is not necessary valid.

Recently a theory was proposed which postulates that various mass extinctions that have occurred during the Earth's geologic past have been caused by showers of comets released from their stable orbits in the Oort Cloud by the gravitational influence of a passing star. Furthermore, the apparent periodicity of 26 million years associated with the repeated extinctions has led some investigators to postulate that the extinction-causing star, dubbed "Nemesis," is actually bound in an orbit about the Sun with an aphelion (maximum distance) of 2.78 light-years and a perihelion (minimum distance) of perhaps 0.01 light years.[14] Since the last mass extinction occurred about 13 million years ago, the hypothetical Nemesis star would today be located between 2.3 and 2.8 light-years away. In order to confirm this theory, investigators in the late 1980s launched a quest to identify the Nemesis star. During the attempt they found that the problem they had set themselves was extraordinarily difficult, much like "looking for a needle in a haystack." That is, finding Nemesis would require searching through images of literally billions of dim objects and then determining the range to each of them using parallax (the apparent shift in position of an interstellar object when viewed from opposite sides of the Earth's orbit) measurements. Beyond the sheer magnitude of this task in terms of volume, a large parallax itself could easily confuse the search by making correlation of two images of the same object difficult. The possibility also existed that Nemesis was insufficiently luminous for imaging with the equipment employed, which would make the entire effort an exercise in futility.

To date, no Nemesis object has been found. Yet the technical problems that surfaced during the search have shown that many objects of stellar mass could easily exist in nearby interstellar space and have not been imaged or, if imaged, have not yet been identified as neighbors.

Beyond the evidence supplied by periodic mass extinctions on Earth, there is other support for the existence of non-imaged stellar objects. Many of the leading current cosmological theories require that the universe contain much more mass than can be accounted for by the total of currently imaged stars, dust, and other matter. Furthermore, it has been observed that many galaxies, including our own, are rotating at too fast a rate to be accounted for by the gravitational attraction of the visible matter that they contain. Many hypotheses have been advanced to account for this discrepancy, including the possible presence of nonluminous matter orbiting the galaxies but outside the plane containing the luminous stars, the existence of hypothetical forms of "nonhadronic" exotic matter that is intrinsically invisible, and the possibilities that galactic rotation is governed by magnetoplasma dynamic effects (i.e., not gravity) and that the various cosmological theories requiring missing matter are just plain wrong. However, also included among the possible explanations are that the missing matter could be composed of nonluminous objects of stellar mass, including black dwarfs (white dwarf stars that have ceased to emit energy), brown dwarfs (proto stars that never ignited), neutron stars, and black holes. If such objects are in fact responsible for the missing mass, they could be quite numerous, since the missing mass is estimated by those who believe in it to outweigh the universe's inventory of visible mass by as much as 10 to 1 or even 100 to 1.

Without passing judgment on the veracity of any of these theories, it is sufficient to observe that there is a significant body of evidence that currently undetected objects of stellar mass may exist in near-solar space. Let's consider what they might be, and how they could assist interstellar flight.

Since no near-solar system stellar mass objects have been detected to date, they must be of such a nature as to not make their presence obvious. In the case of luminous objects, this does not necessarily imply that they are too dim to be imaged, but only that they are so dim as to not provoke further investigation. The apparent motions of the 300,000 brightest stars, down to about the 10th visual magnitude, have been catalogued, so the object must be dimmer than this, and it may be considered unlikely that a large parallax of an object brighter than 11th magnitude could have evaded detection. Dim-type M (red dwarf) and white dwarf stars can have absolute magnitudes of 17 or 18, which at a distance of 2 light-years implies an apparent visual magnitude of 11 to 12. Thus, it may be possible that actual luminous stars exist as close as 2 light-years.

The smallest and dimmest ignited stars that are known to exist have a mass of about 3 percent that of the Sun. Objects smaller and dimmer than this may exist; however, no one has individually assessed each of the billions of the dimmest of undistinguished stellar objects for parallax—circular logic has caused an assumed minimal stellar size to be interpreted as implying that very dim stars cannot be close. However, if we drop that a priori assumption, it may be the case that many such objects can be found. For small red dwarf stars (types K5 through M5) it has been found that the luminosity of the star goes in proportion to the mass of the star squared. Therefore, if we scale from Proxima Centauri, which has an absolute magnitude of 15.45 and a mass of 0.1 solar units, we find that small stellar and substellar mass objects could potentially exist much closer than 1 light-year and shine with less than 12th apparent visual magnitude. The results of such scaling are shown in Table 9.2.

The planet Jupiter has a mass of about 0.001 Suns. Objects with masses between 0.002 and 0.01 solar masses would be brown dwarf stars. These stars either never have ignited or may be regarded as being in such a slow-burning condition that a substantial alteration from their initial composition via nuclear fusion has not occurred. They may, however, be giving off heat through compression (Jupiter is). For the larger brown dwarfs, such thermal emissions may be sufficient to warm close orbiting planets (moons?) to habitable temperatures. As shown in Table 9.2, such systems may exist undetected at distances of less than half a light-year.

If an object is nonluminous, it can be detected either visually due to its

▪ TABLE 9.2 ▪

Possible 12th Apparent Magnitude Objects

Mass (Suns)	Absolute Magnitude	Distance (light-years)
0.1	15.45	6.66
0.05	16.96	3.33
0.02	18.95	1.33
0.01	20.46	0.67
0.005	21.97	0.33
0.002	23.98	0.13

ability to reflect sunlight or by virtue of the effects of its gravity on the solar system. The candidate nonluminous objects—black holes, neutron stars (asteroid size), and black dwarfs (Earth size)—are all quite compact and would be difficult to detect visually at distances greater than 100 AU (0.0016 light-year) from the Sun. Gravitational perturbations caused by their mass would probably give away the fact of their existence, if they were located within 300 AU of the Sun. A summary of possible candidate near-solar objects is given in Table 9.3.

How could such objects aid us in our quest for interstellar flight? In the case of luminous objects, such as white, red, and brown dwarf stars, the answer is obvious—they give us worthwhile destinations to go to that are significantly closer than the known stars, and possibly much more numerous as well. Surprisingly, perhaps, the nonluminous objects may offer us even more. How's that? What can one possibly get from a black hole? Answer: gravity.

We previously explored how spacecraft can exploit the deep gravity wells of Jupiter and the Sun to generate large departure velocities with modest rocket burns. Well, dense objects such as black dwarfs, neutron stars, and black holes offer magnificent gravity wells, with escape velocities of 2 percent, 85 percent, and 100 percent the speed of light, respectively. These can be used as terrific multipliers of rocket effectiveness, allowing even near-term technologies to achieve velocities relevant for interstellar missions.

For example, consider a spacecraft powered by a nuclear thermal rocket

▪ TABLE 9.3 ▪

Candidate Near-Solar Stellar Mass Objects

Object	Mass (Suns)	Radius (Sun)	Distance (light-years)	V_e[a]
Brown dwarf	0.01	0.20	0.67	0.0004*c*
Red dwarf	0.03	0.30	2.0	0.0007*c*
White dwarf	1.0	0.01	2.0	0.022*c*
Black dwarf	1.0	0.008	0.01	0.024*c*
Neutron star	3.0	0.00002	0.01	0.852*c*
Black hole	20.0	0.00008	0.01	1.000*c*
Jupiter	0.001	0.1	0.0001	0.0002*c*
Sun	1.0	1.0	0.00002	0.0022*c*

[a] V_e = escape velocity.

(NTR), such as those demonstrated by NASA in the 1960s. It has an exhaust velocity of 9 km/s. Let's say we give it enough propellant to do a 12 km/s ΔV and enough thrust so that it would have an acceleration at propellant burnout of 10*g*. These are not extraordinary specifications; such a spacecraft could probably be developed within ten years if there were a political decision to do so. Now, let's fly this spacecraft fairly close by a black hole with a mass of 100 Suns to an altitude where the escape velocity is 7,500 km/s, or 2.5 percent lightspeed. We then fire the engine, exerting a rocket ΔV of 12 km/s, but because of the gravity multiplier effect described by Equation 9.1, the spacecraft emerges with a departure velocity of 424 km/s!

The reader may ask why we did not fly deeper into the gravity well, where the escape velocity is higher, and get a bigger multiplier effect. The answer is that we kept our distance to provide the rocket enough time to generate its acceleration during its close (and fast) approach around the object. Our rocket vehicle has a burnout thrust-to-weight ratio of 10. To exploit the benefits of deeper regions of the black hole's gravity well, we need systems that can generate higher accelerations. We can probably build rockets with thrust-to-weight ratios of 100, but if we really want to get a payload up to speed fast, we need to use guns.

Artillery pieces in current use can generate projectile velocities of 2 km/s and accelerations of 40,000*g*. More advanced types of ordnance under current development, including light gas guns, ram accelerators, rail guns, and ram imploders, can generate similar accelerations and muzzle velocities of 10–20 km/s. Let's say we were to fly such a unit to the 100-solar-mass black hole described above and fire it (muzzle velocity = 15 km/s) while passing through an altitude where the escape velocity is 85,000 km/s (28 percent *c*). The result is that our projectile will be sent on its way with a speed of 1,600 km/s (0.5 percent *c*). That's not too shabby for twentieth-century technology.

Of course, given the high acceleration, no such gun-fired payload could be manned. But instruments can be built to survive such accelerations, so such a system might represent a low-cost, low-tech way to send robotic probes to the stars.

The math of how much acceleration can be applied at what altitude over a dense object of a given mass is complicated. Those who are interested can find it all worked out in a paper I wrote in the *Journal of the British Interplanetary Society* several years ago.[15] The bottom-line results, however, are shown in Figures 9.5 and 9.6.

In Figures 9.5 and 9.6, *N* is the mass of the object in Suns and *G* is the

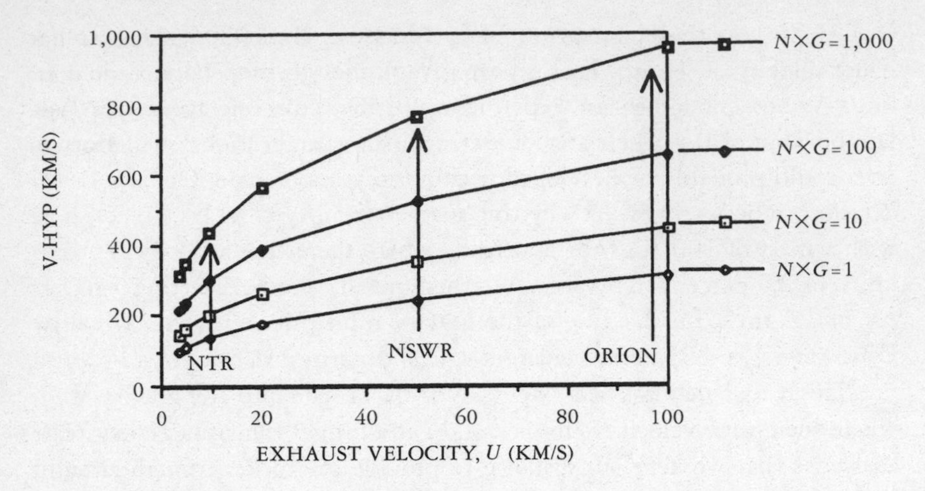

FIGURE 9.5 *Potential performance of near-term rockets in deep gravity wells (V-hyp = departure velocity).*

burnout acceleration (in g's) of the rocket or average acceleration of the gun. The ΔV performed by the rocket system in Figure 9.5 is assumed to be 4/3rds of the exhaust velocity, U. The value of using dense objects to multiply the effects of these systems is apparent.

As rocket technology becomes more advanced, and therefore capable of

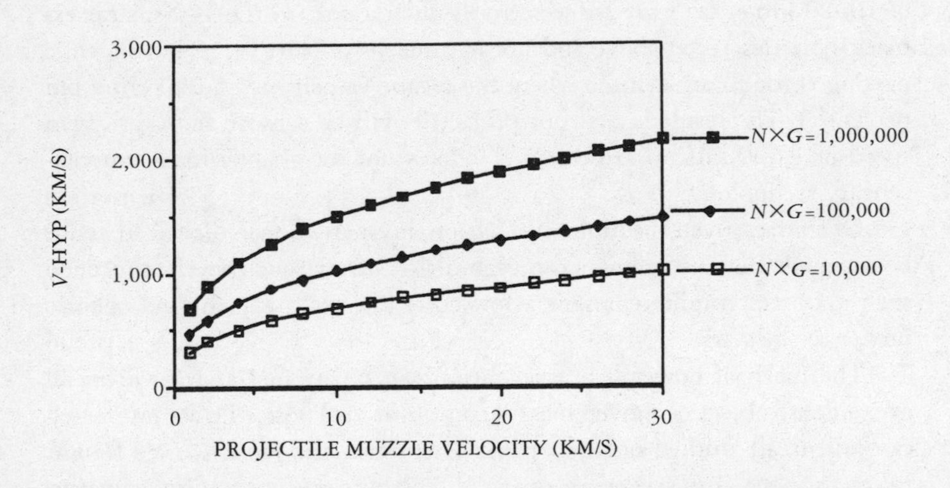

FIGURE 9.6 *Potential performance of guns in deep gravity wells (V-hyp = departure velocity).*

generating larger ΔV's directly, the multiplier effect offered by dense stellar objects decreases. However, even at exhaust velocities of 1 percent *c*, a doubling of system effectiveness can be obtained. It would thus behoove a Type II civilization with interstellar ambitions to mount a significant effort to find one or more of these potential springboards to the stars.

STARFLIGHT AND SPECIES MATURITY

In this chapter we have discussed interstellar travel using mighty systems such as thermonuclear fusion, antimatter rockets, and laser-pushed light sails, all with power ratings in the tens to hundreds of terawatts. It should be obvious to most readers that such systems will be (a) expensive and (b) very dangerous in the hands of minors.

As far as expense is concerned, this will take care of itself. Starflight will not occur until humanity can afford it. But as we have shown, if humanity does develop into a healthy Type II civilization, our resources and power base will continue to expand at a rate that will make even the huge costs associated with interstellar colonization affordable within just a few centuries.

The issue of danger is different. Starflight requires the deployment of vast amounts of energy in compact form. Any system that can dispense such energies is implicitly a weapon of mass destruction with potentials far exceeding the twentieth century's nuclear arsenals.

This brings up an interesting point. Intelligent species, including our own, evolve from aggressive, predatory, highly competitive forebears. It is, in fact, the selection pressures associated with the successful implementation of such a mode of life that call forth the evolution of that adaptation known as intelligence. Moreover, within the history of the species, it is the winners of millennia of tribal conflicts who survived to pass on their genes. It is a nasty but true fact that all of us alive today are descendants of folks who were good at killing people and breaking their stuff. We can be proud of it, we can be ashamed of it, but no matter how we feel about it, we are all the children of warriors.

Wars fought with bows, arrows, and spears are one thing; wars fought with antimatter bombs and planet-frying lasers are quite another. Primitive warfare can be very ugly, but it also carries the redeeming virtue of species selection for intelligence and physical strength. Not so modern war. It is said that Christopher Columbus, observing a Spanish peasant disintegrate a squadron of proud Moorish cavalry with a hand-thrown bomb at the siege of Granada in 1492, commented that "such inventions make war meaningless." The author of humanity's Type I triumph saw far.

We are the children of warriors, but also of loving parents, incurable tinkerers, explorers, and reasoners. We bear the genes, instincts, and capabilities of all these. From the warriors we have inherited not only the instincts that threaten us but the courage to try the unknown. From the explorers we have inherited the drive to take us to the stars; from the tinkerers, the spirit that will give us the tools to get there; and from the lovers and reasoners we have received that which will allow us to use our expanding powers for good instead of evil.

There is no turning back. The spirit of the tinkerers and explorers cannot be suppressed without destroying our humanity. Safety at the cost of doing so would come at too high a price. So there will be Type II civilization with the capability to move comets, and there will be mass-produced plasma explosives, giant lasers, and all the rest of the formidable gear needed to launch interstellar missions. And we will survive the test of their ownership. Because we also have Love and Reason, and when forced to do so, we can and will grow those capacities, too.

People will do anything to survive, even become better.

The heavens will be open to those who deserve them. I think humanity does. We'll need to grow up a bit, but with ingenuity, determination, and species maturity, the stars can be ours.[16]

Ad astra.

FOCUS ON: THE NEARBY STARS

Interstellar voyagers should have a good knowledge of the potential list of nearby destinations. With that in mind, here are two maps (Figures 9.7 and 9.8) and a table (Table 9.4) with the sailing directions to and some key data about the Sun's closest neighbors.[16]

In Table 9.4, the number preceding the Star's name corresponds to the numbered points on the star maps. Distance is given in light-years from the Sun (Sol). The three numbers in parentheses also listed under distance correspond to the star's *X, Y, Z* coordinates, in light-years, with the *X* direction defined as that pointing from Earth to the Sun during the spring equinox (of 1950 to be precise—although the vernal equinox hasn't moved

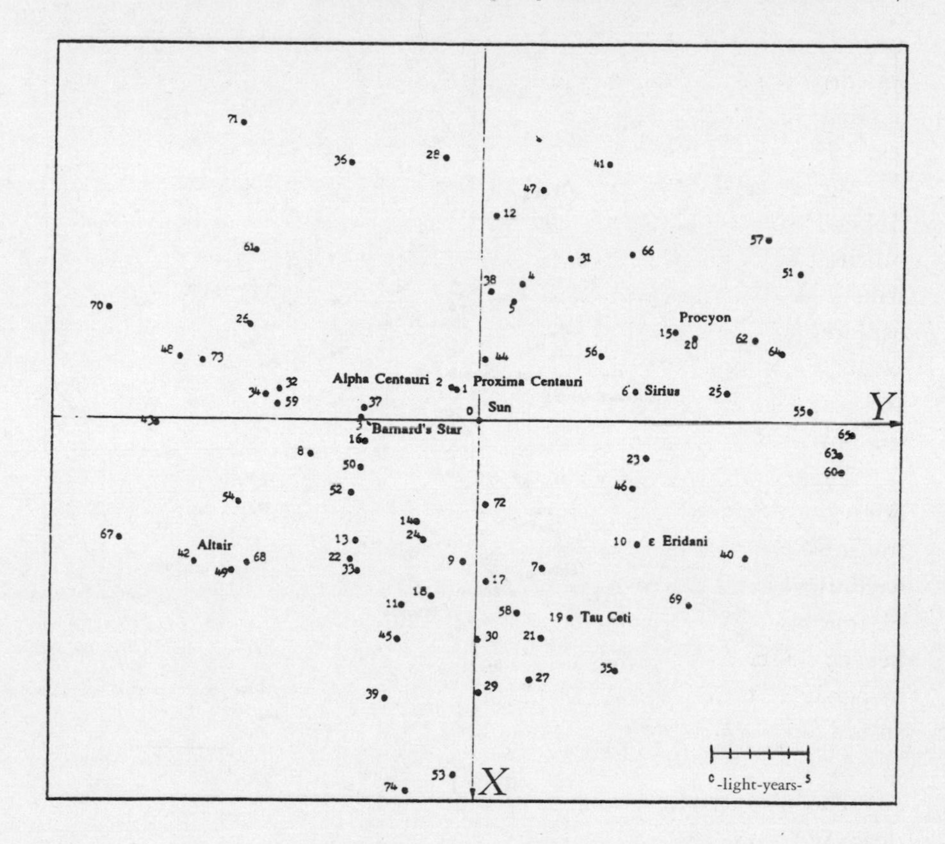

FIGURE 9.7 *Map of nearby interstellar space in the* X-Y *plane (the plane of the equator). (Courtesy of John Wiley and Sons, Inc.)*

much since), the *Z* direction pointing outward from the Earth's North Pole (in spring 1950), and the *Y* direction located in the Earth's equatorial plane and perpendicular to the *X* and *Z* vectors in such a manner that the three *X, Y,* and *Z* vectors form a standard "right-hand rule" coordinate system. The luminosity, mass, and radius for each star are given in the units of Suns. For example, Procyon A's listed luminosity of 7.6 means that it has a brightness 7.6 times that of the Sun. Stars with designations A, B, or C following their names are members of multiple star systems. The main sequence of stars runs through types O, B, A, F, G, K, M, with O0 the brightest and M9 the dimmest. White dwarf stars have spectral types beginning with the letter D. The most suitable stars for colonization are those of type F, G, or K. Bon voyage!

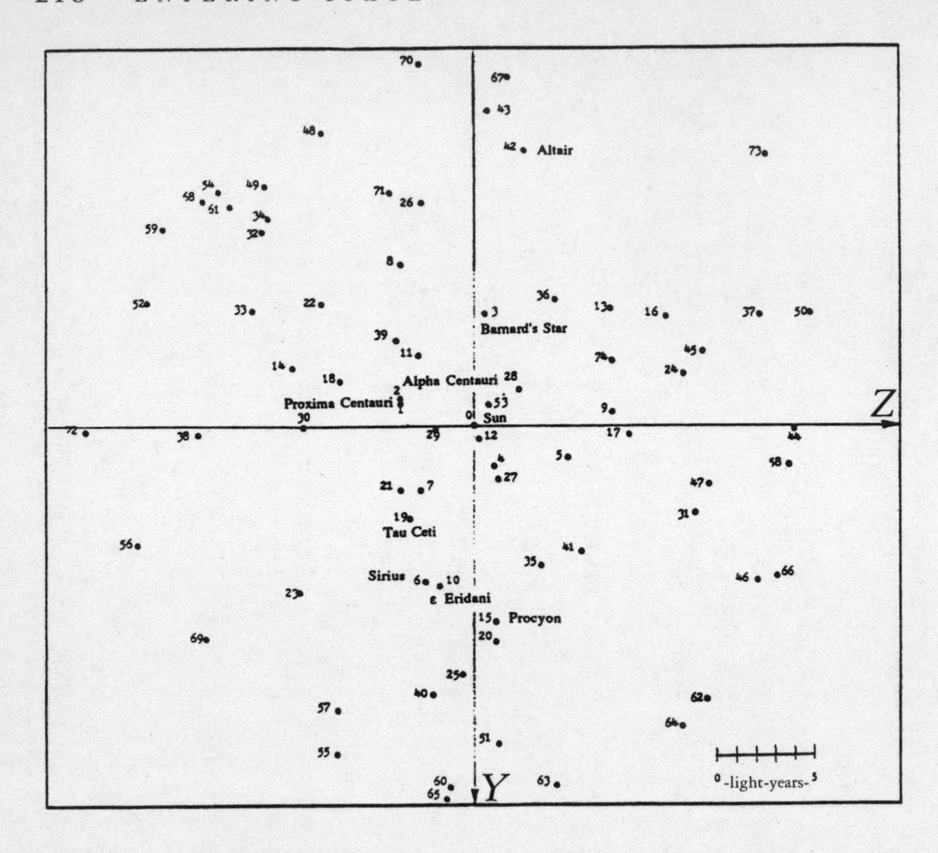

FIGURE 9.8 *Map of nearby interstellar space in the* Y-Z *plane. (Courtesy of John Wiley and Sons, Inc.)*

▪ TABLE 9.4 ▪

The Nearby Stars

Star	Distance (light-years from the Sun)	(*X, Y, Z*) in light-years	Type	Luminosity (Suns)	Mass (Suns)	Radius (Suns)
0. Sol	0	(0.0, 0.0, 0.0)	G2	1.0	1.0	1.0
1. Proxima Centauri	4.3	(−1.6, −1.2, −3.8)	M5e	0.00006	0.1	—
2. α Centauri A	4.4	(−1.7, −1.4, −3.8)	G2	1.3	1.1	1.23
2. α Centauri B	4.4	(−1.7, −1.4, −3.8)	G2	0.36	0.89	0.87
3. Barnard's Star	5.9	(−0.1, −5.9, −0.5)	M5	0.00044	0.15	0.12
4. Wolf 359	7.6	(−7.2, 2.1, 1.0)	M8e	0.00002	0.2	0.04
5. Lalande 21185	8.1	(−6.3, 1.7, 4.8)	M2	0.0052	0.35	0.35
6. Sirius A	8.7	(−1.6, 8.2, −2.5)	A1	23.0	2.31	1.8
6. Sirius B	8.7	(−1.6, 8.2, −2.5)	DA	0.0028	0.98	0.022
7. UV Ceti A	8.9	(7.7, 3.4, −2.8)	M6e	0.00006	0.12	0.05
7. UV Ceti B	8.9	(7.7, 3.4, −2.8)	M6e	0.00004	0.10	0.04
8. Ross 154	9.5	(1.8, −8.5, −3.8)	M5e	0.0004	0.31	0.12
9. Ross 248	10.3	(7.4, −0.7, 7.1)	M6e	0.00011	0.25	0.07
10. ε Eridani	10.7	(6.4, 8.4, −1.8)	K2	0.3	0.8	0.9
11. Luyten 789-6	10.8	(9.7, −3.7, −2.9)	M6	0.00012	0.25	0.08
12. Ross 128	10.8	(−10.8, 0.7, 0.2)	M5	0.00033	0.31	0.1
13. 61 Cygni A	11.2	(6.3, −6.1, 7.0)	K5	0.063	0.59	0.7
13. 61 Cygni B	11.2	(6.3, −6.1, 7.0)	K7	0.040	0.50	0.8

Star	Distance (light-years from the Sun)	(X, Y, Z) in light-years	Type	Luminosity (Suns)	Mass (Suns)	Radius (Suns)
14. ε Indi	11.2	(5.3, –3.0, –9.4)	K5	0.13	0.71	1.0
15. Procyon A	11.4	(–4.7, 10.3, 1.1)	F5	7.6	1.77	1.7
15. Procyon B	11.4	(–4.7, 10.3, 1.1)	DF	0.0005	0.63	0.01
16. +59° 1915 A	11.5	(1.1, –5.7, 9.9)	M4	0.0028	0.4	0.28
16. +59° 1915 B	11.5	(1.1, –5.7, 9.9)	M5	0.0013	0.4	0.20
17. Groombridge 34A	11.5	(8.4, 0.5, 8.0)	M2	0.0058	0.38	0.38
17. Groombridge 34B	11.5	(8.4, 0.5, 8.0)	M4	0.0004	—	0.11
18. Lacaille 9352	11.7	(9.2, –2.3, –6.9)	M2	0.012	0.47	0.57
19. τ Ceti	11.9	(10.3, 4.9, –3.3)	G8	0.44	0.82	1.67
20. Luyten BD	12.2	(–4.4, 11.3, 1.1)	M4	0.0014	0.38	0.16
21. LET 118	12.5	(11.4, 3.4, –3.8)	M5e	—	—	—
22. Lacaille 8760	12.5	(7.3, –6.4, –7.9)	M1	0.025	0.54	0.82
23. Kapteyn's Star	12.7	(1.9, 8.8, –9.0)	M0	0.004	0.44	0.24
24. Kruger 60 A	12.8	(6.3, –2.7, 10.8)	M4	0.0017	0.27	0.51
24. Kruger 60 B	12.8	(6.3, –2.7, 10.8)	M6	0.00044	0.16	—
25. Ross 614 A	13.1	(–1.5, 13.0, –0.6)	M5e	0.0004	0.14	0.14
25. Ross 614 B	13.1	(–1.5, 13.0, –0.6)	—	0.00002	0.08	—
26. BD –12° 4523	13.1	(–5.0, –11.8, –2.8)	M5	0.0013	0.38	0.22
27. Van Maanen's Star	13.9	(13.6, 2.8, 1.2)	DG	0.00017	—	—
28. Wolf 424 A	14.2	(–13.9, –1.9, 2.3)	M6e	0.00014	—	0.09
28. Wolf 424 B	14.2	(–13.9, –1.9, 2.3)	M6e	0.00014	—	0.09

29. G158-27	14.4	(14.3, 0.2, –2.0)	M7	0.00005	—	—
30. CD –37° 15492	14.5	(11.5, 0.1, –8.8)	M3	0.00058	0.39	0.4
31. Groombridge 1616	15.0	(–8.6, 4.6, 11.4)	K7	0.04	0.56	0.5
32. CD –46° 11540	15.1	(–1.6, –10.2, –11.0)	M4	0.003	0.44	0.25
33. CD –49° 13515	15.2	(7.9, –6.0, –11.5)	M3	0.00058	0.37	0.34
34. CD –44° 11909	15.3	(–1.3, –10.9, –10.7)	M5	0.00063	0.34	0.15
35. Luyten 1159-16	15.4	(13.1, 7.3, 3.4)	M8	0.00023	—	—
36. Lalande 25372	15.7	(–13.6, –6.6, 4.1)	M2	0.0076	—	0.40
37. BD +68° 946	15.8	(–0.6, –5.8, 14.7)	M3	0.0044	0.35	0.39
38. Luyten 145-141	15.8	(–6.8, 0.5, –14.3)	DA	0.0008	—	—
39. Ross 780	15.8	(14.6, –4.5, –4.0)	M5	0.0016	0.39	0.23
40. Omicron Eridani A	15.9	(7.1, 14.1, –2.1)	K0	0.33	0.81	0.7
40. Omicron Eridani B	15.9	(7.1, 14.1, –2.1)	DA	0.0027	0.43	0.018
40. Omicron Eridani C	15.9	(7.1, 14.1, –2.1)	M4e	0.00063	0.21	0.43
41. BD +20° 2465	16.1	(–13.6, 6.6, 5.5)	M4	0.0036	0.44	0.28
42. Altair	16.6	(7.4, –14.6, 2.5)	A7	10.0	1.9	1.2
43. 70 Ophiuchi A	16.7	(0.2, –16.7, 0.7)	K1	0.44	0.89	1.3
43. 70 Ophiuchi B	16.7	(0.2, –16.7, 0.7)	K6	0.083	0.68	0.84
44. AC –79° 3888	16.8	(–3.2, 0.2, 16.5)	M4	0.0009	0.35	0.15
45. BD +43° 4305	16.9	(11.5, –3.9, 11.8)	M5e	0.0021	0.26	0.24
46. Stein 2051 A	17.0	(3.5, 8.1, 14.6)	M5	0.0008	—	—
46. Stein 2051 B	17.0	(3.5, 8.1, 14.6)	DC	0.0003	—	—
47. WX Ursa Majoris A	17.5	(–12.2, 3.1, 12.1)	M2	—	—	—
47. WX Ursa Majoris B	17.5	(–12.2, 3.1, 12.1)	M8	—	—	—

Star	Distance (light-years from the Sun)	(X, Y, Z) in light-years	Type	Luminosity (Suns)	Mass (Suns)	Radius (Suns)
48. 36 Ophiuchi A	17.7	(–3.3, –15.5, –7.9)	K2	0.26	0.77	0.90
48. 36 Ophiuchi B	17.7	(–3.3, –15.5, –7.9)	K1	0.26	0.76	0.82
48. 36 Ophiuchi C	17.7	(–3.3, –15.5, –7.9)	K6	0.09	0.63	0.90
49. HR 7703 A	18.4	(7.9, –12.6, –10.9)	K3	0.20	0.76	0.80
49. HR 7703 B	18.4	(7.9, –12.6, –10.9)	M5	0.0008	0.35	0.14
50. σ Draconis	18.5	(2.5, –5.9, 17.3)	K0	0.4	0.82	0.28
51. YZ Canis Minoris	18.5	(–7.8, 16.7, 1.2)	M4	—	—	—
52. δ Pavonis	18.6	(3.8, –6.4, –17.0)	G6	1.0	0.98	1.07
53. 1° 4774	18.6	(18.6, –1.1, 0.7)	M2	0.0001	—	—
54. Luyten 347-14	18.6	(4.3, 12.3, 13.3)	M7	0.0001	0.26	0.08
55. –21° 1377	18.7	(–0.6, 17.3, –7.0)	M1	0.016	0.46	0.59
56. Luyten 97-12	18.9	(–3.4, 6.3, –17.5)	D	0.0003	—	—
57. Luyten 674-15	19.1	(–9.6, 15.0, –7.0)	M	—	—	—
58. η Cassiopeia A	19.2	(10.1, 2.1, 16.2)	G0	1.0	0.85	0.84
58. η Cassiopeia B	19.2	(10.1, 2.1, 16.2)	M0	0.03	0.52	0.07
59. Luyten 205-128	19.2	(–0.8, –10.3, –16.2)	M	0.0002	0.14	—
60. HD 36395	19.2	(2.6, 19.0, –1.2)	M1	0.02	0.51	0.69
61. 40° 9712	19.3	(–8.8, –11.5, –12.7)	M4	0.003	0.44	0.29
62. Ross 986	19.3	(–4.3, 14.4, 12.0)	M5	—	—	—
63. Ross 47	19.4	(1.7, 18.9, 4.2)	M6	0.0008	0.35	0.17
64. Wolf 294	19.4	(–3.6, 15.8, 10.7)	M4	0.008	0.49	0.46

65. LP 658-2	19.6	(0.6, 19.5, –1.4)	DK	—	—	—
66. +53° 1320 A	19.6	(–8.8, 7.9, 15.6)	M0		—	—
66. +53° 1320 B	19.6	(–8.8, 7.9, 15.6)	M0	—	—	—
67. VB10 A	19.6	(6.2, –18.5, 1.7)	M4		0.39	0.43
67. VB10 B	19.6	(6.2, –18.5, 1.7)	M5	0.007	—	0.008
68. –45°13677	19.9	(7.5, –11.8, –14.1)	M0	0.00002	—	—
69. 82 Eridani	20.3	(9.6, 11.2, –13.9)	G5	—	—	—
70. Wolf 630 A	20.3	(–5.8, –19.2, –2.9)	M4		0.38	—
70. Wolf 630 B	20.3	(–5.8, –19.2, –2.9)	M5	—	0.38	—
70. Wolf 630 C	20.3	(–5.8, –19.2, –2.9)	—		—	—
70. Wolf 630 D	20.3	(–5.8, –19.2, –2.9)	M4	—	—	—
71. –11° 3759	20.4	(–15.7, –12.3, –4.4)	M4		—	—
72. β Hydri	20.5	(4.4, 0.4,–20.0)	G1	—	—	1.66
73. +45 Fu46 A	21.0	(–3.1, –14.4, 15.0)	M3		0.31	—
73. +45 Fu46 B	21.0	(–3.1, –14.4, 15.0)	—	—	0.25	—
74. +19° 5116 A	21.0	(19.5, –3.4, 7.1)	M4		—	—
74. +19° 5116 B	21.0	(19.5, –3.4, 7.1)	M6	—	—	—

CHAPTER 10

Extraordinary Engineering

To see it in our power to make a World happy . . .
to exhibit on the theatre of the Universe
a character hitherto unknown—and to have,
as it were, a new creation intrusted to our hands,
are honors that command reflection and can neither
be too highly estimated nor too gratefully received.

—THOMAS PAINE, 1783

No one will be able to look upon it and
not feel prouder to be a man.

—JOHANN ROEBLING, designer of the Brooklyn Bridge, 1868

I LIVE IN Colorado, among the mountains. Not long ago, I had occasion to hike to the top of one of the smaller peaks, whose summit was just above tree line. As I sat, observing the scenery while eating my lunch, an odd question crossed my mind: *How did all these trees get up here?* Conifers lined the slopes nearly to the summit of every peak in sight. *How did a mob of immobile trees ever climb those steep heights?*

As I munched away, pondering this, I noticed a group of chipmunks scurrying about *carrying pinecones.* The answer was thus made apparent. The chipmunks had transported the seeds uphill. Interesting. Every mountain in sight was covered with trees. By moving the seeds upslope, the chipmunks had enormously expanded their "natural habitat." In fact, if by "natural habitat" one means the habitat that would support a chipmunk population that exists prior to and independent of their seed-spreading activity, it's unclear whether any such place exists at all. The chipmunk habitat does not exist "naturally." It exists because the chipmunks (together with a host of other participating species) have *created* it. That's how life works.

A major challenge that humans will face as we become a Type II and then Type III civilization is that of transforming the environments found on other planets to more Earth-like conditions. This must be done because environments friendly to life are a product of the activity of life. Thus, as humans move out into space it is unlikely that we will find environments that perfectly suit our needs. Instead, as life and humanity have done historically on Earth, we will have to improve the natural environments we find to create the worlds we want. Applied to other planets, this process of planetary engineering is termed "terraforming."[1–3]

Some people consider the idea of terraforming other planets heretical—humanity playing God. Others would see in such an accomplishment the most profound vindication of the divine nature of the human spirit—dominion over nature, exercised in its highest form to bring dead worlds to life. Personally, I prefer not to consider such issues in their theological form, but if I had to, my sympathies would definitely be with the latter group. Indeed, I would go further. *I would say that failure to terraform constitutes failure to live up to our human nature and a betrayal of our responsibility as members of the community of life itself.*

These may seem like extreme statements, but they are based on history, about 4 billion years of history. The chronicle of life on Earth is one of terraforming—that's why our beautiful blue planet is as nice as it is. When the Earth was born, it had no oxygen in its atmosphere, only carbon dioxide and nitrogen, and the land was composed of barren rock. It was fortunate that the Sun was only about 70 percent as bright then as it is now, because if the present-day Sun had shined down on that Earth, the thick layer of CO_2 in the atmosphere would have provided enough of a greenhouse effect to turn the planet into a boiling Venus-like hell. Fortunately, however, photosynthetic organisms evolved that transformed the CO_2 in Earth's atmosphere into oxygen, in the process completely changing the surface chemistry of the planet. As a result of this activity, not only was a runaway greenhouse effect on Earth avoided but the evolution of aerobic organisms that use oxygen-based respiration to provide themselves with energetic lifestyles was enabled (though a primeval EPA dedicated to preserving the status quo on the early Earth might have regarded this as a catastrophic act of environmental destruction). This new crowd of critters, known today as animals and plants, then proceeded to modify the Earth still more—colonizing the land, creating soil, and drastically modifying global climate. Life is selfish, so it's not surprising that all of the modifications that life has made to the Earth have contributed to enhancing life's prospects, expanding the biosphere, and accelerating its rate of developing new capabilities to improve the Earth as a home for Life still more.

Humans are the most recent practitioners of this art. Starting with our earliest civilizations, we used irrigation, crop seeding, weeding, domestication of animals, and protection of our herds to enhance the activity of those parts of the biosphere most efficient in supporting human life. In so doing, we have expanded the biospheric basis for human population, which has expanded our numbers and thereby our power to change nature in our interest in a continued cycle of exponential growth. As a result, we have literally remade the Earth into a place that can support billions of people, a substantial fraction of whom have been sufficiently liberated from the need to toil for daily survival that they can now look out into the night sky for new worlds to conquer.

It is fashionable today to bemoan this transformation as destruction of nature. Indeed, there is a tragic dimension to it. Yet it is nothing more than the continuation and acceleration of the process by which nature was created in the first place.

Life is the creator of nature.

Today, the living biosphere has the potential to expand its reach to encompass a whole new world, on Mars, and the Type II interplanetary civilization that develops as a result will have the capability of reaching much further. Humans, with their intelligence and technology, are the unique means that the biosphere has evolved to allow it to blossom across interplanetary and then interstellar space. Countless beings have lived and died to transform the Earth into a place that could give birth to a species with such capabilities. Now it's our turn to do our part.

It's a part that 4 billion years of evolution has prepared us to play. Humans are the stewards and carriers of terrestrial life, and as we spread out, first to Mars and then to the nearby stars, we must and shall bring life to many worlds, and many worlds to life.

It would be unnatural for us not to.

MARS-LIKE WORLDS

Mars is the first extraterrestrial planet that will be terraformed. As discussed in detail in my book *The Case for Mars,* the engineering methods by which this can be done are relatively well understood. The first step will be to re-create the atmosphere of early Mars by setting up factories to produce artificial greenhouse gases, such as perfluoromethane (CF_4) for release into the atmosphere. If CF_4 were produced and released on Mars at the same rate as chlorofluorocarbon (CFC) gases are currently being produced on Earth (about 1,000 tonnes per hour), the average global temperature of the Red Planet would be increased by 10°C within a few decades. This temperature rise would cause vast amounts of carbon dioxide to outgas from the regolith, which would warm the planet further, since CO_2 is a greenhouse gas. The water vapor content of the atmosphere would vastly increase as a result, which would warm the planet still more. These effects could then be further amplified by releasing methanogenic and ammonia-creating bacteria into the now-livable environment, as methane and ammonia are very strong greenhouse gases. The net result of such a program could be the creation of a Mars with acceptable atmospheric pressure and temperature, and liquid water on its surface within fifty years of the start of the program. Even though such an atmosphere would not be breathable by humans, this transformed (essentially rejuvenated) Mars would offer many advantages to settlers: They could now grow crops in the open, space suits would no longer be necessary for outside work (just breathing gear), large supplies of water would be much more accessible, aquatic life could flourish in lakes and ponds oxygenated by algae, and city-sized habitation domes could be constructed as there would be no pressure difference between their interior and the outside world. These short-term advantages would be more than sufficient to motivate Martian settlers to initiate the required terraforming operations to obtain them. In the longer term, plants spreading across the surface of such a partially terraformed Mars would put oxygen in its atmosphere. Using the most efficient plants available today, it would take about a thousand years for enough oxygen to be released to create an atmosphere that humans can breathe. However, future biotechnology may allow the creation of more efficient plants, or other technologies might become available that could accelerate the oxygenation of Mars considerably.

It may be observed that if humans had encountered Mars not in its current condition but in its warm-wet youth, terraforming would have been much simpler. Essentially, we could have skipped the large-scale industrial engineering required to greenhouse the planet with CF_4 and gone directly to the stage of using self-replicating systems such as bacteria and plants to further warm and oxygenate the planet. Thus, a team of interstellar explorers who chance upon a "young Mars" could initiate massive terraforming operations with little more equipment than an appropriate array of bacterial cultures, seeds, and a bioengineering lab.

Mars-like and young Mars-like worlds may be quite common in nearby interstellar space. If so, then the terraforming of Mars will thus give us not only one new world, but the tools for creating many.

WORLDS TOO HOT

Within our own solar system, the other planet that has attracted serious attention from would-be terraformers is Venus.

Venus was once thought to be Earth's sister planet, since it is about 95 percent the diameter of the Earth and has about 88 percent of Earth's gravity. The fact that it orbits the Sun at 72 percent of Earth's distance implied that it would be warmer than Earth, but not necessarily fatally so, and visions of Venus as a world rich with steaming jungles beneath cloudy skies filled astronomy books through the late 1950s. However, in the early 1960s, NASA and Soviet probes reached Venus and discovered that the fair planet of the love goddess not only lacked jungles but was a pure hell, sporting a mean surface temperature of 464°C—hot enough to melt lead. This was especially surprising since Venus is masked by highly reflective clouds—so reflective, in fact, that the planet actually absorbs less solar radiation than the Earth! Based on the amount of sunlight it absorbs, Venus should be colder than Canada. Instead, it is as hot as a self-cleaning oven. The explanation for this paradox was soon found, however; Venus is hot because what heat it does absorb is kept captive due to the greenhouse effect caused by its thick CO_2 atmosphere (this is how the greenhouse effect currently of concern on Earth was first discovered).

Well, then, if the CO_2 atmosphere is baking Venus to death, why not just get rid of it? This was the genesis of Carl Sagan's seminal 1961 proposal to terraform Venus with aerial algae.[4] According to Sagan, Venus could be cooled by dispersing photosynthetic organisms in its atmosphere that could convert it from greenhousing CO_2 to transparent (and breathable) oxygen. This proposal was a landmark in that it was the first serious discussion of terraforming within the world of science and engineering, as opposed to science fiction. However, it would not have worked for a number of reasons.

In the first place, while algae have been found in rainwater, there are no plants that actually live in an aerial habitat. Perhaps they could be engineered, especially for a planet with as thick an atmosphere as Venus. Sagan's proposal faces a bigger problem, however. Though based on the scientific knowledge available at the time, his concept grossly overestimated the amount of water available on Venus. Photosynthesis involves combining water molecules with CO_2 molecules in accordance with:

$$6H_2O + 6CO_2 \rightarrow C_6H_{12}O_6 + 6O_2 \qquad (10.1)$$

As can be observed in equation 10.1, to get rid of a molecule of CO_2 using photosynthesis, you need to use up a molecule of water. On Venus today there is much more CO_2 than water, so if you could actually find organisms to perform reaction 10.1 on a mass scale, you would simply rid Venus of the small amount of water it retains while leaving the large bulk of the CO_2 atmosphere basically untouched. It would take the equivalent of a global ocean 200 meters deep to provide enough water to react away the CO_2 in Venus' atmosphere via photosynthesis; in fact, Venus has only enough water to cover itself with a layer 5 cm deep. Sagan's idea could not have worked because there just isn't enough water on Venus to do the job.

Importing the water isn't an option. It would require moving 92 million iceteroids, each with a mass of a billion tonnes. So, in fact, the only way to cool Venus is to block the Sun with a huge solar sail. If we were to make a sail twice the diameter of Venus out of aluminum 0.1 micron thick, 124 million tonnes of processed materials would be required, to which we would need to add several billion tonnes of ballast. Manufacturing this object out of asteroidal material would be a huge job, but probably not beyond the means of an advanced Type II civilization. Such a sail could be stationed at the position between Venus and the Sun where their gravitational fields, sunlight force, and heliocentric centrifugal force all balance (near the Venus-Sun L1 point, about 1 million kilometers from Venus). In that case it would block over 90 percent of the Sun's rays, leading to the precipitation of Venus' CO_2 atmosphere as dry ice after a cooling-off period of about 200 years. The dry ice could then be buried, and by moving the sail around in a small orbit about the L1 point, we could create an acceptable surface temperature and day-night cycle on Venus. But the planet would still be incredibly dry. (In contrast to Venus's 0.05 meters of water, Mars has 200 meters and Earth 2,000 meters. Only the Moon, at 0.00003 meters, is drier.) All this would tend to imply that terraforming Venus is a very big project offering modest payoff.

However, this was not always so. The young Venus had lots of water, comparable, in fact, to the water inventory of the Earth. According to the "moist greenhouse"[5] theory proposed by planetary scientist James Kasting in 1988 and now generally accepted, the early Venus featured oceans of water at temperatures between 100°C and 200°C. The pressure of the thick overlying atmosphere prevented these oceans from boiling away. The rapid cycling of water in this ultra-tropical environment would have caused most of Venus' CO_2 supply to rain out and react with minerals to form carbonate rocks. This moist greenhouse Venus could exist because the early Sun was only about 70 percent as luminous as the Sun is today. But after a billion years or so, the Sun's luminosity increased to 80 percent of its present value, and temperatures on Venus rose above 374°C, the critical temperature for water. Once this happened, liquid water could no longer exist on Venus and all of its oceans turned to steam. With so much water vapor in the atmosphere, water loss from the planet due to ultraviolet dissociation of upper atmospheric water molecules occurred at a rapid rate. Moreover, once it stopped raining, geologic recycling was able to release the vast supplies of CO_2 stored in Venus's carbonate rocks, thereby creating the hellish runaway greenhouse environment that curses Venus today.

If human explorers had arrived on the scene when Venus was still in its young, moist greenhouse phase, terraforming could have been accomplished by building the Sun shade described above. That would be a significant engineering project, but well worthwhile, since the result would be an Earth-sized planet complete with temperate oceans and a moderate-pressure, nitrogen-dominated atmosphere, fully ready for the rapid propagation of life.

As discussed in *The Case for Mars,* solar sails used as reflectors to increase solar flux will also be a useful auxiliary technique in melting the permafrost and activating the hydrospheres of cold Mars-like planets. A solar sail mirror 15,000 km in radius (190 million tonnes of 0.1 micron aluminum) positioned with Titan at its focal point would increase Titan's solar flux to Mars-like levels, which, together with the strong greenhouse effects produced by Titan's methane atmospheric fraction, should be enough to raise the planet to Earth-like temperatures. A similar mirror would be sufficient to melt surface water ice into oceans and vaporize dry ice to create a CO_2 atmosphere on Callisto. Thus, the ability to manufacture large, thin solar sails in space, a central skill for engineering both Type II interplanetary transportation and emergent Type III starship propulsion, is also a key technology for terraforming Mars-like, Titan-like, Callisto-like, and young-Venus moist greenhouse worlds.

It should be noted that while Sagan's proposal for terraforming Venus with algae would not have worked, there is a class of planet for which it would—young Earths. The early Earth also had a thick CO_2 atmosphere, which was fortunate, because the early Sun was weak. But there could be other young Earths out among the stars that are receiving solar input comparable to what the Earth gets today, but which are too hot to be habitable because of heavy CO_2 atmospheres. In the case of such worlds, appropriately selected or bioengineered photosynthetic organisms might be able to rapidly terraform the planet to fully Earth-like conditions without any additional macroengineering effort. This is important because the Earth has only had its current high oxygen/low CO_2 atmosphere for the past 600 million years, or the most recent 14 percent of the history of the planet. If humans had encountered the Earth during the other 86 percent of its history, we would have needed to adopt Sagan-like terraforming strategies in order to make it habitable.

So Sagan didn't really have the wrong idea—he had the right idea, but applied it to the wrong world. There are undoubtedly many right worlds out there waiting for its application. Based on our own history, young Earths, uninhabitable but ready for Sagan-style biological terraforming, are likely to outnumber already habitable Earths by a considerable margin.[6]

TIDALLY LOCKED WORLDS

The more massive a star is, the faster it burns and the brighter it shines. Astronomers divide the large majority of normal or "main-sequence" stars into classifications by letter, which, starting with the biggest, brightest, and hottest and going to the smallest, dimmest, and coolest, are classes O, B, A, F, G, K, and M (remembered by the mnemonic "oh be a fine girl kiss me"). Each of these classes is further divided into subclasses by number, ranging from 0 to 9, with 0 the brightest and 9 the dimmest. Thus a G0 star is one step below an F9, while a G9 is one step above a K0. Class A stars, such as Sirius, are white hot; F stars, such as Procyon, are yellow-white; G stars, such as the Sun and Alpha Centauri A, are yellow; K stars, such as Epsilon Eridani and Epsilon Indi, are orange; and M stars, such as Proxima Centauri and Barnard's Star, are red. Class O and B stars, blue and blue-white, respectively, burn so hot and fast that they probably do not have time to form habitable planets. But all the rest are candidates for life.

Because the Sun, as a G2 star, falls close to the middle of this classification sequence, people frequently describe it as an "average" star. But this is far from true. If we were to perform a census of stars in our galaxy, we would find that 10 percent of them are white dwarfs (i.e., not on the main sequence described above), 70 percent are type M, 15 percent are type K, 3 percent are type G (with 90 percent of this 3 percent comprising class G3 through G9—i.e., cooler than the Sun), and the remaining 2 percent are divided among types O, B, A, and F. In other words, 97 percent of all stars are dimmer than the Sun, and 95 percent are *much* dimmer. The Sun is a member of the biggest and brightest 3 percent. It is most definitely *not* average.

Massive stars are brighter than little ones not just because they are bigger, but also because they have more gravity, which acts to compress their material and therefore drive fusion reactions faster. As a result, for stars near the Sun's weight class or above, the total power output, and thus luminosity, scales as the fourth power of the star's mass, whereas for much smaller stars luminosity goes as the square of the mass. In either case, the point is that as the mass of a star is decreased, the power output drops a lot faster than the gravity.

As an example, let's consider Barnard's Star, which, as a type M5 red dwarf, really *is* average. It has a mass that is 15 percent that of the Sun, but its luminosity is only 0.044 percent as great. For a planet to get as much light from Barnard's Star as Earth gets from the Sun, it would have to orbit at a distance of 0.02 AU, or 3 million kilometers. But at that distance the gravitational attraction Barnard's Star would exert on the planet would be 375 times as great as that which the Sun exerts on the Earth. As a result, any planet close enough to Barnard's Star to receive enough light for habitability would very likely be tidally locked (the same can be said for any other type M red dwarf or white dwarf). That is, it would rotate with the same face always pointed at its star, just as the Moon always points the same side toward Earth. The only way it could avoid such a fate would be if the planet had a very large moon, or itself was a satellite of a still larger planet.

Now, together, type M red dwarfs and white dwarfs comprise about 80 percent of all stars. So, in eight cases out of ten, any potentially habitable planet we might find will be tidally locked.

Tidal locking is a problem because when it occurs only one side of a planet receives any warmth or illumination from its star. Therefore, one side can become inhospitably hot, or if the planet is far enough from its star that the permanently sunlit side is temperate, then the dark side will become so cold that it could become a trap where the entire atmosphere of the planet freezes out.

There are two ways we could deal with this problem. The simplest is to use orbiting sunshades or mirrors. In the case of a tidally locked planet with a permanently lit, too-hot, sunward-pointing surface, we can just put a sunshade in orbit about the planet with a period equal to the desired day. Such a shade would be relatively close to the planet, so unlike our Venus sunshade at far-off L1, it would only need to be equal in size to the planet (as opposed to twice as big). In the case of the tidally locked planet whose sunlit side is temperate but whose dark side is too cold, we need to position a mirror levitated by sunlight pressure (a "statite,"[7] invented by Robert Forward and discussed in *The Case for Mars*) so that it could always be used to project light onto the cold hemisphere of the planet. In this case, all we need to do is illuminate the dark side of the planet sufficiently so that its temperature is above the freezing point of its coldest freezing gas, probably nitrogen, which would be around 63 K (–210°C). Provided this is done, there will be enough gas (high enough vapor pressure) in the atmosphere that substantial atmospheric convection will even out most of the temperature differences between the sunlit and nightside hemispheres. Now the amount of sunlight required to create a given temperature is proportional to the fourth power of the absolute temperature, so the amount of irradiance needed to generate a surface temperature of 63 K is only 0.002 $[(63/300)^4=0.002]$ as great as the amount of light needed to generate 300 K (27°C). So to warm the dark side of an Earth-sized tidally locked planet whose forward side was at 300 K, we would need a mirror with an area of 0.002 that of the planet, which means a radius of 280 km. If made of 0.1 micron aluminum solar sail material, such a sail mirror would have a mass of 85,000 tonnes. Type I humanity has already produced many oceangoing vessels with masses of this order.

The alternative to using orbiting mirrors and shades is to actually impart rotation to the tidally locked planet. Let's see if this is possible.

The moment of inertia, *I,* of a spherical planet is given by:

$$I=0.33MR^2 \qquad (10.2)$$

where M is the mass of the planet and R is the radius. Let's consider three planets, one Moon-sized, one Mars-sized, and one Earth-sized. The mass, radius, and moment of inertia of each are given in the first three columns of Table 10.1. Now, let's say we go to the outer regions of the planet's solar system and move a 100-km cube (10^{18}kg) iceteroid such that it is sent on a collision course with the planet we intend to spin up. In order to maximize the collision speed and resulting rotation, we send the iceteroid in on a retrograde orbit, hitting the planet's equator with an impact velocity of 80 km/s. Table 10.1 shows the increase in rotational velocity of the planet ($\Delta\omega$) in radians per second as well as the period of rotation that results from such an impact. If more than one such impact is delivered, the length of the period would decrease proportionately.

It can be seen that, if the target planet is only the size of the Moon, a 116-day rotation could be induced with a single hit, which means a 12-day light-dark cycle could be created with ten hits. That's probably good enough for a habitable planet. But if the planet is Mars-sized, the magnitude of the effort would be increased by a factor of 17, whereas an Earth-sized world would need almost 300 times as many such impacts to do the same job.

One-hundred-kilometer iceteroids are probably available in the Kuiper-Belt or Oort Cloud of most target solar systems. If we assume that a 200 m/s ΔV is sufficient to initiate the collision trajectory, and iceteroid steam with an exhaust velocity of 3 km/s (Isp of 300 s) is used as propellant, then only 7 percent of the iceteroid will need to be sacrificed to get it moving. Seven percent of a 100km iceteroid, however, is 70 trillion tonnes. If we assume that the rocket engine employed has a jet power rating of 1,000 TW (1 billion megawatts!), about ten years of continuous firing would be required to perform the ΔV needed to start the maneuver.

▪ TABLE 10.1 ▪

Moments of Inertia of Planets

PLANET	MASS (KG)	RADIUS (M)	INERTIA (KG-M^2)	$\Delta\omega$ RAD/S)	PERIOD AFTER 1 HIT
Moon	7.35 × 1022	1,738,000	2.22×10^{35}	6.26×10^{-7}	116 days
Mars	6.42 × 1023	3,393,000	7.39×10^{36}	3.6×10^{-8}	1,980 days
Earth	5.98 × 1024	6,378,000	2.43×10^{38}	2.0×10^{-9}	34,632 days

While it might be argued that 1,000 TW is "only" four times as much as used by the laser-pushed light sail discussed in chapter 9 as a means of propelling interstellar missions, it should be observed that the aforementioned 240-TW laser would be located in the home solar system of a fully developed Type II or emergent Type III civilization. In the case under discussion, we are dealing with disposing of vaster amounts of power, not at home, but in a far-off stellar system where we are new arrivals and have no resources other than those aboard our ships. It would thus appear that the ability to rotate tidally locked planets is much more difficult than providing compensatory illumination or shading with mirrors and is therefore a skill that will emerge later, rather than sooner, in the course of humanity's efforts among the stars.

ROBOTICS, BIOENGINEERING, NANOTECHNOLOGY, AND PICOTECHNOLOGY

Terraforming will require a lot of work, and the people who attempt it will want helpers. There is thus no doubt that in terraforming, as in all other extraterrestrial engineering projects, robotics will play an important role. No commodity will be in shorter supply in an early Martian colony than human labor time, and as the frontier moves outward among the planets and then to the stars, the labor shortage will grow ever more pressing. The space frontier will thus serve as a pressure cooker for the development of robotics and other forms of labor-saving technologies.

But robots that must be manufactured still demand human labor. This will make them expensive, as space labor will be dear and transportation from Earth will be costly. Expensive robots are acceptable for assisting in certain tasks, such as exploration, where large numbers are not required. But terraforming will need multitudes. The best solution would be robots that make themselves.

Back in the 1940s, the mathematician John Von Neumann proved that self-replicating automatons are possible. That is, he proved that there is no mathematical contradiction that precludes the existence of such systems. But creating them is another issue altogether.

No one today has a clue as to how to do it, but it would not be too big a leap of faith to believe that a machine could be built and programmed so that, if let loose in a room filled with gears, wires, wheels, batteries, computer chips, and all of its other component parts, it could assemble a copy of itself. But who would make the parts? Consider what is necessary to make even a simple part, such as a stainless steel screw.

To make the steel for the screw, iron, coal, and alloying elements from all over the world need to be transported to a steel mill. They need to be transported by rail, ship, truck, or plane, and all of these contrivances must be made in factories or shipyards of great complexity, each of which involves thousands of components shipped in from all over the world, by various devices, made in various facilities, etc. So just supplying the steel for the screw actually involves the work of thousands of factories and millions of workers. If we then consider who made the food, clothing, and housing used by all those workers, who taught them, and who wrote the books that educated them, we find that a large fraction of the present and past human race was involved. And that's just the steel for the screw. If we now consider the processes needed to put the thread on the screw—but I think you get my point. Self-replicating machines cannot exist unless the parts they require are ready-made. This will never be the case for machines built out of factory-produced gadgets.

The only self-replicating complex systems known to exist are living things. Organisms can reproduce themselves because they are made of cells that can reproduce themselves using naturally available molecules as parts for their component structures. Because they can reproduce themselves, bacteria, protozoa, plants, and animals have extraordinary power as terraforming agents; a few of the right kinds, released under the right conditions, can multiply exponentially and radically transform an environment. Of course, for the transformation to be beneficial, some aspect of the organism's self-directed activity must contribute to the terraforming program. As we have seen, in such cases as methanogenic bacteria producing greenhouse gases, or photosynthetic plants eliminating them (while producing useful oxygen), the metabolisms of many forms of life make them natural servants of the terraforming process. This is to be expected because, as discussed earlier in this chapter, life would not exist if it did not terraform.

That said, current bacteria, plants, and animals are not specifically adapted to terraforming virgin planets; their adaptations are focused on terraforming and living on the current Earth. Their ancestors pioneered the early Earth, and they retain some of the necessary skills, but they are by no means the ideal candidates for pioneering new worlds.

However, since the domestication of the dog 20,000 years ago, humans have practiced modification of other species to meet our needs, primarily through the practice of selective breeding. In recent years a series of developments—beginning first with the development of genetics, then the discovery of DNA, and now the actual reading of the genetic code and mastery of recombinant DNA techniques—has enormously expanded our abilities in this area. As a result, it will soon be within our capabilities to design ideal pioneering microorganisms and ultra-efficient plants well suited to transform a wide variety of extraterrestrial environments.

But microorganisms and plants have their limits. They are all based on water/carbon chemistry, which cannot function beyond the temperature boundaries defined by the freezing and boiling points of water. If temperatures are sustained below 0°C, life survives but goes dormant; above the boiling point (100°C at Earth's sea level, 374°C maximum) organisms are destroyed. Many extraterrestrial environments of interest exist beyond these narrow limits.

The question thus arises if it might be possible to develop self-replicating organisms with a fundamental chemistry other than the water/carbon type universal to life as we know it. If, in venturing out into interstellar space, we should discover novel kinds of life, based on silicon or boron, for example, but with their own equivalent of a genetic code that future human bioengineers can master, this problem would be partially solved, as the new chemistry would undoubtedly define a new set of temperature limits. But it is unclear whether such organisms will ever be found or what the extent of their utility might be. From the point of view of the planetary engineer, a more intriguing question is whether we can devise non–water/carbon self-reproducing organisms from scratch.

This is the idea behind "nanotechnology"—the construction of self-replicating, microscopic, programmable automatons out of artificial structures built to design specifications on the molecular level. Why try to build microscopic self-replicating robots when we don't even know how to build human-scale reproducing automatons? The reason, once again, is that parts for large robots need to be manufactured in advance, whereas the molecules used as parts for nanorobots (a nanometer is a billionth of a meter) either come ready-made or can be readily assembled from atoms that do. So while building a nanorobot would unquestionably be more difficult than constructing a normal-sized one, nanorobots are the only kind that hold the promise of being potentially self-replicating.

The vision of nanotechnology is described at length by the field's champion, K. Eric Drexler, in his book *Engines of Creation.*[8] The basic idea is that once we learn how to manipulate individual atoms and molecules, machines can be constructed using small clumps of atoms as gears, rods, wheels, and so on. Because each of these parts would be made out of a precisely assembled group of atoms—just as carbon is arranged in a diamond lattice—they potentially could be very strong. Thus, nanomachines, filled with gears, levers, clockwork, motors, and all kinds of mechanisms, could in principle be built. Energy to drive the units could be obtained from nanophotovoltaic units and stored in nanosprings or nanobatteries. To go from there to nanorobots, we need nanocomputers. Drexler proposes that these could be built out of mechanical nanomachines, along the same principle as the first mechanical computers built by Charles Babbage out of brass gears and wheels in the nineteenth century; such machines could be programmed with punched tape or cards, and presumably nanoscopic analogs for these mechanical software devices could be found as well. Babbage's ingenious mechanical computers don't even remotely compare in capability to modern electronic ones, but the parts used by Drexler's nano-Babbage-machines would be so small that enormous amounts of computing power could be contained in a microscopic speck. So, once we accomplish the admittedly difficult job of building the first nanorobot (with all its necessary nanomechanisms for locomotion and manipulation) and equipping it with a superpowerful version of a programmed and debugged Babbage machine built on the nanoscale, we could set this first "assembler" loose and it would multiply itself through exponential reproduction. The vast horde of assemblers would then turn their attention to accomplishing some task they had been programmed to execute, such as inspect a human body for cancer cells and make appropriate adjustments, manufacture huge solar sails from asteroids, or terraform a planet.

To build macroscopic structures, billions of nanorobots would have to group themselves together to form large robots, perhaps on the human scale or even much bigger. This could lead to the manifestation of systems that would have all of the capabilities of the evil "liquid metal" robot depicted in the movie *Terminator 2,* able to change its shape and disperse and reassemble itself as required. But it actually would be much more powerful, since when it did choose to disperse, each of the billions of its subcomponents could be used as a seed to reassemble an entire unit from dirt. Even Arnold Schwarzenegger would have had a hard time saving the world from one of those!

It certainly sounds like fantasy, but is it? In defense of the nanotechnology thesis, one can advance the statement that it does not defy any known laws of physics, and therefore, given sufficient technological advance, it should become possible. Against it, one can easily point out the enormous technological difficulties that must be mastered before nanotechnology becomes a reality. Furthermore, while nanotechnology may not violate any laws of physics, *controllable* self-replicating robots may well violate the laws of *biology.* Consider that small replicating micromachines will unquestionably undergo random alterations, or mutations, if you will. Those mutations that produce strains that reproduce more rapidly will swiftly outnumber to insignificance those that don't. Clearly, if the goal is to reproduce rapidly, it would be to a nanomachine's advantage not to have to bother with doing work for the benefit of human masters. Instead, evolutionary pressures will dictate that nanorobots attend only to their own needs. Those nanorobots that continue to slave away in obedience to their human-directed programs will not be able to compete with the wild varieties, and will rapidly go extinct. As the saying goes, "Live free or die."

There is another reason to hold nanorobots suspect—we don't observe them. If diamond-geared self-replicating assemblers could be built, they would be ideally suited for dispersal across interstellar space using microscopic solar sails for propulsion. If, in the vast sweep of past time, a single species anywhere in the Milky Way developed such microautomatons, it long since would have been able to use them to colonize the entire galaxy. All life on Earth would be based on nanorobots. But since this is not observed, we are driven toward concluding that either (a) there is no other intelligent life in the galaxy or (b) non-organic nanotechnology of the self-replicating micro-Babbage-robot type described by Drexler is impossible. Since we know that the evolution of intelligent life is possible, but we do not know that nanotechnology is, I must consider (b) the more likely alternative.

It may be observed that bacteria, the organic nanocritters of nature, are also capable of surviving space flight. We would therefore expect that if bacteria had evolved (or been developed) elsewhere in the galaxy they would be the basis for life on our planet. Interestingly, they are. Not only are bacteria the earliest known inhabitants of the Earth, but the higher eukaryotic cells that compose all animals and plants are clearly evolved from symbiotic colonies of bacteria. The possible broader significance of this will be discussed further in the next chapter. For our purposes here, however, it suffices to say that the omnipresence of organic self-replicating nanospacefarers (bacteria) and the absence of non-organic nanoassemblers is strong evidence for doubting the feasibility of Drexler-style nanotechnology.

But maybe nanotechnology isn't impossible; maybe it's just incredibly difficult. Maybe the reason why nobody else has invented it is because they weren't smart enough, or didn't try long and hard enough, or were scared of the consequences of it getting out of control. Maybe there really is a way to initiate and control nanotechnology, and it's just waiting for someone to invent. In every field of endeavor, someone has to be first. Maybe that someone could be us. Maybe.

That's a lot of maybes. But it's worth some speculation, because if the promise ever does pan out, programmable self-replicating nanomachines will offer our descendants powers of creation limited only by the rate at which solar flux provides the energy needed to drive work in a given region. If we continue the vector toward ever-growing technological sophistication that necessarily will accompany our transformation into first a Type II and then Type III civilization, the intricate wizardry required to develop nanotechnology might someday fall within our grasp. And who knows? Perhaps, in the still more distant future, even greater capabilities could become possible—building machines not out of atoms or molecules but from subatomic particles such as atomic nuclei. Operating on a scale thousands of times smaller and faster than even nanomachines, such *picotechnology* might draw its energy not from chemical reactions, but from far faster and more powerful nuclear reactions. The capabilities that such picomachines would make available could only be described today as sheer magic.

In the meantime, however, my bet is on bioengineering. Life offers us a tried-and-true type of self-replicating micromachine, and the programming manual is already in our hands. With our brains and their muscle, human-improved microorganisms will do some very heavy lifting in the hard work required to bring dead worlds to life.

DYSON SPHERES, RINGWORLDS, AND SPACE COLONIES

Princeton professor Freeman Dyson has some very visionary ideas. One of the most remarkable is his suggestion that an advanced Type II civilization, in order to make full use of the power emitted by its Sun, would build a sphere completely enclosing its star with the shell positioned at the correct distance to make the entire inner surface of the sphere habitable. Thus, none of the star's radiated light and heat would be lost to space, and a vast surface area, far exceeding that offered by any planet, would be made available to habitation. For example, if such a "Dyson sphere" were built in our solar system, the shell radius would be at 1 AU and the inner surface would therefore have an area of 283 billion million square kilometers. This is about 553 million times the surface area of the Earth. Dyson suggested that, if such spheres existed, they would radiate strongly in the infrared. Ergo, the search for advanced extraterrestrial civilizations might be conducted by looking for such spheres. To date, no Dyson spheres have been observed.

I'm not surprised. Dyson spheres are impractical. The 1-AU sphere described in the previous paragraph, if given a shell just 1 meter thick, would require the mass from 260 disassembled Earths to build (and 1 meter is much too thin for it to hold together). This much solid material is unavailable in our solar system (Jupiter's hydrogen would be useless for construction), and is unlikely to be available in others. Moreover, while advertising a tremendous nominal surface area, most of the inner surface would really be useless for habitation because, except near the equator, where the rotation of the sphere could supply outward-pointing centrifugal force, the gravity vector almost everywhere on its surface would point in rather inconvenient directions. As a result, there could be no air any significant distance from the equator. Terraforming the rest of the sphere would be impossible, since no matter how much air you made, it would all either flow to the equator or flow to the Sun.

While made famous by its sweeping visionary nature, the implausibility of the Dyson sphere is so extreme that it has even been criticized by science fiction writers. Thus, as an improvement on the Dyson sphere, science fiction novelist Larry Niven offered the concept of a "Ringworld," in which only the equatorial band of the Dyson sphere would be built, and its atmosphere would be contained by rims around the edges of the bands.[9] Since the Earth's atmosphere is only about 50 km high, these rims represent trivial additions to the overall structure of the Ringworld. If positioned in our solar system, the Ringworld might be a band 10,000 km wide, circling and orbiting the Sun at a distance of 1 AU. Such a band would have a surface area of 9.42 trillion square kilometers, or 18,440 times the area of the Earth. If made 100 meters thick, its mass would be about equal to that of Venus, which might be conveniently disassembled to construct it.

Niven's Ringworld is thus clearly a much more practical concept than a Dyson sphere, but it's still obviously a rather big project. To see how big, let's calculate just the energy required to disassemble Venus and then move its mass from its present orbit (at 0.72 AU) to the 1-AU distance required for the Ringworld, to melt it into a solid ring, and then to accelerate the ring so that it spins about the Sun with sufficient velocity to create Earth-normal gravity at its inner surface.

Well, Venus has a mass (M) of 4.87×10^{24} kg and a radius (R) of 6,051,000 m, so it has a total gravitational energy (given by GM^2/R) of 2.6×10^{32} joules (where G is the universal gravitational constant). To move the mass from Venus' orbit to Earth's orbit requires a ΔV of 5,355 m/s, which implies an energy (given by $MV^2/2$) of another 0.7×10^{32} joules. If we assume a megajoule to melt every kilogram, that's another 4.87×10^{30} joules. But then, to spin it up to create Earth-normal gravity, we need to accelerate the ring to a speed of 1,212,435 m/s. This will need an energy of 3.6×10^{36} joules, a requirement that makes all the rest insignificant. To put the matter in perspective, 3.6×10^{36} joules is 11 million billion times the amount of power humanity currently uses in a year; it is equal to the *total* amount of power output from the Sun over a period of three centuries! So short of the development of picotechnology, not even a Ringworld (let alone a Dyson sphere) will ever be constructed.

But there is one kind of space-based macroengineering that a Type III civilization will be able to build: space colonies. The reader will recall that in chapter 4 I discounted the possibility of using space solar energy business plans to drive the near-term construction of billion-tonne O'Neill colonies in Earth orbit. I stand by those arguments. But feats of engineering that are fantasy for a Type I civilization will be reality for Type III. Making use of vast supplies of fusion energy, sophisticated space technology, and very advanced robotics, humans in the era of Type III civilization will hollow out asteroids and Oort Cloud objects to create miniature enclosed worlds with many of the features envisioned by O'Neill. They won't do it to sell space solar power-beaming stock options; humans will be way past that sort of thing by then. Rather, they will do it to create new habitats for life and civilization to develop in novel ways. Some will keep their homes in solar orbit, forming an ever-growing community of diverse societies. Others will set sail for the stars.

LIGHTING STARS

It is better to light a candle than to curse the darkness.

Stars are the sources of life. Enormous engines of nuclear fusion, they pour light out into the cosmos, warming the dead cold of space, and provide the antientropic power needed for the self-organization of matter. Starry nights have a mystic beauty, but when considered from a scientific standpoint they are even more beautiful than they look. For the million specks of light that adorn the black velvet of a dark night sky are, in fact, nothing less than a million fountains of life.

Without question, there are numerous worlds too far from any star to support life. In our own solar system, we find world-sized moons of Jupiter and Saturn that can be terraformed only with the aid of giant reflectors, and worlds beyond, such as Neptune's giant moon Triton, for which the huge efforts required for such an expedient make it difficult even to contemplate.

As mentioned earlier, our Sun is actually among the brightest of stars. Most of the stars we know are much dimmer type M red dwarfs, and there are likely legions dimmer still—brown dwarfs, too small to ignite fusion, whose dead planets therefore orbit endlessly in frozen darkness.

What if we could light their fires?

If the object in question is an actual luminous star, such as a type M red dwarf, we could amplify its power by using solar sails to reflect back a small portion of the star's output. The rate at which thermonuclear fusion proceeds in a star goes as a strong power of its temperature. For proton-proton fusion in a star the size and temperature of our Sun, reaction rates scale as the fourth power of temperature, whereas for cooler stars the temperature dependence is stronger. If the CNO cycle is being used by the star to catalyze fusion, then the reaction rate will increase in proportion to T^{20}(!!!). So even increasing the temperature of a star by a small amount through reheating with reflected light can cause a large increase in power generated. This increased output will cause the star's temperature to rise further, which will amplify output yet again.

But what if the "star" is a brown dwarf, or a gas giant planet, incapable of generating sustained fusion energy? In this case, astroengineers will have to make use of black holes.

According to British cosmologist Stephen Hawking, large numbers of primordial black holes would have been formed during the Big Bang event that is generally believed to have initiated our universe.[10] The smallest of these, with masses under a billion tonnes, would have evaporated by now due to Hawking radiation, but those useful for lighting stars would have a mass at least a million times greater than this and would therefore have experienced very little mass change since the beginning of the universe. As noted by British scientist Martyn Fogg, the motion of such black holes could be directed by charging them with an ion or electron beam and then pushing with an electric field or by applying thrust periodically to an object in orbit about the hole. Alternatively, the hole could be pushed by shining a powerful laser or firing projectiles at it; the light or projectiles would be absorbed but their momentum would be conserved. The best way, however, would be to find a way to use the energy released by the hole itself when matter falls into it. Conceivably, this might be done by positioning a large light reflector equipped with a magsail in orbit around the hole and then feeding matter into the system to be annihilated when the reflector is on the correct side to exert thrust. Since the reflector is gravitationally bound to the hole, it will pull it along when the emitted light and plasma push it.

Now let's say we discover a brown dwarf somewhere in interstellar space with an accompanying family of stillborn frozen planets. If we also can find (or more speculatively, make) a black hole of appropriate size and direct it into the brown dwarf, we can turn it into a star!

This no doubt sounds impossible, as intuition suggests that the whole brown dwarf would pour itself nearly instantly down the bottomless pit of the black hole, never to be seen again. But this is incorrect. In fact, if a black hole were placed within a brown dwarf (or Jupiter, or any other massive object) it would *start* swallowing mass, but in the process it would produce a huge output of radiation that would push back against the infalling material. This would very quickly regulate the rate of infall to a steady state known as the Eddington limit,[11] which is given by

$$L\,(\text{watts}) = \varepsilon R_b c^2 = 4\pi g M_b m_p c/\sigma = 6.39\, M_b\,(\text{kg}) \qquad (10.3)$$

Here L is the luminosity of the radiation given off by the black hole in watts, ε is the efficiency at which infalling mass is turned into energy (between 0.06 and 0.4 depending on the rotation of the hole), R_b is the rate at which mass is accreted by the black hole, c is the speed of light, M_b is the mass of the black hole in kilograms, m_p is the mass of a proton (1.67×10^{-27}kg), and σ is the electron-scattering cross section (6.57×10^{-29} m^2). This equation has a lot of fancy terms, but for our purposes the first and last parts are what is important; the luminosity of the hole (in watts) will be six times its mass (in kilograms).

Our Sun has a luminosity of 3.9×10^{26} W. Let's say we found a brown dwarf with a nice-sized planet orbiting at a distance of 1.5 million kilometers, a bit farther out than the 1.2 million kilometer distance that Titan orbits Saturn. Since 1.5 million kilometers is 1/100th of an AU, and since solar flux falls as the square of the distance, this means we would need a luminosity of 1/10,000th that of the Sun, or 3.9×10^{22} W, to create Earth-like temperatures on the planet. Using equation 10.3, we see that this implies that we need a black hole with a mass of 6.1×10^{21} kg to do the job. This would be an object about 8 percent the mass of the Earth's moon (or about 1/1,000th that of the Earth). While a formidable task, setting such an object in motion might not be beyond the capability of a Type III civilization. For example, to give it a ΔV of 10 km/s to send it on its way would involve an energy of 3×10^{29} J. Under ordinary circumstances that would be too much energy (9.5 billion TW-years!) to expect even a very advanced Type III civilization to wield. But we've got ourselves a black hole here, in this case one with a potential power output of 3.9×10^{22} W. At that power level, it could generate the required energy to move itself in 90 days. Assuming the black hole could convert mass to releasable energy with an efficiency of 10 percent, to generate the required 3×10^{29} J we would need to feed it with 30 billion tonnes of mass, which is to say we would need to be able to move a 2-km asteroid—a modest task for a Type III civilization.

How long would our new-created star last? Not as long as our Sun, but long enough to make the project worthwhile. If we manipulate equation 10.3 we obtain

$$M_b/R_b = \tau = \varepsilon c^2/6.39 = \varepsilon(447 \text{ million years}) \qquad (10.4)$$

The term on the left-hand side of equation 10.4 is the mass of the black hole divided by the rate at which its mass changes. This quotient is therefore equal to a time, which we call τ. This characteristic time, τ, is how long it will take for the black hole to grow to 2.72 times its initial mass and power output. If efficiency, ε, equals 0.1 (a typical value,) this time will be about 45 million years. In 1 million years, the output power will increase only by 2 percent. In 10 million years, luminosity will rise 25 percent, which, if nothing is done, will cause temperatures on the planet illuminated by the artificial star to rise by 5.7 percent.

But the inhabitants of the planet around that star will be advanced Type III people, equipped with the full grab bag of terraforming anti-greenhouse techniques. They should be able to handle the situation without difficulty.

In his book, *The Starmaker,*[12] written in the 1930s, British philosopher Olaf Stapledon compared the starmaker to God.

Gods we'll never be. But starmaking *is* a very noble profession.

If we learn to light stars, we will become capable of bringing not only planets, but whole solar systems to life. That's not too shabby for the children of tree rats.

In the early universe, nearly all matter was hydrogen or helium. The heavier elements, including the carbon, oxygen, and nitrogen vital to organic chemistry and life, were all made in stars, and spread through the cosmos by stars in nova and supernova explosions. We are stardust, warmed to life by a mother star, and now ready to leave the nest to seek our fortune and make our mark among her siblings.

Ex astra, ad astra.

Stars have made life. Life should therefore make stars.

CHAPTER 11

Meeting ET

Where are they?

—ENRICO FERMI, Los Alamos, 1950

TWO THOUSAND YEARS ago the Roman philosopher Metrodorus, in contemplating the question of life in the universe, asked, "Is it reasonable to suppose that in a large field, that only one shaft of wheat should grow, and in an infinite Universe, to have only one living world?" The question is well put.

We now know that the universe is a fertile field. The possibility that this could be the case was suspected by some, such as the Italian Renaissance humanist Giordano Bruno, who guessed the existence of other solar systems, even before the telescope revealed the stars to be suns like our own. Until the present decade it was still possible for pedants to argue that our planetary system could be a unique phenomenon, as no others had been observed. However, during the 1990s, over a dozen extra-solar planetary systems have been detected, thereby proving not merely the existence of these particular planets, but the fact that the processes that lead to the formation of solar systems must be common, and probably intrinsic to the process of star formation itself. Much more could be said on this subject, but suffice it to say that, on the basis of current scientific knowledge, it's an excellent bet that the majority of stars have planets.

We now have fossil evidence for life on Earth going back 3.8 billion years—practically to the immediate time following the conclusion of the early solar system's massive meteorite bombardments that life *could* have existed here. The recent discovery of evidence of possible past life on Mars—dating back 3.5 billion years—gives additional strong support to the contention that the development of life on planets is a common event. Furthermore, the entire history of life on Earth shows clearly that, once life starts, it exhibits a continual tendency toward development of greater complexity, activity, and intelligence. In other words, based on everything that we know, life and intelligence should be common in the universe. If we add to this the fact that, even at our relatively primitive level of technology, we today can conceive of how interstellar voyages can be performed with currently understood engineering, and then consider the evolutionary advantages accruing to any species that engages in such, the conclusion becomes inescapable: Our galaxy is almost certainly currently inhabited by large numbers of starfaring species.

When we go out among the stars, we may meet them.

WHERE ARE THEY?

There are 400 billion stars in our galaxy, and it's been around for 10 billion years. Even adopting the most pessimistic assumptions, it would appear obvious that numerous starfaring civilizations should have already appeared. Our galaxy is 100,000 light-years in diameter. If an advanced civilization were to adopt an expansion program and move out at a pitiful 0.5 percent of the speed of light, it would take at most 20 million years to occupy the whole place. As long as this might seem, it is only 0.2 percent the age of the galaxy. In other words, by rights, extraterrestrials should already be here! It was a calculation of this sort that led the great physicist Enrico Fermi to pose his celebrated question at a 1950 Los Alamos National Laboratory lunchtime meeting; if all that is so, then, "Where are they?"

This question is known as the "Fermi Paradox."

In 1961, radio astronomer Frank Drake (one of the pioneers in the search for extraterrestrial intelligence, of which more will be said later) developed a pedagogy for analyzing the question of the frequency of extraterrestrial civilizations. According to Drake, in steady state, the rate at which new civilizations form should equal the rate at which they pass away, and therefore we can write:

$$\text{rate of demise} = N/L = R_* f_p n_e f_l f_i f_c = \text{rate of formation} \qquad (11.1)$$

Equation 11.1 is therefore known as the Drake Equation.[1] In it, N is the number of technological civilizations in our galaxy and L is the average lifetime of a technological civilization, so the left-hand side term, N/L, is the rate at which such civilizations are disappearing from the galaxy. On the right-hand side, we have R_*, the rate of star formation in our galaxy; f_p, the fraction of these stars that have planetary systems; n_e, the mean number of planets in each system that have environments favorable to life; f_l, the fraction of these that actually developed life; f_i, the fraction of these that evolved intelligent species; and f_c, the fraction of intelligent species that developed sufficient technology for interstellar communication. (In other words, the Drake equation defines a "civilization" as a species possessing radiotelescopes. By this definition, civilization did not appear on Earth until the 1930s.)

Well, N, the number of civilizations, is a quantity of great interest, and so some people have attempted to use the Drake equation to compute it. For example, if we estimate that $L = 50{,}000$ years (ten times recorded history), $R_* = 10$ stars per year, $f_p = 0.5$, and each of the other four factors, n_e, f_l, f_i, and f_c, equals 0.2, we would calculate that the total number of technological civilizations in our galaxy, N, equals 400.

Four hundred civilizations in our galaxy may seem like a lot, but scattered among the Milky Way's 400 billion stars, it would represent a very tiny fraction: just one in a billion to be precise. In our own region of the galaxy, (known) stars occur with a density of about 1 in every 320 cubic light-years. If the calculation in the previous paragraph were correct, it would therefore indicate that the nearest extraterrestrial civilization is likely to be about 4,300 light-years away.

But Drake developed his equation only as a pedagogical tool for delineating various factors influencing the probable frequency of extraterrestrial intelligence.[2] He never intended his equation to be used as a computational tool, and if used as such it is patently incorrect. For example, the equation assumes that life, intelligence, and civilization can evolve in a given solar system only *once*. This is manifestly untrue. Stars evolve on a time scale of billions of years, species of millions of years, and civilizations of thousands of years. Current human civilization could knock itself out with a thermonuclear war, but unless humanity drove itself into complete extinction there is little doubt that 1,000 years later global civilization would be fully reestablished. An asteroid impact on the scale of the K-T event that eliminated the dinosaurs might well wipe out humanity completely. But 5 million years after the K-T impact the biosphere had fully recovered and was sporting the early Cenozoic's promising array of novel mammals, birds, and reptiles. Similarly, 5 million years after a K-T class event drove humanity and most of the other land species to extinction, the world would be repopulated with new species, including probably many types of advanced mammals descended from current nocturnal or aquatic varieties. Human ancestors 30 million years ago were no more intelligent than otters. It is unlikely that the biosphere would require significantly longer than that to recreate our capabilities in a new species. This is much faster than the 4 billion years required by nature to produce a brand new biosphere in a new solar system.

PANSPERMIA

The Drake equation also ignores the possibility that both life and civilization might propagate across interstellar space. At the turn of the century, the Swedish chemist Svante Arrhenius proposed the theory that life on Earth may have originated from bacterial spores transported through space by light pressure. This theory, known as *panspermia,* was later championed by British astronomer Fred Hoyle but then became unfashionable after Hoyle's stock went down with the refutation of his "Steady State" cosmological theory by Big Bang advocates in the 1960s. However, the recent discovery of possible bacterial traces in ancient Martian meteorites has caused it to be reexamined.

As well it should—recovery of live bacteria (of terrestrial origin) from Surveyor spacecraft cameras left on the Moon has proved without doubt that bacteria can survive in the hard vacuum of space and, provided a few microns of dust or ice are available as shielding, can survive the cosmic ultraviolet environment as well. Bacteria are also terrifically radiation resistant—it takes over 10 million rads to completely sterilize a culture of *Micrococcus radiodurans,* for example. By contrast, 1,000 rads will kill a human. Spores of bacteria have been recovered from amber and Permian salt deposits and revived to life after periods of dormancy of 90 and 230 million years(!!!), respectively. In other words, bacteria, which appeared on our planet almost as soon as it was habitable for them, possess all the adaptations required for surviving space flight. Think about that for a minute; life frequently preserves unnecessary adaptations, but it never goes out of its way to create them. *If bacteria originated indigenously in the clement ponds and oceans of Earth, why are they adapted to survive the hard vacuum, the temperature extremes, the radiation, and the extended periods of dormancy required for interstellar space flight?*

If archeologists were to discover that the first inhabitants of some island wore sailor suits, would they not conclude that they came from the sea?

How could bacteria develop adaptations for space flight? There are only two possibilities—natural selection or design.

Let's consider the most conservative explanation, natural selection of microorganisms originating on Earth. Suppose that on the early Earth, many types of microbes developed, and a few of these, by chance, possessed characteristics that would allow them to survive space flight. The early Earth was subject to massive, frequent asteroid impacts, up to and including the world-shattering projectile that created the Moon. These impacts were so bad that many of them may have completely sterilized the planet but, in the process, kicked debris-containing microorganisms into space. This debris would then orbit about the solar system for thousands and millions of years, until some of it reencountered the Earth. However, the only microbes that would survive the trip, and thus live to recolonize the Earth with life, would be those that chance had preadapted for space. Their descendants, though, would share their astronautical capabilities.

But if this could have happened on Earth, then it could have happened on other planets. And there are likely billions of other planets where it could have happened first and, therefore, in all probability, did.

An alternative speculation is that bacteria were created by design, by an intelligent species somewhere interested in propagating life through the cosmos. In other words, bacteria *are* nanorobots. Consider, if we had nanorobots, one suggestion for their use might be to send them to other solar systems and then reconstruct people, thereby allowing us to colonize without spending the huge amounts of energy required to transport humans across interstellar space. But humans as we know them might not be properly adapted to live in the environments of the destination planets. A more flexible approach might be to send bacteria, perhaps with an implicit code written into their genetic structure determining that eventually they would evolve toward intelligence, but allowing them the flexibility to develop an array of plant, animal, and sentient species appropriate for the new world. The mentality that would conceive of such a program would have to take a much vaster and long-term view of things than we do, but perhaps that is what real intelligence is. If so, then our species has a long way to go to reach such maturity.

It would be interesting if microbiologists could conceive of some test that would allow us to determine whether the natural or nanotechnological explanation for the origin of spacefaring bacteria is correct.

The theory of panspermia is intrinsically unsatisfying to investigators of the origin of life because it basically ducks the fundamental question of how non-living matter complexified itself into the first cells. As an answer to the mystery of the origin of life, it's virtually worthless—it just moves the problem elsewhere. But for an investigation into the probable frequency of life, it is fundamental. In fact, even if somehow panspermia was not operative 3.8 billion years ago when life appeared on our planet, it is almost certainly the case that it is under way now—originating, if nowhere else, from the Earth! For the past 3.8 billion years, every significant asteroid impact into our planet has spewed debris loaded with microorganisms into interplanetary space. Once in space, light pressure from the Sun and gravity assists from close encounters with various planets have propelled innumerable numbers of these spores out of our solar system in every direction. Traveling at characteristic speeds on the order of 30 km/s (the Earth's velocity about the Sun), they could reach nearby stars in 50,000 years, during which time the bacteria would receive a cosmic-ray dose of about 1 million rads—considerably less than the 10 million rads required to kill them. As they spread, each new planet colonized would become a source point for additional propagation. In about a billion years, they could colonize the entire galaxy.

The first life on Earth were bacteria, and all subsequent life here evolved from them. It is not surprising that the Earth's first inhabitants were capable of surviving space flight. Quite the contrary, logically it would be expected that the first species anywhere in the galaxy that evolved adaptations for space flight would become the basis for life nearly everywhere else.

Bacteria are everywhere. This means that nearly every planet whose prebiotic environment was acceptable to bacteria probably developed a biosphere. Once a biosphere develops to a certain point, it becomes capable of contributing to the regulation of planetary conditions through various forms of feedback (if a planet gets too hot and rich in CO_2, for example, plant activity accelerates, reducing CO_2 greenhousing activity and cooling the surface through water evapotranspiration from increased leaf area). By taking control of a planet in this way, a biosphere increases its chances for long-term survival, during which time it can evolve ever more complex species capable of greater degrees of activity and therefore promote further expansion of the biosphere from oceans to land, to arid, torrid, and frigid regions, to mountains, and ultimately multiply it by colonizing new and previously uninhabitable planets across space.

In a word, by ignoring the capability of both microorganisms and intelligent life to propagate across interstellar distances, as well as the capability of life to regenerate intelligence within a planetary system many times within the life of that system, the Drake equation drastically underestimates the probable frequency of life and civilization in the galaxy. So let's reconsider the question.

CALCULATING THE GALACTIC POPULATION

There are 400 billion stars in our galaxy, and about 10 percent of them are good G- and K-type stars that are not part of multiple stellar systems. Almost all of these probably have planets, and it's a fair guess that 10 percent of these planetary systems feature a world with an active biosphere, probably half of which have been living and evolving for as long as the Earth. So that leaves us with 2 billion active, well-developed biospheres filled with complex plants and animals, capable of generating technological species on time scales of somewhere between 10 million and 40 million years. As a middle value, let's choose 20 million years as the "regeneration time" t_r. Then we have

$$\text{rate of demise} = N/L = n_s f_g f_b f_m / t_r = n_b / t_r = \text{rate of creation} \qquad (11.2)$$

where N and L are defined as in the Drake equation; n_s is the number of stars in the galaxy (400 billion), f_g is the fraction of them that are "good" (single G and K) stars (about 0.1), f_b is the fraction of those with planets with active biospheres (we estimate 0.1), f_m is the fraction of those biospheres that are "mature" (estimate 0.5), and n_b, the product of these last four factors, is the number of active mature biospheres in the galaxy.

If we stick with our previous estimate that the lifetime, L, of an average technological civilization is 50,000 years and plug in the rest of the above numbers, equation 11.2 says that there are probably about *5 million* technological civilizations active in the galaxy right now. That's a lot more than suggested by the Drake equation; it indicates that 1 out of every 80,000 stars warms the home world of a technological society. Given the local density of stars in our own region of the galaxy, this implies that the nearest *center* of extraterrestrial civilization could be expected at a distance of about 185 light-years.

But technological civilizations, if they last any time at all, will become starfaring. In our own case (and our own case is the only basis we have for most of these estimations), the gap between development of radiotelescopes and the achievement of interstellar flight is unlikely to span more than a couple of centuries, which is insignificant when measured against L = 50,000 years. So once a civilization gets started, it's likely to spread. As we saw in chapter 9, propulsion systems capable of generating spacecraft velocities on the order of 5 percent the speed of light appear to be possible. However, interstellar colonists will probably target nearby stars, with further colonization efforts originating in the frontier stellar systems once civilization becomes sufficiently well established there to launch such expeditions. In our own region of the galaxy, the typical distance between stars is 5 or 6 light-years. So, if we guess that it might take 1,000 years to consolidate and develop a new stellar system to the point where it is ready to launch missions of its own, this would suggest that the speed at which a settlement wave spreads through the galaxy might be on the order of 0.5 percent the speed of light. However, the period of expansion of a civilization is not necessarily the same as the lifetime of the civilization; it can't be more, and it could be considerably less. If we assume that the expansion period might be half the lifetime, then the *average* rate of expansion, V, would be half the speed of the settlement wave, or 0.25 percent the speed of light.

We have assumed that the average life span, L, of a technological species is 50,000 years, and if that is true, then the average age of one is half of this, or 25,000 years. If a typical civilization has been spreading out at the above estimated rate for this amount of time, the radius of its settlement zone would be 62.5 light-years ($VL/2$ = 62.5 light-years), and its domain would include about 3,000 stars. If we multiply this domain size by the number of expected civilizations calculated above, we find that about 15 billion stars, or 3.75 percent of the galactic population, would be expected to lie within somebody's sphere of influence. If 10 percent of these are actually settled, this implies that there are about 1.5 billion civilized stellar systems within our galaxy. Furthermore, we find that the nearest *outpost* of extraterrestrial civilization could be expected at a distance of 185 – 62.5 = 122.5 light-years.

The above calculation represents my best guess as to the shape of things, but there's obviously a lot of uncertainty in the calculation. The biggest uncertainty revolves around the value of *L;* we have very little data to estimate this number and the value we pick for it strongly influences the results of the calculation. The value of *V* is also rather uncertain, although less so than *L,* as engineering knowledge can provide some guide. In Table 11.1 we show how the answers might change if we take alternative values for *L* and *V* while keeping the other assumptions we have adopted constant.

In Table 11.1, *N* is the number of technological civilizations in the

▪ TABLE 11.1 ▪

The Number and Distribution of Galactic Civilizations

	$V = 0.005c$	$V = 0.0025c$	$V = 0.001c$
L = 10,000 years			
N (number of civilizations)	1 million	1 million	1 million
C (number of civilized stars)	19.5 million	2.4 million	1 million
R (radius of domain)	25 light-years	12.5 light-years	5 light-years
S (separation between civilizations)	316 light-years	316 light-years	316 light-years
D (distance to nearest outpost)	291 light-years	304 light-years	311 light-years
F (fraction of stars within domains)	0.048%	0.006%	0.0025%
L = 50,000 years			
N (number of civilizations)	5 million	5 million	5 million
C (number of civilized stars)	12 billion	1.5 billion	98 million
R (radius of domain)	125 light-years	62.5 light-years	25 light-years
S (separation between civilizations)	185 light-years	185 light-years	185 light-years
D (distance to nearest outpost)	60 light-years	122.5 light-years	160 light-years
F (fraction of stars within domains)	30%	3.75%	0.245%
L = 200,000 years			
N (number of civilizations)	20 million	20 million	20 million
C (number of civilized stars)	40 billion	40 billion	18 billion
R (radius of domain)	500 light-years	250 light-years	100 light-years
S (separation between civilizations)	131 light-years	131 light-years	131 light-years
D (distance to nearest outpost)	0	0	31 light-years
F (fraction of stars within domains)	100%	100%	44%

galaxy (5 million in the previous calculation), *C* is the number of stellar systems that some civilization has settled (1.5 billion above), *R* is the radius of a typical domain (62.5 light-years above), *S* is the separation distance between the centers of civilization (185 light-years above), *D* is the probable distance to the nearest extraterrestrial outpost (122.5 light-years above), and *F* is the fraction of the stars in the galaxy that are within someone's sphere of influence (3.75 percent above).

Examining the numbers in Table 11.1, we can see how the value of *L* completely dominates our picture of the galaxy. If *L* is "short" (10,000 years or less), then interstellar civilizations are few and far between, and direct contact would almost never occur. If *L* is "medium" (~50,000 years), then the radius of domains is likely to be smaller than the distance between civilizations, but not much smaller, and so contact could be expected to happen occasionally (remember, *L, V,* and *S* are *averages;* particular civilizations in various localities could vary in their values for these quantities). If *L* is large (200,000 years or more, then civilizations are closely packed, and contact should occur frequently. (These relationships between *L* and the density of civilizations apply in our region of the galaxy. In the core, stars are packed tighter, so smaller values of *L* are needed to produce the same "packing fraction," but the same general trends apply.)

THE LIMITS OF *L*

It may be asked, why should *L* have any limits? After all, while a Type I civilization can be wiped out by an asteroid or an epidemic, and a Type II civilization could be eliminated by an interplanetary war or having its star go nova, what could possibly destroy a Type III civilization once it has expanded to 50 light-years or so and colonized several hundred stars?

I can think of several things. One would be another Type III civilization. Another would be bad ideas.

The scientist Richard Dawkins has postulated a theory that views ideas, like genes, as self-interested entities dedicated to the purpose of furthering their own reproduction.[3] An idea, or *meme,* as he calls it, spreads when it has appropriate features for infecting minds, which will then give it hosts and more bases from which to venture forth to infect other minds. Those memes best styled for rapid reproduction crowd out the slower ones and therefore come to predominate. Memes may carry useful ideas and therefore spread because they are valuable to their hosts (i.e., intelligent organisms like us). But usefulness or truth is not a criterion for the spread of a meme, just capability to penetrate the brains of their hosts and then reproduce. And clearly, some memes can be pathological.

If we look at human history, we can see that most of our worst catastrophes were caused by bad memes that spread like contagions. Such mental diseases have included the messianic nihilism of the Mongol horde, Muslim fundamentalism, which destroyed the potential of Islam as a major world civilization, many other types of fundamentalism, the Pol Pot madness, and Nazism, which nearly led to the autocannibalization of European civilization within our own century.

One bad meme can wreck a society. And memes can spread by radio. A 100-planet interstellar empire built over 50 millennia can be destroyed in 50 years by a radio-transmitted cult spreading at the speed of light.

In order to have had time to grow in the first place, a society must have significant internal resistance to bad memes. But complete bulletproofing is impossible, as intelligence and civilization themselves by their very nature require the ability to assimilate and spread new ideas.

And new ideas will be generated, because conditions will change. To take one obvious example, as the frontier of a civilization radiates outward into space, the interior worlds of a domain, which will constitute an ever greater majority of the whole, will have no physical frontier of their own. They will no longer be able to launch missions to new and virgin stars but must content themselves with the challenge of developing the interior. For how long will this be sufficient to prevent stagnation and the ensuing establishment of pathological forms of social organization with their concurrent deadly pathological memes? It's impossible to say, but clearly the potential exists, and, given sufficient time, anything that can happen will. Sooner or later the ax must fall, thereby putting a finite limit to L.

A third factor that may put a boundary on L is biological evolution. We tend to think of ourselves as permanent, but *Homo sapiens* has been around for only 200,000 years, and there is no reason to believe that evolution has stopped. Quite the contrary, as humans gain control of their genome and move into novel environments in space, there is every reason to believe that evolution will accelerate dramatically. We will become something else, probably many new somethings else. The same would hold true for other species that master the scientific method, with its accompanying gifts of genetic knowledge and space-flight technology. Everyone will change, and it is unclear whether the beings we change into will require or desire the continuation of interstellar expansion or even technological civilization itself. Could a species that has created for itself an ideal environment then degenerate into a group of lotus eaters, needless of effort or thought? Might such beings then prove helpless to stop their own extinction when conditions changed? It does not seem impossible. Certainly we see many parallels to such cycles of heroic age, golden age, and collapse in human history.

So, one way or another, even the grandest of interstellar empires must come to an end. We know *L* is finite; if it wasn't, the Earth and every other planet would be crawling with aliens. But the human race has already existed for 200,000 years, and so *L* values on this order seem entirely possible. If that is the case, then the nearest extraterrestrial outposts could be quite nearby. If *L* is smaller, we may have a fair space of open range. But either way, one thing seems rather certain: They're out there. Plenty of them.

THE SEARCH FOR EXTRATERRESTRIAL INTELLIGENCE

The finitude of *L* thus provides one kind of answer to Fermi's Paradox. They are not here because they are not yet everywhere. But we are still left with the very interesting literal question: Where are they?

The first practical proposal to try to detect extraterrestrial civilizations at interstellar distances was advanced by Cornell physicists Giuseppe Cocconi and Phillip Morrison in 1959. In a calculation published in the prestigious science journal *Nature,* Cocconi and Morrison pointed out that existing radiotelescopes were capable of projecting a signal that could be detected by similar gear nearly at the distance to Alpha Centauri. Better equipment, already on the horizon (and which we possess today), could communicate much farther. Thus, while interstellar travel might require speculative advanced technology, interstellar communication does not and therefore might well be under way, and therefore detectable, in the radio bands.

Of course, when listening for radio signals, one must know what frequency to tune in on, as there are millions possible. Cocconi and Morrison argued that the frequency of choice for interstellar communication would be near 1,420 MHz (21-cm wavelength) as the emissions of hydrogen gas at that frequency make it the most important and most listened-to band in radio astronomy.

This latter argument has some logic, although to me it has always seemed a bit like, "We have a hammer, therefore this must be a nail." ("Our radiotelescopes are tuned in on 1,420 MHz, therefore that is where ET must be signaling.") In the late 1950s and early 1960s, frequencies near 1,420 MHz (L band and S band) were popular for spacecraft communication, and this gave added credence to the idea. However, since that time, we have already gone to higher frequencies that support higher data rates, which suggests that the idea of 1,420 MHz as a universal communication band for advanced civilizations may be considered a form of temporal chauvinism, much like nineteenth-century astronomers trying to detect life on Mars by searching for the telltale emissions of Martian gaslights.

However, whatever its weaknesses, the Cocconi-Morrison suggestion represented a clear and *implementable* strategy for attempting the detection of extraterrestrials. In 1960, Frank Drake, then a young post-doc staff member at the National Radio Astronomy Observatory (NRAO) in Green Bank, West Virginia, decided to put it to the test. In an experiment called Project Ozma, Drake pointed the Green Bank 26-meter-diameter radio dish at Tau Ceti and Epsilon Eridani, two nearby "good" G and K (respectively) stars, and searched the radio waves around 1,420 MHz for signs of intelligent communication. Except for a false alarm caused by a signal from a high-altitude military aircraft, the results of the search were nil. Nevertheless, the Ozma experiment caused a sensation, and over the years that followed numerous other investigators with better equipment and more observatory time have greatly expanded on Drake's work.[4]

In addition to innumerable informal efforts by radio astronomers with some extra instrument time on their hands, follow-on Search for Extraterrestrial Intelligence (SETI) activities have included the Big Ear program, run at Ohio State University from 1973 to 1998, the META and BETA programs sponsored by the Planetary Society (currently operating from Argentina and Massachusetts, respectively), and the SERENDIP and SETI Australia programs, which piggyback special SETI signal analyzers on whatever is being observed by the powerful Arecibo, Puerto Rico (300-m diameter), and Parkes, New South Wales (94-m diameter), radio telescopes, respectively. The most powerful current SETI search is Project Phoenix funded by the SETI Institute (now headed by Frank Drake) with private money after the U.S. Congress killed the program's budget.

In his 1960 search for radio signals from Tau Ceti and Epsilon Eridani, Drake had a motor turn the radio dial to examine one frequency at a time. Thus, only a few frequencies out of the hundreds of millions possible near

1,420 MHz were briefly examined. In contrast, Phoenix, which is scheduled to study the thousand nearest Sun-like stars, can listen to 28 million channels at once. In the course of a day, Phoenix points Green Bank's 43-meter-diameter radio telescope at a given star and listens successively to one block of 28 million channels after another for five or ten minutes at a time. At day's end, 2 billion frequencies between 1,000 and 3,000 MHz have been studied. Then the next star is studied in the same way, then the next, then the next. All told, it is estimated that the Phoenix study is 100 trillion times as comprehensive as Drake's original Project Ozma.

However, despite the improved equipment, all SETI searches conducted to date have been completely unsuccessful.

There are many possible reasons for the lack of success. For example, even the Phoenix only looks at a given star for a few minutes in a given frequency. Planets rotate. What if at the particular moment Phoenix is looking at an *inhabited* planet in the *correct* frequency, ET's radio transmitter is (or rather was) pointed in the wrong direction? We'd miss the show. We could be failing simply because the search is not intensive enough, with much more time devoted to each star of interest.

An alternative explanation for the failure is technological. The Cocconi-Morrison 1,420-MHz recommendation that still focuses most SETI work was convenient to radio astronomers and appeared to have some validity since it mirrored technology that was then state-of-the-art for spacecraft communication. But even forty years into the space age, such frequencies are already obsolescent. The amount of data that a spacecraft can send with an antenna and power of a given size is proportional to the square of the frequency. Thus, the 8,418-MHz (3.5-cm) X-band radios in common use on interplanetary spacecraft today can support a data rate 35 times greater than equivalent systems transmitting at 1,420 MHz (L band). Using radio in the millimeter region could push the data rate up by another factor of a thousand, while going to optical frequencies (i.e., laser communication at 0.5 microns) would raise it millions of times higher still. The advantage of using these technologies for interstellar communication is shown in Table 11.2, which compares the data rate possible for each assuming a 10-MW transmitter and a target receiver at a distance of 10 light-years.

Examining Table 11.2, it should be apparent why the assumption that extraterrestrials will choose to use 21-cm radio as the preferred medium of interstellar communication is highly questionable. Millimeter radio provides 40,000 times the data capacity for the same power and transmitter

▪ TABLE 11.2 ▪

Data Transmission Possible with 10 MW to 10 Light-Years

Technology	Wavelength	Transmitter/Receiver	Data Rate	Beam Diameter
L band	21 cm	70-m dish	1.4 b/s	1,998 AU
X band	3.5 cm	70-m dish	50 b/s	333 AU
Millimeter radio	1 mm	70-m dish	61,000 b/s	9.5 AU
Laser optical	0.5 micron	5-m telescope	6,378,000 b/s	0.066 AU

dish size, while offering a beam diameter of 9.5 AU at 10 light-years—more than enough to make tracking the receiver planet unnecessary.

The higher-frequency, shorter-wavelength transmitter systems achieve their higher data rates because they can be focused with a much tighter beam. Some SETI researchers frown on them for this reason—unless they are pointed directly at us, we won't see them. But that is precisely the reason they are probably preferred! Few people design their communication systems for the benefit of eavesdroppers.

The answer that the L-band SETI researchers have is that ET may be broadcasting in order to try to advertise its existence to potential correspondents. Well, maybe, but why do it with an inefficient ephemeral communication technology? But I think it is more likely that they will want to choose their pen pals themselves. It would be irrational for any civilization to give away its position and other vital intelligence to a universe filled with potentially hostile competitors by broadcasting powerful radio signals indiscriminately in all directions. Indeed, the survival value of such a custom could be so negative that any civilization which adopts it may not last to do it very long. Even if the large majority of ET societies are benevolent and "civilized," it takes only one bunch of paranoid-aggressive emergent Type III savages armed with weapons of mass destruction to launch a preemptive strike of catastrophic proportions. Caution in opening communications may thus well be the rule.

Ground-based radio SETI research is a long shot, but it is cheap, and the potential payoff is enormous. It should therefore be continued and in fact expanded to include higher frequencies right up through the optical. But if we want a real chance of finding ET, we may have to use an alternative approach.

DETECTING STARSHIPS

A Type III extraterrestrial civilization operating within 50 light-years of our solar system could probably determine that there was life on Earth by astronomical means using large space-based optical interferometers, as the presence of free oxygen in our atmosphere is strongly indicative of the presence of photosynthetic plants. In order to monitor developments here, they therefore might choose to send robotic probes into our solar system to set up observation stations. Such instruments may thus have been stationed within our solar system for a long time, quietly sending back a tight beam of reconnaissance data to their owners. If we could find such objects, it would obviously be of great scientific interest, as they would not only prove the existence of extraterrestrial intelligence but possibly reveal significant information about the nature, location, and technological acumen of at least one ET variety.

It must be assumed, however, that any such craft would be designed and operated to minimize its chance for detection. It certainly would not transmit constantly back to its home world with an omnidirectional broadcast, or wide beam of any sort, not would it ever make a transmission capable of being intercepted by an observer on Earth. Rather, any transmissions would be very brief and on a tight beam pointing well away from our planet. The infrared signal of the device would probably be masked, visual camouflage might be used to give it as natural an appearance as possible, and stealth technology might be employed to minimize the chances of detection by radar.

In other words, short of establishing a wide-scale human presence throughout the solar system affording numerous off-Earth points of observation to pick up an outward pointing signal and/or plenty of explorers on extraterrestrial bodies where such systems might be placed, it is unlikely that such a device will ever be found. Detecting alien robotic reconnaissance probes will therefore probably have to wait until the establishment of a mature Type II civilization in our solar system.

Even then, there is another problem with looking for probes—they might not be here. We like to think of our solar system as prime real estate, because it is our home. But actually our Sun, as a type G2 star, is near the extreme bright, hot, and ultraviolet-intense end of the spectrum of stars where it is possible for a biosphere to evolve. The large majority (more than

90 percent) of life-supporting planetary systems probably surround dimmer, redder, type G3–G9 and type K stars. For most everybody else, therefore, *those* kinds of stars are home, and while an advanced Type III civilization could undoubtedly find some terraforming expedient to deal with the Sun's excessive ultraviolet radiation, why bother when there is so much more territory around that can be had without the hassle?

In any case, if our strategy is to try to find extraterrestrials by looking for their artifacts, we would be much better off searching for the spectral signature created by those activities that are vital to their propagation as a species but that necessarily entail a large unmasked release of energy into the universe. The most obvious such activity is crewed interstellar travel. Propelling large starships across interstellar distances at relevant speeds requires a great deal of power, and each of the types of propulsion systems that might be employed—antimatter, fusion, magnetic sails—would release this power into the universe in a characteristic way that could be readily distinguished from natural astrophysical phenomena. Furthermore, unlike communication that is governed by a fairly arbitrary selection of technology and mutually agreed on conventions, transportation systems are governed much more stringently by the laws of physics. No understanding of alien psychology is necessary to detect a starship. Thus, the best way to find ET may not be by trying to overhear his conversations or searching for his robotic scouts but by listening for the sound of his engines.[5]

In chapter 9, we discussed various types of interstellar propulsion systems from the point of view of the engineer seeking performance. Now let's reexamine three of the most promising—antimatter, fusion, and magsails—from the viewpoint of an astronomer attempting detection.

Depending on the propulsion system employed, a starship could reveal itself via various forms of radiation. If antimatter is employed, after several intermediate but very short time scale reactions, about 40 percent of the total energy will be released in the form of very hard gamma rays with energies between 130 and 350 MeV. It would be both difficult and undesirable to attempt to block all of these rays from escaping the starship structure, and thus the primitive proton-proton annihilation spectrum could be expected to be radiated into space. To obtain the high specific impulse necessary for interstellar flight, the antimatter could be used either to heat a plasma, presumably magnetically confined, or to heat a radiator to produce thrust in the form of photons. If a plasma confinement system is used, there will be both cyclotron and bremsstrahlung radiation, which will be broadcast into space. In order to obtain the maximum specific impulse in an antimatter-fed plasma drive, the plasma will be heated to several million electron volts and will thus produce bremsstrahlung gamma rays in this energy range. The frequency of the cyclotron radiation would be determined solely by the strength of the magnetic field used. If the field strength is 5 Tesla, then there will be electron cyclotron radiation at 140 GHz (high-frequency radio) and higher harmonics, along with ion cyclotron radiation at 80 MHz and higher harmonics. If photon propulsion is used, about half of the hard gamma radiation plus all of the rest of the annihilation energy will be thermalized to heat, which will be radiated to space by a set of radiators. Because the amount of power that can be radiated goes as temperature to the fourth power, it is highly advantageous to run the radiator as hot as possible. The maximum temperature of the system is governed by the long-duration temperature limits of materials, which, based on our current knowledge, would be about 2,800 K (~2,500°C for graphite or tungsten). Radiators operating at this temperature will emit strongly in both the visible and infrared portions of the spectrum. The light they emit could be distinguished from that given off by stars because it would not have hydrogen lines in its spectrum. As discussed in chapter 9, in order to maximize the useful thrust, reflectors will be used to channel the emitted photons into as small a cone angle as possible.

If thermonuclear fusion power is employed, there will be cyclotron radiation and bremsstrahlung, whose frequency will be governed by the plasma temperature spectrum. As we have seen, the optimum fusion reaction for interstellar rocket propulsion may well be D-He3, since nearly all of the energy it releases is in the form of charged particles whose momentum can be converted to thrust. The products of this reaction, a proton and an alpha particle, are released with energies of 18 and 4.5 MeV, respectively, and thus some gamma rays may be expected with energies in this range. However, the optimum power/mass-ratio fusion reactor using this fuel will be realized if the plasma temperature is kept at about 60 keV. The bremsstrahlung emittance from such a reactor will thus be dominated by x rays in this frequency range.

As I noted earlier, a magnetic sail (or "magsail") would be of unique value to an interstellar spacecraft because of its ability to decelerate a ship without the use of propellant. It could be used to decelerate a ship using any propulsion system for acceleration, including antimatter, fusion, or the laser-pushed light sail. During deceleration, the magnetosphere of the magsail will create a standoff shock wave, which will heat the interstellar medium it encounters from hundreds of keV to a few MeV, depending on the ship's velocity. The plasma so created will then encounter the magnetic field of the magsail, where it will emit cyclotron radiation.

In the 1989 paper I co-authored with Dana Andrews,[6] we showed that the characteristic trajectory of a magsail-decelerated spacecraft will always have the form

$$W = Wo/(1 + kWo^{1/3}t)^3 \qquad (11.3)$$

where W is the velocity of the spaceship at a given time t after braking begins, Wo is the velocity at the start of the deceleration maneuver, and k is a constant. Since the strength of the magnetic field created in the magsail's bowshock will be directly proportional to the ship's speed, and the frequency of the cyclotron radiation emitted by the plasma caught in the bowshock will be proportional to the magnetic field, by measuring the frequency of the emitted cyclotron radiation we can measure the speed of the ship. If we discover that the way this velocity changes over time matches equation 11.3, it's a dead giveaway that we are looking at a magsail.

To emphasize the above point: The cyclotron frequency emitted by a magsail is not a function of spacecraft design but only of the ship's velocity and the (fixed, known) density of ions in the interstellar medium. At a density of 1 ion/cc, a ship traveling at $0.1c$ would produce electron cyclotron radiation at a frequency of about 12 kHz (similar to marine mobile radio).

The magnitude of the radiation that a starship will emit is a function of the magnitude of the power of its rocket engine, as well as of the engine design. If we assume, somewhat arbitrarily, that the dry mass of an interstellar spacecraft is a million tonnes, then the estimated magnitudes of various types of radiation emitted by the different propulsion systems are given in Table 11.3. For some types of engines, the fraction of jet power emitted as certain types of radiation can be calculated accurately. Where such information is lacking, I have assumed that the magnitude of a major type of radiation generic to an engine is 10 percent of the jet power.

In Table 11.3 it is assumed that the ship is on a 6-light-year mission and that the ship has a big enough engine so that the acceleration from zero velocity to maximum flight speed equals ¼ of the trip time. The maximum velocity, W, for a given propulsion system is chosen to be at the upper end of what the engine's exhaust velocity, U, can reasonably be expected to generate.

In Table 11.3, A is the maximum ship acceleration required, P_{jet} is the power of the engine in terawatts, and P_{rad} is the power of the radiation emitted by the engines at various energies. It can be seen that the antimatter

▪ TABLE 11.3 ▪

Characteristic Power Levels of Interstellar Propulsion Systems

TECHNOLOGY	U/c	W/c	A (*g*'s)	P_{jet}(TW)	P_{rad}(TW)
Fusion plasma	0.08	0.1	0.005	600	60 @ 1–100 KeV, 60 cyclotron
Antimatter plasma	0.2	0.2	0.02	6,000	4,000 @ 200 MeV, 600 @ 20 MeV, 600 cyclotron
Antimatter photon	1.0	0.4	0.08	120,000	40,000 @ 200 MeV, 120,000 visible/IR
Magsail fusion	—	0.1	0.026	780	80 @ 2 MeV, 80 @ 12 kHz
Magsail antimatter plasma	—	0.2	0.0065	4,000	400 @ 8 MeV, 400 @ 24 kHz
Magsail antimatter photon	—	0.4	0.0166	20,000	2,000 @ 32 MeV, 2,000 @ 48 kHz

rockets emit the most radiation, with large amounts of it in the form of 200-MeV gamma rays. Unfortunately, since each gamma ray carries a lot of energy, the photon count created by such systems at interstellar distances is quite small and thus extremely difficult to distinguish from instrument noise and background radiation. For example, a starship emitting 10,000 TW of 200-MeV gamma radiation at a distance of 1 light-year from Earth will cause 7.5 photons *per year* to impact on a 1-square-meter collection device. This would obviously be undetectable. Unlike gamma rays, x rays can be collected to some extent with a kind of grazing incidence telescope. However, even with a telescope that could focus a 1-meter aperture down to a 1-cm collection plate, the x-ray emissions (kilovolt-class radiation) of the various starships in Table 11.3 would be undetectable beyond 10 light-years.

However, while the gamma-ray emissions from the engine of an antimatter photon rocket would be undetectable, the visible radiation composing its exhaust is another story. If we consider the sample photon rocket in Table 11.3 with a jet power of 120,000 TW, and assume that it uses a reflective nozzle to focus the emitted light to a half angle of 30 degrees, then it will shine in the direction of its exhaust with an effective irradiated power of 1,800,000 TW. Such an object at a distance of 1 light-year would be seen from Earth as a 17th-magnitude light source and could be detected on film by a first-class *amateur* telescope. The 200-inch telescope on Mount Palomar could image it at 20 light-years, and the Hubble Space Telescope would record it at a distance of about 300 light-years. The implications of this are displayed in Figure 11.1, which shows curves of apparent magnitude versus distance, as well as the number of stellar systems ($n = R^3/80$) within range. Since, at least for the upper-end telescopes considered, the number of stellar systems within range is significant (100,000 stars are within 200 light-years of Earth), this approach offers some hope for a successful search. The light from the photon rocket could be distinguished from that of a dim star by the lack of hydrogen lines in the rocket's emissions.

The other kind of starship radiation that we might hope to detect is radio waves emitted as a result of plasma interaction with the deceleration field of a magnetic sail. Magsails produce electron cyclotron radiation with frequencies of tens of kilohertz. Magsail radiation is thus below the cutoff frequency for passage through the Earth's ionosphere and could be detected only by antennas positioned in space.

Because the frequency of magsail radiation would be low (much lower, for example, than the gigahertz-class cyclotron radiation given off by the plasmas in fusion reactors), the required receiver bandwidth needed to cap-

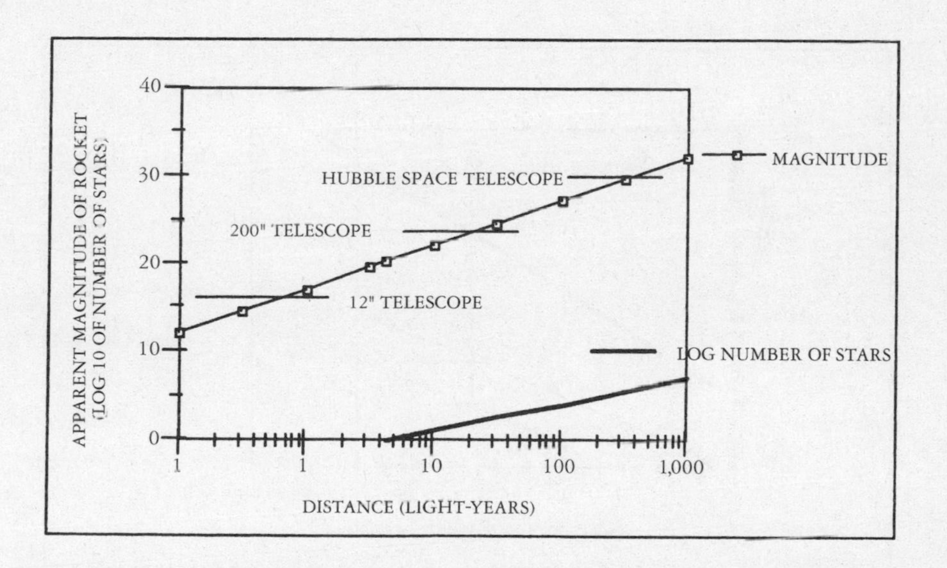

FIGURE 11.1 *The light from antimatter photon rockets could be detected at interstellar distances by existing telescopes.*

ture most of it can be much narrower. Now, since the signal-to-noise ratio (SNR) of a radio receiver is inversely proportional to its bandwidth, this means that, with the same size antenna and power source, magsail radiation can produce an SNR 6 orders of magnitude greater than that possible from a plasma drive. Furthermore, since the low-frequency magsail radiation has very long wavelengths (12 kHz = 25 km wavelength), huge collection areas can be created with very little mass by orbiting dishes or antennas made of sparsely placed wires or crossed tethers. For these reasons, magsail cyclotron radiation would be much easier to detect than that from plasma engines.

If we assume an SNR of 2 and a bandwidth of 1 kHz, sufficient to capture a significant fraction of electron cyclotron radiation emitted by a magsail (we assume 10 percent of the emitted electron cyclotron radiation within this bandwidth), and orbiting antennas with effective equivalent radii of 6 km and 30 km, respectively, then the power that needs to be emitted by a magsail for it to be detectable on Earth is shown in Figure 11.2.

It can be seen that the magsail radiation of a characteristic fusion starship being decelerated from a cruise velocity of 0.1*c* could be detected by a 6-km orbiting antenna from a distance of 400 light-years, while that emitted by a characteristic antimatter photon rocket in its deceleration phase could be seen as far away as 2,000 light-years. There are about 100

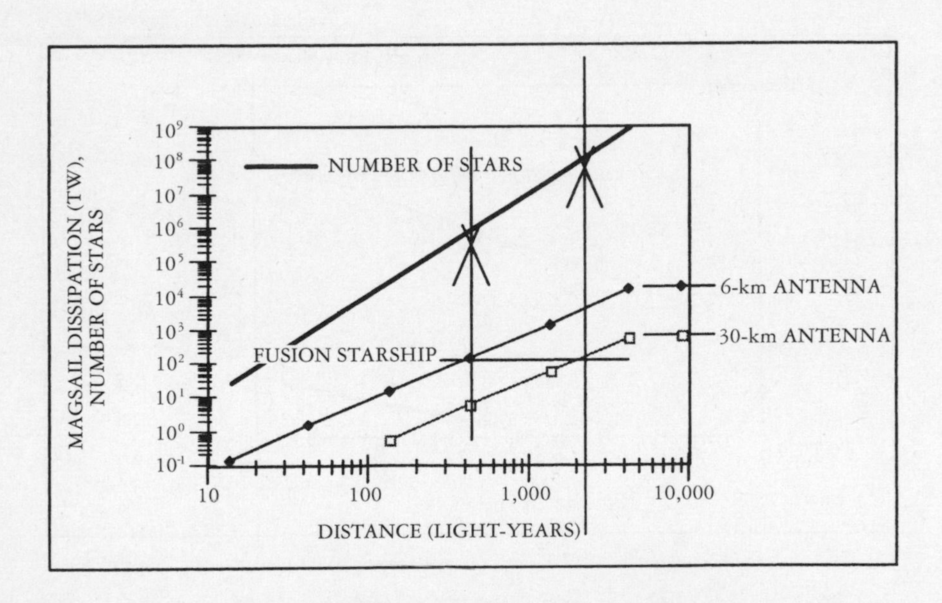

FIGURE 11.2 *Detectability of magsails over interstellar distances.*

million stellar systems to be found within the latter distance. This extended range detection capability combined with magsail radiation's unique time-dependent frequency spectrum appears to make a search for magsail radiation the most promising option for extraterrestrial starship detection.

It may be noted that our estimate of starship mass is a speculative guess. However, since the signal detectability is proportional to the mass of the ship divided by the square of the distance, a decrease of ship mass by 2 orders of magnitude results only in a decrease in detectability distance by 1 order of magnitude. Thus, even if the true characteristic mass of starships is 10,000 tonnes, and not the 1,000,000 tonnes we have postulated as a baseline, magsail radiation would still be detectable by the 6-km antenna at 40 light-years and by the 30-km antenna at 200 light-years.

It's worth a try.

GALACTIC CIVILIZATION

While the assumption of a moderate (50,000 years or less) value for L readily explains the absence of extraterrestrials on Earth in the present, unless very low values for L and V are chosen, it is difficult to understand how it could have been inaccessible to outsiders at all times in the past. The solar system is moving with respect to other nearby stars with a relative velocity of about 10 km/s. So even in most cases where the galaxy is only partially settled, the Earth should have drifted through somebody's interstellar empire at some time in its history. How many times depends on all the factors we have considered so far, but mostly on L and V. Keeping all the other factors (f_g, f_b, f_m, and t_r) considered in equation 11.2 constant with the same values we assumed earlier but varying L and V, we can calculate the average probable time between such drift-through encounters. (Such a calculation is done by computing the average path length between settled domains and dividing that distance by the solar system's relative interstellar drift speed of 10 km/s.) The results are shown in Table 11.4.

The Earth has been habitable for creatures like ourselves for about 600 million years and readily terraformable by anybody with a good bioengineering capability for 3.6 billion years. If we take our previous "best-guess" case of L = 50,000 years, V = 0.0025c, the calculations given in Table 11.4 show that our solar system, while unlikely to be within or very near a settled region today, has probably drifted through settled domains of inter-

• TABLE 11.4 •

Average Time between Earth/ET Domain Drift-Through Encounters (millions of years)

L (YEARS)	V = 0.005c	V = 0.0025c	V = 0.001c
5,000	15,645	62,582	391,139
10,000	1,955	7,822	48,892
20,000	244	977	6,111
30,000	71	289	1,810
40,000	29	121	763
50,000	14	62	390
60,000	6.8	35	226
70,000	3.1	21	142
80,000	0.8	14	95
90,000	0	9.0	66
100,000	0	5.9	48
120,000	0	2.3	27
140,000	0	0.2	17
160,000	0	0	11
180,000	0	0	7.0
200,000	0	0	4.6

stellar space about once every 62 million years. So under those assumptions, it would appear that the solar system has been inside of someone else's sphere of influence about 10 times during the 600 million years that Earth has been prime real estate. Lower assumptions for *L* and *V* lead to less frequent encounters, but, all in all, it seems that the odds are high that someone has stopped by.

For some reason, they didn't stay.

As mentioned earlier, one explanation could be that they were orange K-star people (i.e., members of the cosmic majority) and regarded the excessive ultraviolet environment of our yellow G-star world with disdain. But any Type III civilization worth its salt should have been able to handle that problem.

A more intriguing explanation is that they chose not to invade because they thought it would be wrong to interfere with the development of a promising biosphere, or more interesting and rewarding to wait to see what it might bring forth.

This may seem like a bizarre idea—cosmic environmentalism, or, perhaps, cosmic husbandry or stewardship. But recall our discussion of the destructive potential of Type III technologies. No species can last long as a Type III civilization if it is not also wise. In most matters, the long view is the wise view.

True, as discussed in conjunction with the SETI search, there could be anomalies, occasional examples of aggressive Type III societies, sufficient to make others cautious. But it is not true that the "noninterference" hypothesis requires *every* civilization in the galaxy to be well behaved to explain the apparent lack of extraterrestrial intrusion—just our immediate neighbors.

Intelligence, I would argue, can take innumerable distinct forms, each of which can offer unique insights into problems scientific, technological, artistic, ethical, philosophical, and so on. The development of such a new type of intelligence thus potentially offers much more in the long run to neighboring civilizations than the benefits associated with gaining possession of one more piece of cosmic real estate. Furthermore, aside from potential long-range benefits and aesthetic, moral, or scientific appeal, the maintenance of a noninterference policy toward developing biospheres by a Type III civilization may have short-term positive survival value. Remember, Type III civilizations are intrinsically capable of inflicting destruction on an astronomical scale. Therefore, should any new Type III civilization show itself to be aggressive, it would be a menace whose existence could not be tolerated. Its older starfaring neighbors might therefore combine to wipe it out.

Bacteria are everywhere. So the subsistence of bacteria on a near-sterile planet like Mars over billions of years without further development does not constitute a true biosphere. Such planets should be terraformed to make them a real home for life. But a true biosphere, filled with multiplicities of complex plants and animals in the process of evolving toward higher forms, is something entirely different. It is unique; it is precious; it holds unfathomable promise. No sane species would destroy one for the sake of a little extra living space. The same is even more true for the case of a world, or a solar system, sporting indigenous intelligent life.

In this book we have discussed humanity expanding to become a Type I global, Type II interplanetary, and ultimately Type III starfaring civilization. But Type III is qualitatively different from the others in this important respect—no one species will ever expand to mastery of the entire galactic theater. The distances involved are too great, the required time span for the activity too long, the number of indigenous civilizations too vast, and the numerical and tactical advantages naturally accruing to defenders of a solar system against an interstellar invasion force so manifest that no campaign of galactic conquest could ever succeed. Galactic civilization will not be an empire spread from a single Type III, but a society of many Type III civilizations linked together.

So the question is, when we meet the galactic club, will we be fit to join it?

MEETING THE UNIVERSE

When I was a child, I spake as a child, I understood as a child, I thought as a child: but when I became a man, I put away childish things. For now we see through a glass, darkly; but then face to face: now I know in part; but then shall I know even as also I am known.

— SAINT PAUL, I Corinthians 13:11-12

In my view, to be successful in joining the club we will need to have attained a degree of both technological and ethical maturity that can only come with having reached Type III status ourselves. In fact, far from being beneficial, there is good reason to believe that encountering a Type III civilization will be psychologically devastating for humanity, if we have consigned ourselves to a Type I role. One only has to look at human history to see the results when more primitive cultures have come into contact with the West. Humans require dignity. We cannot survive with the thought that we are beings of a lesser order. Turning off the radio telescopes to perform a kind of cosmic withdrawal is not an option, as it amounts to conceding inferiority in advance. The only solution is to grow up.

Since time immemorial, we humans have prided ourselves upon the conceit that we are the universe's sole vessels of sentient life. If we maintain a scientific outlook, that conceit is likely to be destroyed. We will not be able to base our pride and dignity on the childlike notion that there are no others. Instead, we will have to grasp a deeper and more mature idea, that we are worthwhile not because we are alone, but because we are unique and precious representatives of something far bigger than ourselves. Something to which we can contribute—*as equals.*

We must *make* ourselves fit to join the galactic club.

On the other hand, if there is no galactic club, if the galaxy is really a Hobbesian jungle war of all against all, then technological maturity of Type III level is required for survival. Species with little empathy can still be allies; witness the strategic maneuvers of terrestrial geopolitics. But to survive in such a universe, we would need to become "worthwhile as friends, undesirable as enemies."

Finally, even if there are no extraterrestrials, making ourselves worthy to meet them is still what we need to do for our own sakes.

Type III calls.

CHAPTER 12

North to the Stars

There is grandeur in this view of life, with its several powers, having been originally breathed by the Creator into a few forms or into one; and that . . . from so simple a beginning endless forms most beautiful and most wonderful have been and are being evolved.

—CHARLES DARWIN, *On the Origin of Species,* 1859

It is possible to believe that all the past is but the beginning of a beginning, and that all that is and has been is but the twilight of the dawn. It is possible to believe that all the human mind has ever accomplished is but the dream before the awakening, . . . We are creatures of the twilight. But it is out of our race and lineage that all minds will spring . . . that will reach forward fearlessly to comprehend this future that defeats our eyes. All this world is heavy with the promise of greater things. . . .

—H. G. WELLS, *"The Discovery of the Future"*[1]

THE EXTRATERRESTRIAL IMPERATIVE

Humans are the descendants of explorers. Four hundred million years ago, our distant ancestors forsook the aquatic environment in which they had evolved to explore and colonize the alien world above the shoreline. It is remarkable when you think about it, sacrificing the security of the waters for the hazards of the land. Fish are superbly adapted to life in the marine habitat. Compared to their crawling arthropod, molluscan, and echinoderm competitors, fish underwater are like gods, flying effortlessly through their liquid firmament. In contrast, fish out of water—even ones with stubby fins that allow them to drag themselves across the sand—are hapless, helpless creatures, running imminent risk of death by desiccation, asphyxiation, or predatory attack by crabs or other low-life.

But on land, animals could raise their body temperatures higher, first through accessing direct sunlight and later through the evolution of warm-bloodedness, thereby more readily increasing their level of activity and brain development than would have been possible had they stayed in the ocean. While the land habitat of the Earth is actually spatially smaller than the oceans, it offers much more variety, and far better possibilities for separate, independent development, than does the sea. It also offers greater challenges, including those associated with rapid seasonal, climatic, and other forms of environmental change. Therefore, on land, the possibilities and driving forces for evolution have been much greater, and it has occurred faster. In the 400 million years since some fish left the ocean, their compatriots who stayed behind have not changed much, but the descendants of the emigrants have evolved legs and wings, feathers and fur, far-seeing eyes, nimble hands, and clever brains.

On land it is possible to build fires. On land it is possible to see the stars.

Out of the oceans to the land. Out of Africa to the north. Out of Earth and into space. The future is best served not by those who remain behind to walk in the footsteps or swim endlessly in the wakes of their ancestors, but by those who dare the unknown, willing to risk all to take on the challenge of the new.

The human desire to explore is thus one of our primary adaptations. We have a fundamental need to see what is on the other side of the hill, because our ancestors did, and we are alive because they did. And, therefore, I am firmly convinced that humanity will enter space. We would be less than human if we didn't.

In this book I have mentioned the need for humanity to mature if we are to join in the grand enterprise of Type III civilization. By maturing, I mean leaving behind the infantile selfishness and violence of our species' childhood. But I do not mean become old. Quite the opposite, we must preserve and expand our youthful characteristic of curiosity. Juvenile animals explore. Humans explore as adults. When we cease exploring, we die—first in soul, then in body. To be really alive as individuals, to be really alive as a civilization and as a species, we must be willing to continually seek out and experience the new.

The human expansion into space will not be without cost, and that cost will be measured in units far more dear than dollars. The Oregon Trail was lined with graves. Not all of our missions will be successful, for in enterprising the unknown, the possibilities for disasters abound. Progress was never without risk.

The day may come when humanity no longer has a need for wars. But we will always have a need for heroes.

SPACE COLONIZATION AND HUMAN DIVERSITY

In biology, an animal type is generally considered to be successful if it manages to diversify itself into many species inhabiting a wide array of habitats. The reason for this is that a single species with a single mode of life is a slender thread whose line to the future can be easily cut if conditions change adversely. Viewed from this point of view, the genus *Homo* cannot currently be considered very successful; certainly we are not safe, as for the past 20,000 years (since the extinction of our Neanderthal cousins) we have been limited to a single species. Of course, our species is of global extent and has featured many widely differing cultures. However, in the twentieth century, with the advent of global communications and jet aircraft, the potential for terrestrial geographic barriers to maintain cultural diversity has been eradicated. In consequence, the world is now rapidly being homogenized to a single culture, and in the twenty-first century, this tendency will only accelerate. Were we to remain Earth-bound, we would soon not be only one species, but one culture. If studies of biology and evolution are any guide, that is a prescription for disaster.

However, fortunately, the human expansion into space should generate conditions for the rapid regeneration of both cultural and biological diversity. In nature, large interconnected gene pools are very slow to evolve, as it takes a very long time for any new trait to become dominant. In contrast, the generation of new species is favored when a small group becomes isolated from the main stock of its kind and put in a new environment where it is offered new opportunities and subject to novel adaptive stresses. Under such conditions, the generation of new traits is called forth, and since the genes for the new trait are not being constantly washed out by interbreeding with the primary population, new varieties and new species rapidly result. Analogous processes of innovation-in-isolation are necessary for the generation of substantially new cultures by human populations.

As humans expand to Mars, the asteroids, the outer solar system, and ultimately the stars, precisely such conditions for, first, cultural and, ultimately, genetic diversification will obtain. In fact, due to the intrinsic enormous differences in environment from one extraterrestrial habitat environment to another (down to things as fundamental as the gravitational field within which a civilization exists), it is certain that both culture and heredity will be driven fast and hard in many diverse directions.

It might be maintained that in the future, the increasing human ability to control heredity will impede this process. I would argue the contrary. In fact, since cultural evolution occurs normally on a much faster time scale than genetic evolution, as soon as human beings gain control of the genetic code (i.e., culture gains control of heredity), biological evolution will occur at a greatly accelerated pace. It might be the case that in one locality or another, governments will act to suppress this kind of self-directed evolution. However, enforcement of any sort of government edicts across interstellar space is likely to be impossible. Therefore, among the far-flung culturally diverse civilizations, some might choose to suppress change, but others will drive it. The difference of opinion on this score will thus serve only to accelerate the process of multiple speciation.

One of the results of these programs of self-modification will no doubt be a drastic extension of the individual human life span. The aging process itself may well be defeated. If this is the case, however, the necessity for space expansion will be greatly accentuated, as the younger generation will face old worlds in which the determining roles have already been assigned. Humans need to matter. In the age of immortality, new generations will need new worlds to give their lives immortal purpose. Fortunately, long-lived people will be able to undertake long voyages. Thus, the interstellar diaspora, and its production of ever more diversity, will be driven even further.

In science fiction television and film, such as *Star Trek,* the galaxy is frequently depicted as being inhabited by numerous species of "aliens" who are humanoid in all respects except for minor differences, such as skin color or ear shape, and who can sometimes interbreed. Humans are the process of 4 billion years of terrestrial evolution, and while for reasons of convergent evolution it is possible that aliens might be found with a general form similar to humans (i.e., two arms, two legs, head-on-top is a fairly practical design plan), they would obviously differ enormously from us internally at every level, including organs, tissues, and cellular structure. Interbreeding would thus clearly be an impossibility. However, if the process of human diversification alluded to in the previous paragraph were to go forward, then it is highly probable that several hundred thousand years from now, interstellar space in this region of the galaxy will be populated by numerous human-descended intelligent species which will differ from each other in appearance, emotional makeup, and other characteristics to a considerably greater extent than the cosmopolitan species that populate the *Star Trek* universe. Local stylistic fads could well create races with green skin or pointed ears; more serious considerations such as gravity differences might drive the development of outlandishly tall-thin (low gravity) or short-powerful (high gravity) varieties. Many of them, as a result of their self-directed evolution, will be far more intelligent, sensitive, healthy, long-lived, athletic, graceful, and (to themselves, anyway) beautiful than we.

But they will all be human.

HUMANITY AND COSMIC EVOLUTION

Before the evolution of life . . . the portals of the future remain wide open.

—HENRI BERGSON

The universe is evolving. In the beginning, there was nothing but energy, which condensed to form electrons and protons, which then formed hydrogen atoms and molecules, which then separated into giant rotating clouds of gas that formed the basis for the galaxies. Within these galaxies, the gas condensed further into first-generation stars, which then proceeded to produce heavier elements through the process of hydrogen nuclear fusion. The larger of these stars burned out quickly and exploded, spreading their elemental products throughout the galaxy. With the availability of these heavier components, such as oxygen, carbon, nitrogen, iron, and silicon, second-generation stars could form, and with them families of solid planets capable of hosting the development of life. On our planet at least, and probably many, many others, life then evolved, complexified itself, and developed intelligence sufficient to understand physical law and expand through space.

In relation to the above-described processes, and to its future, the universe is still young. According to the generally accepted Big Bang hypothesis, the universe itself is only 12 billion years old; it is expected to last at least several trillion years into the future. By comparison, our Milky Way galaxy is 10 billion years old, our Sun is 4.6 billion years old, and life on Earth is 3.8 billion years old. We've arrived on the scene just recently, but the show had been going on only a short time before us—over 99 percent is yet to go.

In the history of the universe to date, all of its creations—hydrogen gas, galaxies, first-generation stars, heavy elements, planets, and life—have each played a role in determining and enabling the subsequent phase of development. Gas created galaxies, galaxies created first-generation stars, those stars created the heavy elements, the heavy elements created planets, the planets created life, and life, ever advancing, created mind.

Mind has yet to play its hand. Except for the outside possibility that intelligent life may be responsible for the development of bacteria for use in propagation of life through space, there is no evidence of extraterrestrial intelligence playing any role in the creation or development of our biosphere, or anything else for that matter.

This fact, that Mind has not yet, at least in any large or apparent way, created something else fundamentally new in the universe, or substantially affected its development, suggests to me that Mind is still immature—that intelligence, while no doubt existing in innumerable locations, has just begun to extend itself outward to link itself together on a cosmic scale. We have not yet met ET and ET has not yet met us. Could we be the only people out of the loop? Not likely. Rather, it seems to me more probable that the galactic club has yet to get itself organized and everybody connected. And the intergalactic club has even further to go. Life's children are newborns, awakening in the early dawn of the first day of the cosmic spring. It will be a while before we all meet.

Evolutionary theory tells us that living things evolve and behave in the way that they do in order to survive and perpetuate their species. That is undoubtedly true; yet there is a playfulness to Life, in the delightfully sporting way that it goes about achieving this very serious purpose.

Creativity is joy, and Life and Mind both love to play. Life takes over planets to further life's development, and to open new theaters for evolution's frolic of innovation. Mind can similarly affect planets and, as we have seen, probably stars too. It can dream of far more. But can it do more? Will Mind someday be able to influence developments on a cosmic scale?

Type I civilizations have control over the resources of their planet, Type II of their solar system, and Type III of their galaxy. Can there be something greater, a Type IV civilization, one that has mastery over the development of its universe?

If it is to have any hope for a say in such matters, Mind will need to become connected on a galactic and intergalactic basis. Within the scope of physics as we know it, there are limits to how effectively this can be done. The speed of light mandates that communication from the center to the edge of our own galaxy cannot be accomplished in less than 50,000 years. That's not too bad—50,000 years is a blink on the cosmic time scale. It's less than some reasonable estimates for *L,* the average lifetime for technological civilization. Conceivably, if a galactic communication exchange were set up, that very fact might extend the *L* of the participants. But signaling to our nearest galactic neighbor, Andromeda, would take a million years, and to our farthest galactic relatives, over 10 billion.

But perhaps, if the knowledge and insights of millions of Type III civilizations within our galaxy can be put together, something new will emerge. Perhaps, if so many pieces of the puzzle can be joined, a deeper understanding of nature and its laws can be achieved, and new powers over nature thereby attained.

We really don't know a lot about physics today. We know how a variety of phenomena work, but we don't know why. We don't know why matter exists or has mass, or why mass has inertia, or bends spacetime to exert gravity. We don't know why all mass is positive (as opposed to negative or imaginary) or if indeed it is. We don't know why charge exists, or why the charges of the fundamental particles are what they are, or why like charges repel but unlike charges attract, and why the magnitude of these forces have the particular values that they do. We don't know why mass-energy or charge is conserved, or if it really is under all circumstances. We don't know why fundamental particles with a given self-repelling charge don't blow themselves to pieces. We know that the ratio of electric and magnetic forces determines the speed of light, but since we don't know why these forces have the ratio they do, we really don't know why light travels at the speed it does. We don't really know what space, or time, or spacetime, is, or where it comes from, or why it is continuous. We don't know why there are four, and only four, fundamental forces of nature, or if indeed there are only four. We don't know what caused the Big Bang, or time, or causality for that matter. We don't know why time runs forward but not backward or sideways. We don't know if our universe is unique, or if millions or trillions, or even infinite numbers also exist. By definition, if something is in another universe, it is not in ours, and we cannot interact with it. But was this always so, or must it always be so? We don't know why the laws of the universe follow the geometric relations they do, or why they should have any relation to geometry at all. We don't know why the fundamental constants governing the magnitude of forces in our universe are constant, or if they are. These are only a few examples of our ignorance. The amount we have yet to learn is immense. We can use all the help we can get.

If answers to any of the above questions were found, the technological value could be extraordinary. To take just one example, if we could learn how to find or make negative matter (not antimatter, but matter with negative mass), we could use it to negate the inertia of ordinary matter and achieve starflight at the speed of light. But perhaps we could go even further.

The universe and its physical laws did not always exist. What were the processes that brought them into being? Can humans or other intelligent species duplicate those processes, perhaps altering them to advantage? Today humanity is in the position of someone learning how to play chess on a single chess set. But if the accumulated knowledge of the galaxy were combined, could we learn to build new chess sets? Could we learn how to write new rules? Could we change the laws of the universe? Or design new universes, with better laws?

Penn State University physicist Lee Smolin has written a very interesting book, entitled *The Life of the Cosmos,* in which he speculates on why the laws of the universe appear to be so finely tuned to favor the existence of life.[2] Indeed, it can be readily shown that if any one of many crucial physical constants were adjusted a bit higher or lower, not only intelligence and life but even stars would be impossible. Smolin suggests that universes are born within black holes, which are formed from stars, and that the universes born within black holes have laws that are similar, but not identical, to those of their parent stars. According to Smolin, if the daughter universes have better laws for producing stars (and thus black holes), they will multiply faster than those that produce few or no stars. Therefore, by a kind of natural selection, universes favoring stars, and thus life, would come to predominate, and our own life-friendly universe, far from being an unlikely anomaly, would be the odds-on-favorite.

Smolin's theory can be regarded as very speculative, on any number of grounds. For example, it is unclear whether there are any other universes, and there seems little basis to either support or refute his notion that the laws of nature within a black hole daughter universe should be "just a little bit different" from those of the parent universe. But let us speculate even further. Let's say that intelligence has evolved within a given universe, and it has linked itself together, and grown in potency and knowledge to the point where it understands the laws of creation and is capable of using them. Might it not then seek to create new universes, even more favorable for the development of life and mind? What if our universe is one of those? And if so, as the beneficiaries of such work, could we not participate in bringing the process forward another step? Could it be that it is not stars, but intelligence, that is responsible for the propagation and self-perfection of universes?

These are wild speculations. I indulge in them only because the true ultimate potential of a fully interlinked Type III galactic civilization is unknowable at this time. It may be that compared to any one of its component members, the whole may be something entirely new. The body of an intelligent organism is not just an assembly of cells; it is a phenomenon of a higher order altogether. Mind may be immanent in the cell, but it is certainly not predictable. We may be on a path to something truly astounding.

Ideas like unified intelligence of the galaxy can sound pretty scary, as they can be suggestive of some kind of loss of individuality. Actually, nothing could be further from the truth. Establishment of communication within a society is what empowers individuals; only through communication can a single mind affect society at large. A mind whose thoughts are unheard or unread is doomed to oblivion. Similarly, if the human race, in its finite span, is to make a contribution that goes beyond itself, it must link up in communication with others. Should we do that, then the thoughts of a single individual may one day be able to reach an entire galaxy.

And should we wish for more than a finite span, then we must have progeny.

In any case, if intelligence is going to have a role in cosmic evolution, I think that humanity should be part of it. But to do so, we will have to learn much, and not only about physics, but about purpose.

I suggest that we go forth and find out.

CONCLUSION

The human race first became technological (i.e., truly human in its relationship with nature) when it left Africa to take on the challenge of the north. Later, as humans became maritime, it was the stars—with poetic truth, the North Star—that gave us the guidance we needed to become a truly global species.

Today the stars beckon again, but this time not to new continents, but new solar systems. Multitudes of new worlds yet unknown await, filled with menaces to be faced, challenges to be overcome, wonders to be discovered, and history to be made. The first chapter of the human saga has been written, but vast volumes lying out among the stars are still blank, ready for the pens of new peoples with new thoughts, new tongues, astonishing and beautiful creations, and epic deeds.

The tree of life has many branches. No branch lasts long as a single twig. Rather, each limb of the tree grows for a while and then forks out into multiple directions, with most of the resulting twigs terminating in extinction. No individual twig has much chance of lasting long; survival of the line is assured only by multitudes of branchings.

Today the human race is a single twig on the tree of life, a single species on a single planet. Our condition can thus only be described as extremely fragile, endangered by forces of nature currently beyond our control, our own mistakes, and other branches of the wildly blossoming tree itself. Looked at this way, we can then pose the question of the future of humanity on Earth, in the solar system, and in the galaxy from the standpoint of both evolutionary biology and human nature. The conclusion is straightforward: Our choice is to grow, branch, spread, and develop, or stagnate and die.

Yet grow we can, because as fragile as it is, our twig is currently on the edge of the tree, where we can see the light and reach for the sky.

APPENDIX: FOUNDING DECLARATION OF THE MARS SOCIETY

The time has come for humanity to journey to Mars.

We're ready. Though Mars is distant, we are far better prepared today to send humans to Mars than we were to travel to the Moon at the commencement of the space age. Given the will, we could have our first teams on Mars within a decade.

The reasons for going to Mars are powerful.

We must go for the knowledge of Mars. Our robotic probes have revealed that Mars was once a warm and wet planet, suitable for hosting life's origin. But did it? A search for fossils on the Martian surface or microbes in groundwater below could provide the answer. If found, they would show that the origin of life is not unique to the Earth, and, by implication, reveal a universe that is filled with life and probably intelligence as well. From the point of view of learning our true place in the universe, this would be the most important scientific enlightenment since Copernicus.

We must go for the knowledge of Earth. As we begin the twenty-first century, we have evidence that we are changing the Earth's atmosphere and environment in significant ways. It has become a critical matter for us better to understand all aspects of our environment. In this project, comparative planetology is a very powerful tool, a fact already shown by the role Venusian atmospheric studies played in our discovery of the potential threat of global warming by greenhouse gases. Mars, the planet most like Earth, will have even more to teach us about our home world. The knowledge we gain could be key to our survival.

We must go for the challenge. Civilizations, like people, thrive on challenge and decay without it. The time is past for human societies to use war as a driving stress for technological progress. As the world moves towards unity, we must join together, not in mutual passivity, but in common enterprise, facing outward to embrace a greater and nobler challenge than that which we previously posed to each other. Pioneering Mars will provide such a challenge. Furthermore, a cooperative international exploration of Mars would serve as an example of how the same joint action could work on Earth in other ventures.

We must go for the youth. The spirit of youth demands adventure. A humans-to-Mars program would challenge young people everywhere to develop their minds to participate in the pioneering of a new world. If a Mars program were to inspire just a single extra percent of today's youth to scientific educations, the net result would be tens of millions more scientists, engineers, inventors, medical researchers, and doctors. These people will make innovations that create new industries, find new medical cures, increase income, and benefit the world in innumerable ways to provide a return that will utterly dwarf the expenditures of the Mars program.

We must go for the opportunity. The settling of the Martian New World is an opportunity for a noble experiment in which humanity has another chance to shed old baggage and begin the world anew, carrying forward as much of the best of our heritage as possible and leaving the worst behind. Such chances do not come often and are not to be disdained lightly.

We must go for our humanity. Human beings are more than merely another kind of animal—we are life's messenger. Alone of the creatures of the Earth, we have the ability to continue the work of creation by bringing life to Mars, and Mars to life. In doing so, we shall make a profound statement as to the precious worth of the human race and every member of it.

We must go for the future. Mars is not just a scientific curiosity; it is a world with a surface area equal to all the continents of Earth combined, possessing all the elements that are needed to support not only life, but technological society. It is a New World, filled with history waiting to be made by a new and youthful branch of human civilization that is waiting to be born. We must go to Mars to make that potential a reality. We must go, not for us, but for a people who are yet to be. We must do it for the Martians.

Believing therefore that the exploration and settlement of Mars is one of the greatest human endeavors possible in our time, we have gathered to found this Mars Society, understanding that even the best ideas for human action are never inevitable, but must be planned, advocated, and achieved by hard work. We call upon all other individuals and organizations of like-minded people to join with us in furthering this great enterprise. No nobler cause has ever been. We shall not rest until it succeeds.

The above declaration was signed and ratified by the 700 attendees of the Mars Society Founding Convention, held August 13–16, 1998, in Boulder, Colorado. Further information is available at www.marssociety.org or by writing the Mars Society, P.O. Box 273, Indian Hills, CO 80454.

GLOSSARY

kb/s: Kilobits per second.

keV: Kilo-electron volts.

kHz: Kilohertz, a measure of frequency used in radio; 1 kHz equals 1,000 cycles per second.

km/s: Kilometers per second.

kW: Kilowatts.

kWe: Kilowatts of electricity.

kWe-h: The total amount of energy associated with the use of 1 kilowatt of electricity for 1 hour.

kWh: The total amount of energy associated with the use of 1 kilowatt for 1 hour.

MHz: Megahertz, a measure of frequency used in radio; 1 MHz equals 1,000,000 cycles per second.

m/s: meters per second.

MWe: Megawatts of electricity.

MWt: Megawatts of heat; 1 megawatt equals 1,000 kilowatts.

TW: Terawatt; 1 terawatt equals 1,000,000 megawatts. Human civilization today uses about 14 TW.

TW-year: The total amount of energy associated with the use of 1 terawatt for 1 year.

W/kg: Watts per kilogram.

W rf: Watts of radiated power.

ΔV: See delta-*V*.

aerobraking: A spacecraft maneuver using friction with a planetary atmosphere to decelerate from an interplanetary orbit to one about a planet.

aeroshell: A heat shield used to protect a spacecraft from atmospheric heating during aerobraking.

apogee: The highest point in an orbit about a planet.

atmospheric pressure: The pressure an atmosphere exerts. On Earth at sea level, the atmospheric pressure is 14.7 pounds per square inch. This amount of pressure is therefore known as one "atmosphere" or one "bar."

bipropellant: A rocket propellant combination including both a fuel and an oxidizer. Examples include methane/oxygen, hydrogen/oxygen, kerosene/hydrogen peroxide, etc.

buffer gas: An effectively inert gas that is used to dilute the oxygen required to support breathing or combustion. On Earth, the 80 percent nitrogen found in air serves as a buffer gas.

cosmic ray: A particle, such as an atomic nucleus, traveling through space at very high velocity. Cosmic rays originate outside of our solar system. They typically have energies of billions of volts and require meters of solid shielding to stop.

cryogenic: Ultra cold. Liquid oxygen and hydrogen are both cryogenic fluids as they require temperatures of -180°C and -250°C, respectively, for storage.

DC-X: An experimental Reusable Launch Vehicle build by the Strategic Defense Initiative Organization.

Delta 2: An expendable launch vehicle manufactured by McDonnell Douglas, capable of throwing 1,000 kg on a direct trajectory from Earth to Mars.

delta-*V*: The velocity change required to move a spacecraft from one orbit to another. A typical delta-*V* (also written ΔV) required to go from low Earth orbit to a trans-Mars trajectory would be about 4 km/s.

departure velocity: The velocity of a spacecraft relative to a planet after effectively leaving the planet's gravitational field. Also known as **hyperbolic velocity.**

direct entry: A maneuver in which a spacecraft enters a planet's atmosphere and uses it to decelerate and land without going into orbit.

direct launch: A maneuver in which a spacecraft is launched directly from one planet to another without being assembled in orbit.

electrolysis: The use of electricity to split a chemical compound into its elemental components. Electrolysis of water splits it into hydrogen and oxygen.

electron density: The number of electrons per cubic centimeter. The higher the electron density of an ionosphere, the better it reflects radio waves.

endothermic: A chemical reaction requiring the addition of energy to occur.

equilibrium constant: A number that characterizes the degree to which a chemical reaction will proceed to completion. A very high equilibrium constant implies near complete reaction.

ERV: Earth Return Vehicle.

ET: External tank.

EVA: Extravehicular activity.

exhaust velocity: The speed of the gases emitted from a rocket nozzle.

exothermic: A chemical reaction that releases energy when it occurs.

fairing: The protective streamlined shell containing a payload that sits on top of a launch vehicle.

free return trajectory: A trajectory which, after departing Earth, will eventually return to the Earth without any additional propulsive maneuvers.

GCMS: Gas chromatograph mass spectrometer.

geothermal energy: Energy produced by using naturally hot underground materials to heat a fluid, which can then be expanded in a turbine generator to produce electricity.

gravity assist: A maneuver in which a spacecraft flying by a planet uses that planet's gravity to create a slingshot effect that adds to the spacecraft's velocity without any requirement for the use of rocket propellant.

heliocentric: Centered about the Sun. By a heliocentric orbit what is meant is an orbit that transverses interplanetary space and is not bound to the Earth or any other planet.

Hohmann transfer orbit: An elliptical orbit one of whose ends is tangent to the orbit of the planet of departure and whose other end is tangent to the orbit of the planet of destination. The Hohmann transfer orbit is the purest incarnation of the conjunction-class orbit and, as such, is the lowest energy path from one planet to another.

hydrazine: A rocket propellant whose formula is N_2H_4. Hydrazine is a monopropellant, which means that it can release energy by decomposing, without any additional oxidizer required for combustion.

hyperbolic velocity: The velocity of a spacecraft relative to a planet before entering, or after effectively leaving, the planet's gravitational field. Also known as approach velocity or **departure velocity.**

hypersonic: A speed many times the speed of sound; in common usage, Mach 5 or greater.

ionosphere: The upper layer of a planet's atmosphere in which a significant fraction of the gas atoms have split into free positively charged ions and negatively charged electrons. Because of the presence of freely moving charged particles, an ionosphere can reflect radio waves.

Isp: A commonly used abbreviation for **specific impulse.**

ISPP: In situ propellant production.

JSC: Johnson Space Center.

JPL: Jet Propulsion Laboratory.

Kelvin degrees: The Kelvin or "absolute" scale is a method of measuring temperature that starts with its zero point set at "absolute zero," the temperature at which a body in fact possesses no heat. 273 degrees Kelvin is the same temperature as 0 degrees Celsius, the freezing point of water. Each additional degree Kelvin corresponds to one additional degree Celsius.

LEO: Low Earth orbit.

LOR: Lunar Orbit Rendezvous.

LOX: Liquid oxygen.

magnetic sail: A device for propelling spacecraft using the pushing force of plasma on a magnetic field.

magsail: A **magnetic sail.**

MAV: Mars Ascent Vehicle.

methanation reaction: A chemical reaction forming methane. In the Mars Direct mission, the methanation reaction is the Sabatier reaction in which hydrogen is combined with carbon dioxide to produce methane and water.

millirem: 1/1,000th of a **rem.**

minimum energy trajectory: The trajectory between two planets requiring the least amount of rocket propellant to attain (see **Hohmann transfer orbit**).

MOR: Mars Orbit Rendezvous.

MSR-ISPP: Mars sample return using in situ propellant production.

NEP: Nuclear electric propulsion.

NIMF: Nuclear rocket using indigenous Martian fuel.

NTR: Nuclear thermal rocket.

perigee: The lowest point in an orbit around a planet.

pyrolyze: The use of heat to split a compound into its elemental constituents.

regolith: What most commonly refer to as dirt.

rem: The measure of radiation dose most commonly used in the United States. 100 rem equals 1 Sievert, the European unit. It is estimated that radiation doses of about 60 or 80 rem are sufficient to increase a person's probability of fatal cancer at some time later in life by 1 percent. Typical background radiation on Earth is about 0.2 rem/year.

RLV: Reusable launch vehicle.

RTG: Radioisotope thermoelectric generator.

RWGS: reverse water gas shift reaction.

Sabatier reaction: A reaction in which hydrogen and carbon dioxide are combined to produce methane and water. The Sabatier reaction is exothermic, with a high **equilibrium constant.**

Saturn V: The heavy-lift launch vehicle used to send the Apollo astronauts to the Moon. The Saturn V could lift about 140 tonnes to LEO.

SEI: Space Exploration Initiative.

SNC meteorites: Named for the locations where the first three were found (Shergotty, Nakhla, and Chassigny), SNC meteorites are believed on the basis of very strong chemical, geologic, and isotopic evidence to be debris thrown off of Mars by impacting meteorites.

Sol: One Martian day.

solar flare: A sudden eruption on the surface of the Sun that can deliver immense amounts of radiation across vast stretches of space.

solar sail: A device for propelling a spacecraft by utilizing the pushing force of sunlight.

specific impulse: The specific impulse of a rocket engine is the number of seconds it can make a pound of propellant deliver a pound of thrust. If you multiply the specific impulse of a rocket engine, given in seconds, by 9.8, you will obtain the engine's exhaust velocity in units of meter/second. Specific impulse is generally viewed as the most important factor in judging a rocket engine's performance. Frequently abbreviated "Isp."

SRB: Solid rocket booster.

SSME: Space shuttle main engine.

SSTO: Single stage to orbit.

stable equilibrium: An equilibrium condition that, if displaced by some external force, will return on its own to its original state. A ball on the flat surface on top of a hill is in unstable equilibrium, because if pushed it will roll away, accelerating itself from its original position. A ball on a flat surface at the bottom of a bowl is in stable equilibrium, because if pushed it will roll back to its starting point.

STR: Solar thermal rocket.

telerobotic operation: Remote control of some device, such as a small Mars rover equipped with TV cameras, by human operators at a significant distance away.

thrust: The amount of force a rocket engine can exert to accelerate a spacecraft.

Titan IV: An expendable launch vehicle manufactured by the Lockheed Martin Corporation capable of delivering 20,000 kg to LEO or 5,000 kg to a minimum energy trans-Mars trajectory.

TMI: Trans-Mars injection, a maneuver that places a payload or spacecraft on a trajectory to Mars.

unstable equilibrium: See **stable equilibrium.**

vapor pressure: The pressure exerted by the gas emitted by a substance at a certain temperature. At 100°C, the vapor pressure of water is greater than Earth's atmospheric pressure and so it will boil.

REFERENCES

Chapter 1: On the Threshold of the Universe

1. Christopher Stringer and Robin McKie, *African Exodus: The Origins of Modern Humanity* (New York: Henry Holt, 1997).
2. James Shreve, *The Neanderthal Enigma: Solving the Mystery of Modern Human Origins* (New York: Avon Books, 1995).
3. William McNeill, *The Rise of the West* (New York: Mentor Books, 1965).
4. Frederick Jackson Turner, *The Frontier in American History* (New York: H. Holt & Co., 1920).
5. David Ehrenstein, "Immortality Gene Discovered," *Science,* January 9, 1998, p. 177.
6. Carroll Quigley, *The Evolution of Civilizations* (Indianapolis: Liberty Fund, 1961).
7. Francis Fukuyama, *The End of History* (New York: Free Press, 1992).
8. James Horgan, *The End of Science* (New York: Broadway Books, 1997).
9. Daniel Boorstin, *The Discoverers* (New York: Random House, 1983).

Chapter 2: The Age of Dinosaurs

1. S. Isakowitz, *Space Launch Systems* (Washington, D.C.: American Institute of Aeronautics and Astronautics, 1995).
2. Hans Mark, *The Space Station: A Personal Journey* (Durham, NC: Duke University Press, 1987).
3. G. Harry Stine, *Halfway to Anywhere* (New York: Evans and Company, 1996).

Chapter 3: The New Space Race

1. Ben Iannotta, "Rocket Revolutionary," *New Scientist,* no. 2145, August 1, 1998.
2. R. Zubrin and M. Clapp, "An Examination of the Feasibility of Winged SSTO Vehicles Utilizing Aerial Propellant Transfer," AIAA 94-2923 30th AIAA/ASME Joint Propulsion Conference, June 27–29, 1994, Indianapolis, IN.
3. F. W. Smith, "Practical Applications of Hypersonic Flight: Possibilities for Air Express" (Columbus, OH: Battelle Institute, 1986).

Chapter 4: Doing Business on Orbit

1. Charles D. O'Dale, "Establishing an Infrastructure for Commercial Space," *JBIS,* February 1997, pp. 43–50.
2. Diana Gallagher, "A Reconsideration of Anatomical Docking Maneuvers in a Zero-Gravity Environment," Firebird Arts and Music, Portland, OR.
3. P. Collins, R. Stockmans, and M. Maita, "Demand for Space Tourism in America and Japan, and Its Implications for Future Space Activities," Japanese National Space Laboratory, 1995.
4. B. Sherwood and C. R. Fowler, "Feasibility of Commercial Resort Hotels in Low Earth Orbit," The Boeing Company, 1991.
5. Gerard O'Neill, *The High Frontier* (New York: Bantam Books, 1978).

Chapter 5: The View from the Moon

1. W. Mendell, "Lunar Bases and Space Activities of the 21st Century," Lunar and Planetary Institute, Houston, TX, 1995.
2. L. Haskin and P. Warren, "Lunar Chemistry," Chapter 8 of *Lunar Sourcebook,* ed. G. Heiken, D. Vaniman, and B. French (Cambridge: Cambridge University Press, 1991).
3. J. Pearson, "The Orbital Tower: A Spacecraft Launcher Using the Earth's Rotational Energy," *Acta Astronautica* 2 (1975), p. 785.

Chapter 6: Mars: The New World

1. M. Carr, *The Surface of Mars* (New Haven: Yale University Press, 1981).
2. R. Zubrin with Richard Wagner, *The Case for Mars: The Plan to Settle the Red Planet and Why We Must* (New York: Free Press, 1996; Simon and Schuster, 1997).
3. C. Stoker and C. Emmart, *Strategies for Mars* (San Diego: Univelt, 1996).
4. R. Zubrin, *From Imagination to Reality: Mars Exploration Studies of the Journal of the British Interplanetary Society* (San Diego: Univelt, 1997).

Chapter 7: Asteroids for Good and Evil

1. D. Cox and J. Chestek, *Doomsday Asteroid: Can We Survive?* (Amherst, NY: Prometheus Books, 1996).
2. W. Alvarez, *T. Rex and the Crater of Doom* (Princeton, NJ: Princeton University Press, 1997).
3. C. Sagan, *Pale Blue Dot* (New York: Random House, 1994).
4. J. Lewis and R. Lewis, *Space Resources: Breaking the Bonds of Earth* (New York: Columbia University Press, 1987).

Chapter 8: Settling the Outer Solar System

1. J. Beatty and A. Chaikin, eds., *The New Solar System* (Cambridge, MA: Sky Publishing Corporation, 1990).
2. W. Burrows, *Exploring Space: Voyages in the Solar System and Beyond* (New York: Random House, 1990).
3. R. Terra, "Islands in the Sky: Human Exploration and Settlement of the Oort Cloud," chapter 6 of *Islands in the Sky,* ed. S. Schmidt and R. Zubrin (New York: Wiley, 1995).
4. B. Finney and E. Jones, eds., *Interstellar Migration and the Human Experience* (Berkeley and Los Angeles: University of California Press, 1985).

Chapter 9: The Challenge of Interstellar Travel

1. R. G. Ragsdale, "To Mars in 30 Days by Gas Core Nuclear Rockets," *Astronautics and Aeronautics* 65, January 1972.
2. R. G. Ragsdale, "High Specific Impulse Gas Core Reactors," NASA TM X-2243, NASA Lewis Research Center, March 1971.
3. T. Latham and C. Joyner, "Summary of Nuclear Light Bulb Development Status," AIAA 91-3512, AIAA/NASA/OAI Conference on Advanced SEI Technologies, Cleveland, OH, September 1991.
4. A. Martin and A. Bond, "Nuclear Pulse Propulsion: A Historical Review of an Advanced Propulsion Concept," *Journal of the British Interplanetary Society* 32 (1979), pp. 283–310.
5. R. Zubrin, "Nuclear Salt Water Rockets: High Thrust at 10,000 sec Isp," *Journal of the British Interplanetary Society* (1991), pp. 371–376.
6. S. K. Borowski, "A Comparison of Fusion/Antiproton Propulsion Systems for Interplanetary Travel," AIAA-87-1814, 23rd AIAA/ASME Joint Propulsion Conference, San Diego, CA, June 29–July 2, 1987.
7. R. Forward and Joel Davis, *Mirror Matter: Pioneering Antimatter Physics* (New York: Wiley Science Editions, 1988).
8. L. Friedman, *Starsailing: Solar Sails and Interstellar Travel* (New York: John Wiley and Sons, 1988).
9. R. Forward, "Roundtrip Interstellar Travel Using Laser Pushed Lightsails," *Journal of Spacecraft and Rockets* (1984), pp. 187–195.
10. R. Bussard, "Galactic Matter and Interstellar Flight," *Acta Astronautica* (1960), pp. 179–196.

11. R. Zubrin and D. Andrews, "Magnetic Sails and Interplanetary Travel," AIAA 89-2441, AIAA/ASME Joint Propulsion Conference, Monterey, CA, July 10–12, 1989. Reprinted in *Journal of Rockets and Spaceflight,* April 1991.

12. R. Zubrin, "The Magnetic Sail," chapter 12 of *Islands in the Sky,* ed. S. Schmidt and R. Zubrin (New York: Wiley, 1995).

13. D. Andrews and R. Zubrin, "MagOrion," AIAA/ASME Joint Propulsion Conference, Seattle, WA, July 7–9, 1997.

14. Richard A. Muller, *Nemesis: The Death Star* (New York: Weidenfeld and Nicolson, 1988).

15. R. Zubrin, "The Use of Currently Unknown Near-Solar Stellar Objects in Facilitating Interstellar Missions," *Journal of the British Interplanetary Society* (1995), pp. 467-474.

16. Eugene Mallove and Gregory Matloff, *The Starflight Handbook* (New York: Wiley Science Editions, 1989).

Chapter 10: Extraordinary Engineering

1. James Oberg, *New Earths* (Harrisburg, PA: Stackpole Books, 1981).

2. Martyn Fogg, *Terraforming: Engineering Planetary Environments,* Society of Automotive Engineers, Warrendale, PA, 1995.

3. J. Pollack and C. Sagan, "Planetary Engineering," in *Resources of Near Earth Space,* ed. J. Lewis, M. Matthews, and M. Guerreri (Tucson: University of Arizona Press, 1993), pp. 921–950.

4. C. Sagan, "The Planet Venus," *Science* (1961), pp. 849–858.

5. J. Kasting, "Runaway and Moist Greenhouse Atmospheres and the Evolution of Earth and Venus," *Icarus* (1988), pp. 472–494.

6. M. Fogg, "A Planet Dweller's Dreams," in *Islands in the Sky,* ed. S. Schmidt and R. Zubrin (New York: Wiley, 1995).

7. R. Forward, "The Statite: A Non-Orbiting Spacecraft," AIAA 89-2546, AIAA/ASME 25th Joint Propulsion Conference, Monterey, CA, July 1989.

8. K. Eric Drexler, *Engines of Creation* (New York: Anchor Books/Doubleday, 1987).

9. Larry Niven, *Ringworld* (New York: Ballantine, 1990).

10. Stephen Hawking, "Gravitationally Collapsed Objects of Very Low Mass," *Monthly Notes of the Royal Astronomical Society* (1971), pp. 75–78.

11. Martyn Fogg, "Terraforming: Engineering Planetary Environments" (Warrendale, PA: Society of Automotive Engineers, 1995).

12. Olaf Stapledon, *The Starmaker,* 1937. Republished by Dover, New York, 1968.

Chapter 11: Meeting ET

1. I. S. Shklovskii and Carl Sagan, *Intelligent Life in the Universe* (New York: Delta Books, 1966).

2. Frank Drake, personal communication, 1997.

3. Richard Dawkins, *The Extended Phenotype* (New York: Oxford University Press, 1982).

4. Seth Shostak, *Sharing the Universe* (Berkeley: Berkeley Hills Books, 1998).

5. R. Zubrin, "Detection of Extraterrestrial Civilizations via the Spectral Signature of Advanced Interstellar Spacecraft," in *Progress in the Search for Extraterrestrial Life: Proceedings of the 1993 Bioastronomy Symposium, Santa Cruz, CA, August 16–20, 1993,* ed. Seth Shostak, Astronomical Society of the Pacific Conference Series (San Francisco: Astronomical Society of the Pacific, 1996), vol. 74, pp. 487–496.

6. R. Zubrin and D. Andrews, "Magnetic Sails and Interplanetary Travel," AIAA 89-2441, AIAA/ASME Joint Propulsion Conference, Monterey, CA, July 10–12, 1989. Republished in *Journal of Spacecraft and Rockets,* April 1991.

Chapter 12: North to the Stars

1. H. G. Wells, "The Discovery of the Future," *Nature* 65 (1950), pp. 326–331.

2. Lee Smolin, *The Life of the Cosmos* (New York: Oxford University Press, 1997).

INDEX

Page numbers in *italics* refer to tables, figures, and illustrations

Ackerman, James, 56
Adonis, 134
Advent CAC-1 Passenger Rocket, 56
Advent Launch Services, 54, 56
aerial refueling, 45–46, 47, 53
AeroAstro, 56
aerobraking technology, 181
Aerojet, 153
aerospace corporations, 24, 33–35, 54
aerospike rocket engine, 32, 44
Africa, xiii–xiv, 4–5
aging, 8, 277
agriculture:
  on Mars, 101, 105–6
  on Moon, 81–82
  on Titan, 166
Air Force, U.S., 23, 25, 27–28, 45, 46, 53, 180
Alan Hills (ALH) 84001, 112
Alpha Centauri, 187, 188, 190, 191, 194, 201, 204, 209, 258
Alvarez, Luis and Walter, 130–31, 151
American Institute of Aeronautics and Astronautics (AIAA), 44
Anderson, Walt, 44
Andrews, Dana, 205–6, 265
antimatter, 208
  propulsion, 197–201, *199*, 263–64, *267*, 268
Apollo, 134
Apollo program, xv, 9, 10, 11, 14, 25, 30, 42, 102, 143, 174
"Apollo 25 Years Later" (Sidey), 11
Arctic, 121
Ares rocket, 141
Ariane rocket, 25, 142
Arizona, 129–30
Arizona, University of, 146, 154, 156
Arrhenius, Svante, 250
art, 5
artificial gravity research, 142
Artsutanov, Y. N., 98–99, 100
Ash, Robert, 154
Asia, xiii, 5
asteroids, xvi, 107, 108, 128–56, 158, 242
  brightness of, 170
  colonization of, 150, 180–81, 184
  composition of, 136, 139–40, *139*, 146, 156, 170
  deflection of, 134–39
  discovery of, 133–34
  exploration of, 139–41
  Gaiashield mission and, 141–46, *143*, *144*, *145*
  iceteroids, 170–72, 182–84, 207–8, 234
  impact energy of, 128–29, 130, 131, 134
  Life and, 150–52
  Main Belt of, 133–34, 147, 170, 171, 172, 181, 184
  mass extinctions caused by, 16, 17, 39, 130–32, 142, 151–52, 209, 249–50, 256
  mining of, 146–50, 156, 171
  near-Earth, *132*, 134–35, 139–40, 141, 147
  number of, 133, 134, 147
  panspermia role of, 251, 252
  past major impacts of, 128–30, *130*
  relocating of, 171–72, *171*
  rocket propellant made from ice on, 137–38
  size of, 133, 134, 140, 146
astronomical observatory base, Moon as, 96–98
Astronomical Unit (AU), definition of, 133, 187
astronomy, 133
  early solar system discoveries in, 173
  legacy of, 98

Athena, 115–16, 117, 120
Atlas rocket, 22, 23, 24, 25, 26, 27, 33
atomic (fission) bombs, 195
for fusion propulsion, 191–94, *192, 193*
Atomic Energy Commission, U.S., 192
Augustine, Norm, 32, 33
Australia, 150, 259
aviation prizes, 55–57

Babbage, Charles, 238, 239
backpack gas thrusters, 143
bacteria, 239–40, 250–53
Ballistic Missile Defense Organization (BMDO), 31
Barnard's Star, 232
Beal Aerospace, 54
beanstalks:
lunar, 98–100
on Mars, 108
Beggs, James, 28
bell-nozzle rocket engine, 32
Benford, Greg, 120
Bennett, Gary, 176
Bergson, Henri, 278
BETA program, 259
Big Bang, 250, 279, 281
Big Ear Program, 259
Bill of Rights, 124–25
Binder, Alan, 93
bioengineering, 240
biological evolution, 257–58
Black Colt, 47–49, 53
black dwarfs, 210, 212
black holes, 210, 212–13, 243–46, 281–82
Black Horse, 45–47, 48, 50, 53
Boeing, 34, 35, 45, 50, 205
Boorstin, Daniel, 20
bremsstrahlung radiation, 263–64
Brown, George, 29
brown dwarfs, 210, 212, 243, 245
Bruckner, Adam, 155
Bruno, Giordano, 247
"Building America's Bridges to the 21st Century," 180
Bush, George, 7, 8, 104
business parks, in space, 64–65
Bussard, Robert, 204, 205

Callisto, 167, 168, 170, 173, 231
Cape Canaveral, 102, 103, 176
carbon dioxide, 108–9, 132, 138, 151, 225, 227, 228–31, 252
carbon-nitrogen-oxygen (CNO) catalytic fusion cycle, 205, 243
Carbotek Corporation, 95, 153
carbothermal reduction, 153
Case for Mars, The (Zubrin and Wagner), xvii, 10, 102, 108, 119, 120, 138, 149, 227, 230, 233
Cassini, Giovanni, 110, 158, 173
Cassini probe, 144, 163, 176, 179
catalyzed D-D fusion, 195
catapult, electromagnetic, 108
Celestri, 40
cell phones, 39
Ceres, 133, 148, 149
CERN, 198
Challenger, 27–28
chemical propulsion, 189–90
chemosynthesis, 175
Cheng Ho, 18–19, 20
Chicxulub, 131
China, Imperial, 6, 15
global exploration by, xiv–xv, 18–20
chipmunks, "natural habitat" expanded by, 224–25
Chiron, 170, 173
chlorofluorocarbon gases, 109, 227
Christendom, medieval, 15, 18
Chu, Ching Wu (Paul), 15, 206
Church, 173
City College of New York, 176
Civil Space Transport Study (CSTS), 50
Clancy, Tom, 44
Clapp, Mitchell Burnside, 45, 46, 49
Clarke, Arthur C., 11, 71, 99, 202
Clementine, 92–93, 94
Clementine II, 92
Clinton, Bill, 116
Clinton administration, 32, 180
CNO (carbon-nitrogen-oxygen) catalytic fusion cycle, 205, 243
Cocconi, Giuseppe, 258, 260
Cohen, Aaron, 42
Cold War, 6, 7, 8, 11, 12, 13, 23, 51, 118
Collins, Patrick, 63
Columbus, Christopher, 123, 181, 188, 216
comets, 158, 182–83, 201
composition of, 91
detection and deflection of, 183
lunar water deposits from, 91
origin of, 171
communication, long-distance, 177–79, *179,* 188
communications satellites, 39–40, 43, 54–55, 58, 59, 69, 75–76

Congress, U.S., 27, 104, 134, 259
Constitution, U.S., 124
Coons, Steve, 155
Coriolis forces, 142
cosmic evolution, 278–83
cosmic radiation, 144–45, 146, 166
cosmic rays, 197
cosmology, 250, 281
"cost plus" government contracts, 24
Cousteau, Jacques, 121
Criswell, David, 82–83
critical mass, 195
crystal spheres, in heavens, 146
CSTS (Civil Space Transport Study), 50

Dace, Harry, 56
Dactyl, 140
Dante Alighieri, 1, 185
dark matter, 210, 211–12
Darwin, Sir Charles, 274
data transmission:
  from outer solar system, 177–79, *179*
  to 10 light years, *261*
Dawkins, Richard, 256
DC-X program, 16, 31–32, 33, 40, 41, 43, 45
dead-stick landings, 48, 54
Declaration of Independence, U.S., 124–25
deep gravity wells, 212–15, *214*
Deep Space Network, 188
Deep Space 1, 141
Delta Clipper, 43
Delta rocket, 25, 27, 35, 40, 116
Design Reference Mission, xvii, 104
deuterium, 107–8, 183, 194–95
deuterium-helium-3 (D-He3) fusion, 90, 160, 161, 193, 205, 264
deuterium-tritium (D-T) fusion reactors, 86–87, 88
Diamandis, Peter, 56
dinosaurs, extinction of, 16, 39, 130–32, 249–50
Discovery mission, 141
"Discovery of the Future, The" (Wells), 274
diversity, human, 276–78
Drake, Frank, 248, 259, 260
Drake Equation, 249, 250, 253, 254
Drexler, K. Eric, 238, 239
Dyson, Freeman, 182, 183, 192, 240–41
Dyson spheres, 241

Eannes, Gil, 19
Earth:
  age of life on, 279
  Life's transformation of, 151, 225–26
Earth Return Vehicle (ERV), 102, 103
East Africa, xiii–xiv, 4–5
ECCO/CCI, 40
ecliptic, 182
economy, U.S., space programs and, 10
Eddington limit, 244
Ehricke, Krafft, 79
"eighth continent," Moon as, 79
Einstein, Albert, 197, 198, 201
electricity, satellite generating of, 70–74
Ellipsat, 40
Emerson, Ralph Waldo, 39
Endeavour, 66
End of History, The (Fukuyama), 8
End of Science, The (Horgan), 8–9
energy, Moon as source of, 82–90
Engines of Creation (Drexler), 238
entrepreneurial space initiatives, 74–76
Epsilon Eridani, 259
Eros, 134, 140
ERV (Earth Return Vehicle), 102, 103
Europa, 16, 167, 173, 174–75
Europe, xiii, xv, 5, 7, 15, 104, 118
evolution:
  biological, 257–58
  cosmic, 278–83
Evolution of Civilizations, The (Quigley), 8
exotic physical phenomena, 188
extraterrestrial imperative, 275–76
extraterrestrial life, xvii, 247–73
  detecting starships of, 262–69, *266, 267, 268*
  on Europa, 174–75
  on Mars, xv, 16, 17, 110–13, 117, 248
  panspermia and, 250–53
  population and distribution of, 253–56, *255*
  probability and frequency of, 17, 113, 248–50
  search for, 248, 258–62, *261*
  time between encounters with, *270*
  see also Life
extra-vehicular activities (EVA), 63

Faerie Queene, The (Spenser), 157
fast package delivery, 51–52
Federal Aviation Administration (FAA), 64
Federal Express, 50
Fermi, Enrico, 247, 248
Fermilab, 198
Fermi's Paradox, 248, 258
Feynman, Richard, 56

fission propulsion, 191–94, *192, 193*
Fleeter, Rick, 56
"flying carpet" vehicle, 41, 42
Fogg, Martyn, 244
F-125 jet engine, 47, 49
Forward, Robert, 203, 233
fossils, 116–17, 247–48
Fountains of Paradise, The (Clarke), 99
Frankie, Brian, 154
free-fall, 53
Friedman, Louis, 202
"frontier shock," 6, 7
Fukuyama, Francis, 8, 9, 127, 133
fundamental forces, 9, 281
fusion, proton-proton, 204–5, 264
fusion plasmas, 84–88, *85,* 161
fusion propulsion, xvii, 15, 90, 180, 194–97, *196,* 206, 263, 264
fusion reactors, 80, 84, 107, 183–84

Gaiashield mission, 141–46, *143, 144, 145*
galactic civilization, *see* Type III (galactic) civilization
galaxies, rotation of, 210
Galileo Galilei, 173
Galileo probe, 16, 140, 144, 158, 174–75, 176, 177, 178
Gama, Vasco da, 19–20
gamma-ray spectrometer (GRS), 116
Ganymede, 167, 173
gas chromatograph-mass spectrometer (GCMS), 111
Gaspra, 140
Gates, Bill, 40
Gaubatz, William, 31, 33
Gemini program, 10, 26
General Dynamics, 34, 50
General Electric Astrospace, 34
Genesis, xiv
geostationary beanstalk, 98–100
geosynchronous Earth orbit (GEO), 71, 72, 73, 74, 83, 98
Gibson, Everett, 112
Glenn, John, 26
global civilization, *see* Type I (global) civilization
global passenger service, 52–55
Global Surveyor probe, 16
Goddard, Robert, 11
gold, 61
Goldin, Dan, 28, 104, 115
Goldstone Deep Space Communications Complex, 140
Gore, Al, 29
government space initiatives:
  entrepreneurial initiatives compared with, 74–76
  national and international Mars initiatives, 113–19
gravity assists, 168–70, 171–72, *171,* 174, 187, 189–90, 212
gravity wells, 212–15, *214*
Great Britain, 149–50
greenhouse gases, 109, 226, 227, 228–31, 252
Griffin, Mike, 104, 120
ground-penetrating radar (GPR), 94
GRS (gamma-ray spectrometer), 116
Grumman, 34

Hale-Bopp comet, 182–83
Harrison, John, 56
Haughton Impact Crater, 121
Hawking, Stephen, 244
heavy-lift launch vehicles, 28–29
Hegel, G. W. F., 127, 133
Heinlein, Robert, 31, 38
helium-3, 80, 81, 84, 87, 88, 90, 97, 159–60, 161–63, 166, 170, 180, 181, 183–84
Henry the Navigator, 19
Herschel, Sir William, 110, 133, 173
high-temperature superconductivity, xv, 15, 206
high-vacuum, 59, 60
Hildebrand, Alan, 131
history, "end" of, 8–9, 118, 127
HMX Engineering, 43–44
Hohmann transfer elliptical trajectory, 143
Homo erectus, xiii, 4
Homo sapiens, xiii–xiv, 4–5, 258
Homo technologicus, xiv, 5
Hong Kong, 41
Horgan, James, 8–9
horizontal takeoff/horizontal landing (HTHL) vehicles, 45–49
hotels, orbital, 62–64
House Space Subcommittee, 29, 73
Houston, University of, 206
Houston-Clear Lake, University of, 82
Hoyle, Fred, 250
Hubble Space Telescope, 66, 67, 68, 97, 267
Hudson, Gary, 31, 43–44
human diversity, 276–78
Hunter, Max, 31, 43
Huygens, Christiaan, 173

ICBMs (intercontinental ballistic missiles), 23, 26, 50
Ice Age, xiii, 5

iceteroids, 170–72, 182–84, 207–8, 234
Ida, 140
Inferno (Dante Alighieri), 1
intelligence, 17, 113, 279–80, 282–83
intercontinental ballistic missiles (ICBMs), 23, 26, 50
International Space Station (ISS), 9, 14, 59, 65, 75, 142
  cost of, 68, 116
  design of, 28–29
  servicing of, 67–68
  zero gravity research and, 60–61
Internet, 7, 40, 52, 126
interstellar travel, xvii, 187–223, 254
  antimatter propulsion for, 197–201, *199*
  characteristic power levels for, *266*
  chemical propulsion and, 189–90
  distances involved in, 187, 188
  fission propulsion for, 191–94, *192, 193*
  fusion propulsion for, 194–97, *196,* 206
  light sails, 201–4, *203*
  magnetic sails for, 15, 204–7, 263, 264–65, 267–69, *268*
  obstacles to, 187–89
  species maturity and, 215–16
  unknown stars in, 209–15, *211, 212*
  using Oort Cloud objects, 207–9
Io, 173, 174
iridium (element), 39, 130–31
Iridium (satellites), 39–40, 49
Iridium 2, 40
ISAS, 141
ISP (specific impulse), definition of, 35

Japan, 63, 104, 141
Jefferson, Thomas, 128
Jet Propulsion Lab (JPL), 16, 91, 114, 116, 140, 141, 154, 174
JET tokamak, 86
Johns Hopkins Applied Physics Laboratory, 140
Johnson Space Center, 42, 104
Journal of the British Interplanetary Society, 213
Juno, 133
Jupiter, 134, 140, 147, 172, 173, 176, 187, 189–90, 212, 241, 243
  departure velocity from, 162–63, 168–70, *169*
  moons of, 16, 166–70, *167,* 173, 174–75
Justice Department, U.S., 34

Kaku, Michio, 176, 177
Kardashev, Nikolai S., xiv, xvi
Kasting, James, 230
Kelly, Bob, 53–54
Kelly Space Technologies, 54
Kennedy, John F., xv, 11–12, 14
Kepler, Johannes, 201
Kevlar, 106
Kistler, Walter, 40–42, 43
Kistler Aerospace, 41–43, 54
Kito, Tomiko, 154
Komsomolskaya Pravda, 98
Kowal, Charles, 173
Kuiper, Gerard, 173
Kuiper Belt, 170, 182–84, 207
Kulcinski, Jerry, 84

L (life span), of civilizations, 253–58, *255*
LaGrange points, 72, 73, 100
"Land People," 7
laser fusion, 195
lasers, 207, 235
  light sails pushed by, 203, 206
Lassell, William, 173
Lauer, Chuck, 49, 64
launch costs, 21–24, 71
  effects of high levels of, 23–24, 67
  to low Earth orbit, 61, 89–90
  reasons for high levels of, 23–24, 26, 34, 49, 69
  of Space Shuttle, 21–22, 27
  of Titan rocket, 25, 27
launch vehicles:
  heavy-lift, 28–29
  see also reusable launch vehicles; rockets
Lawrence Livermore Laboratory, 92, 135, 195
Lawson parameter, 85, 86, 88
Lebedev, Peter N., 201
Lee, Pascal, 121
LEO (low Earth orbit), 59–60, 61, 72, 89–90, 97
Lewis, John, 146, 147, 156
Life, 280
  asteroids and, 150–52
  Earth transformed by, 151, 225–26
  potential benefits from discovering nature of, 117–18
  see also extraterrestrial life
Life of the Cosmos, The (Smolin), 281–82
light, speed of, 191, 198, 199
light (solar) sails, 201–4, *203,* 229–30
Lindbergh, Charles, 55
"liquid metal," 238
LMLV, 33
Lockheed, 50
Lockheed Martin, xvii
  corporate mergers and, 34

Lockheed Martin (*cont.*)
"cost plus" contracts and overhead structure of, 24–25
launch vehicles designed by, 33
X-33 program and, 32–33, 34
longitude, determination of, 56, 173
Loral Corporation, 40, 54
Los Alamos National Laboratory, 135, 192, 248
low Earth orbit (LEO), 59–60, 61, 72, 89–90, 97
Lowell, Percival, 110–11
Lunar Prospector, 16, 93–94

Macartney, Lord, 20
McCaw, Craig, 40
McDonnell Douglas, 31, 32, 33, 34, 35, 40, 41, 43, 50
McFadden, Lucy Ann, 140
McKay, Chris, 120
McKay, Dave, 112
McKinney, Bevan, 43
magnetic fusion, 84–88, *85*
magnetic sails (magsails), 15, 204–7, 263, 264–65, 267–69, *268*
magnetic traps, 84–85
MagOrion sail, 206, 207
magsails, *see* magnetic sails
mapping rates, 179
Marie Curie, 120
Mariner missions, 10, 111, 114
Mark, Hans, 28
Marlowe, Christopher, 58
Mars, xvi, xviii, 14, 26, 30, 49, 101–27, 134, 147, 180, 188
agriculture on, 101, 105–6
asteroids and, 148–49
atmosphere of, 101, 102, 105, 108–9, 110, 115, 138, 153–56, 172
"canals" of, 110–11
colonization of, xvii, 101, 105–10, 118
degree of tilt of, 110
hypothetical cost of ticket to, 22
importance of missions to, 118, 122–24
length of day on, 110
life on, xv, 16, 17, 110–13, 117, 248
manned mission proposed for, 102–5
mineral resources on, 101, 107–8
Moon as base for flight to, 96
moon of, 108
NASA's missions to, 114–16, 119, 120, 138, 144; *see also* Mariner missions; Viking missions
NASA's Web site for, 119, 121
need for national and international initiatives for missions to, 113–19
orbital distance of, 187
relocating asteroids to, 171–72, *171*
rights on, 124–27
terraforming of, 108–10, 171, 227–28
underground hydrothermal reservoirs on, 106
water on, 101, 106, 108, 115
Mars Direct plan, xvii, 102–5, 106, 141
Mars Global Surveyor (MGS) orbiter, 114, 115
Marshall Space Flight Center, 42, 48
Mars Observer, 92
Mars Sample Return (MSR) mission, 116
Mars Society, 120–22
Founding Declaration of, 285–86
Mars Underground, 120
Martin Marietta, 34, 45, 48–49, 50, 53, 67–68, 153
Martin Marietta Astronautics, 40
Massachusetts Institute of Technology (MIT), 28–29
"mass drivers," 72
mass-energy equivalency equation, 197, 198
mass extinctions, asteroids and, 16, 17, 39, 130–32, 142, 151–52, 209, 249–50, 256
matter, four states of, 84
memes, 256–57
Mercury, 91–92, 163, 173
Mercury program, 10, 26
META program, 259
meteorites, 129, 139, 248
Metrodorus, 247
Mexico, 131
Microcosm, 54
Milky Way, 216–23
age of, 248, 279
maps of, *217, 218*
nearby stars of, *219–23*
number of stars in, 113, 248, 253
see also extraterrestrial life
Mind, 279
Ming Empire, xiv–xv, 18–20
Mir space station, 68
molecular-scale engineering, 56
moments of inertia of planets, 233–34, *234*
Moon, xv, xvi, 9, 11, 12, 14, 30, 79–100, 102, 106, 180, 232
agriculture on, 81–82
as astronomical observation base, 96–98
bacteria on, 250
beanstalk to, 98–100
chemical and mineral resources on, 80–81, *80,* 152–53, 181

energy from, 82–90
establishing a base on, 94–96
NASA's missions to, *see* Apollo program
water on, 16, 91–95
Morrison, Phillip, 258, 260
Motorola, 39–40, 43, 54
Mount Palomar, 267
MSLS, 33
MSR (Mars Sample Return) mission, 116
Mstar, 40
Mueller, George, 42
Muncy, Jim, 73
Murray, Bruce, 91
MUSES C, 141

nanobacteria, 112, 251–52
nanotechnology, 237–40, 252
National Aeronautics and Space Administration (NASA), xvii, 16, 42, 49, 59, 61, 67–68, 112, 121, 142, 175, 178, 180, 191, 213, 228
asteroid programs of, 135, 141
budget of, 10, 116
Clementine and *Lunar Prospector* and, 92–94
Design Reference Mission for Mars of, xvii, 104
early accomplishments of, 10
greatest accomplishments of, 174
lunar missions of, *see* Apollo program
Mars missions of, 114–16, 119, 120, 138, 144; *see also* Mariner missions; Viking missions
Mars Web site of, 119, 121
misallocated priorities of, 116
1989 "90 Day Report" of, 104, 144
Nixon-era cutbacks and, xv
Space Shuttle program and, 27–29, 30
viewed as idle bureaucracy, 30, 34
X-33 program and, 32–33
X-34 project of, 48
National Aerospace Laboratory, 63
National Ignition Facility (NIF), 195
National Radio Astronomy Observatory (NRAO), 259
National Security, 12
National Space Society, 12, 120
"natural habitat," 225
Nature, 258
navigation prize, 56
Neanderthals, xiii, 5
Near Earth Asteroid Rendezvous mission (NEAR), 140, 141
Near-Earth Object Program Office (NEOPO), 135
near-Earth objects (NEOs), *132,* 134–35, 139–40, 141
near-solar stellar mass objects, *212*
Nemesis, 209–10
NEP (nuclear electric propulsion) systems, 161, 181, 191
Neptune, 26, 163, 170, 173, 174, 176, 179, 182, 183, 187
Nereus, 141
NERVA, 138
neutron stars, 210, 212
"New Millenium" program, 141
Newton's law of reaction, 35
"New World Order," 7, 8
New York Times, 16
NIF (National Ignition Facility), 195
NIFT (Nuclear Indigenous Fueled Transatmospheric) vehicles, 162, 163, 164
Night Thoughts on Life, Death and Immortality (Young), 187
"NIMF" vehicles, 138
Niven, Larry, 241–42
Nixon administration, xv, 30
NK-31 rocket engine, 47, 49
Nobel Prize, 201
"noninterference" hypothesis, 270–71
nonluminous matter, 210, 211–12
North America, 7, 149–50
Northrop, 34
Nozette, Stu, 92
NRAO (National Radio Astronomy Observatory), 259
nuclear electric propulsion (NEP) systems, 161, 181, 191
Nuclear Indigenous Fueled Transatmospheric (NIFT) vehicles, 162, 163, 164
nuclear power, 188
and outer solar system exploration, 176–80
nuclear salt water rocket (NSWR), 193–94, *193*
nuclear thermal rockets (NTRs), 137–38, 171, 181, 191, 212–13
"nuclear winter," asteroid-caused, 130, 131–32

Oberth, Hermann, 3, 11
Office of Management and Budget, 174
oil, 161
Olbers, Heinrich, 133
O'Neill, Gerard, 72, 73, 74–75, 82, 96, 190, 242
On the Beach, 118
On the Origin of Species (Darwin), 274
Oort Cloud, 170–71, 181, 182–84, 207–8, 209, 242
orbital hotels, 62–64
orbital industries, 61–62
Orbital Properties, 64–65

orbital research labs, 59–61
Orbital Sciences, 40
Orion, Project, 192–93, 194, 197
Orteig, Raymond, 55
Orteig Prize, 55, 57
Ostro, Steve, 140
Outer Space Treaty (1967), 12–14
Owen, Henry, 12–13
Ozma, Project, 259, 260

Pacific American Launch Systems, 43
Pacific Ocean, 1990 asteroid crash in, 128–29
package delivery, 51–52
PAHs (polycyclic aromatic hydrocarbons), 112
Paine, Thomas, 101, 224
Pallas, 133
Palomar, Mount, 267
panspermia, 250–53
Paradiso (Dante Alighieri), 185
parallax, 209
Paris Gun, 26
passenger service, global, 52–55
Pathfinder (Mars landing vehicle), 16, 114–15, 119, 121
Pathfinder (rocketplane), 49, 50, 53, 57
Paul, Saint, 272
Pax Mundana, 8, 14, 15, 17, 126
Pax Romana, 8
Pearson, Jerome, 99
Penning traps, 198
Penn State University, 281
Phobos, 108
Phoenix, 43, 44
Phoenix, Project, 259–60
photoelectric effect, 201
photon rocket, 199–200, *199,* 267, 268
photosynthesis, 229
physics, limits to laws and knowledge of, xviii, 280–81
Piazzi, Giuseppe, 133
picotechnology, 240
Pioneer Astronautics, 49, 154
Pioneer missions, 10, 158, 166, 173–74, 176
Pioneer Rocketplane, 49, 54, 56, 64
Planetary Society, 120, 202, 259
planets:
  moments of inertia of, 233–34, *234*
  outside the solar system, xv, 15–17, 113, 247
plasma, 84, 205
plasma bombs, 206
plasmas, fusion, 84–88, *85,* 198–99
Pluto, 173, 187
plutonium-238, 176–77
polycyclic aromatic hydrocarbons (PAHs), 112
population growth, 159
Portugal, 19–20
power, for exploring outer solar system, 158–60, *159, 160*
power relay satellite (PRS), 74
"Prairie" (Sandburg), 21
Pratt and Whitney F-100 jet engines, 49
prizes, 55–57
Progress module, 68
Project Orion, 192–93, 194, 197
Project Ozma, 259, 260
Project Phoenix, 259–60
propellant cast, 72
propellant types:
  antimatter, 197–201, *199,* 263–64, *267,* 268
  characteristic power levels of, *266*
  chemical, 189–90
  CH_4/O_2, 148–49
  fission, 191–94, *192, 193*
  fusion, 194–97, *196,* 206, 263, 264
  hydrogen/oxygen, 30, 42, 89, 94
  hydrogen peroxide, 41, 48
  hydrogen peroxide/kerosene, 46
  from iceteroids, 171–72
  kerosene/oxygen, 22–23, 42, 44, 47, 49
  liquid oxygen, 48
  made from Martian atmosphere, xvii, 101–2, 105, 153
  methane/oxygen, 101–2
  for nuclear thermal rockets, 137–38
Proteus, 56
proton-proton fusion, 204–5, 264
Proton rocket, 142
proximity operations vehicle ("proxops"), 68
PRS (power relay satellite), 74

Quigley, Carroll, 8

radar and radio occultation science, 178–79
radioisotope heating units (RHUs), 176, 177
radioisotope thermoelectric generators (RTGs), 176, 177
radio telescopes, 98, 254, 258–61
Ramohali, Kumar, 154
Ranger missions, 10
RD-120 rocket engine, 49
Reagan, Ronald, 28
Reagan administration, 28
"rectennas," 70
red dwarfs, 210, 211, 212, 232–33, 243
Redstone rocket, 23

regeneration time, 253
research labs, orbital, 59–61
Reusable Aerospace Vehicle (RASV), 34, 45
reusable launch vehicles, xv, 16, 26–35, 58
  DC-X program, 16, 31–32, 33, 40, 41, 43, 45
  disadvantages of, 26–27
  keys to economic feasibility of, 30
  rocketplanes, 49–55
  Space Shuttle as, 27, 29–30
  SSTO, 30–33, 41–49, 72, 149
  X-33 program, 32–33, 34, 41
RHUs (radioisotope heating units), 176, 177
Rice University, 12, 14
Rift Valley, xiii–xiv, xv
ringworlds, 241–42
Robinson, Kim Stanley, 120
robotic satellites, 58, 69
robotic space missions, 9
robots, self-replicating, 235–40
"Rocket Equation," 35
rocketplanes, xvii, 49–55
  fast package delivery using, 50–51
  global passenger service using, 52–55
  military applications for, 51–52
rocketry and spaceflight, fundamentals of, 35–38, *37*
rockets:
  launch costs for, 21–24, 26
  technological stagnation in development of, 24–25
  weaponry origins of, 26
  see also propellant types; reusable launch vehicles; *specific rockets*
Rockwell, 32, 33, 34, 50
Roebling, Johann, 224
Rohrabacher, Dana, 73
Romanek, Chris, 112
Rosenberg, Sanders, 153
Rostow, Walt W., 12
Rotary Rocket Company, 44
Roton space-launch helicopter, 43–44
RTGs (radioisotope thermoelectric generators), 176, 177
Rusk, Dean, 13
Russia, 9, 67
Rutan, Burt, 56

Sabatier reaction, 153, 154
Sagan, Carl, xiii, 128, 228–29, 231
sails:
  light (solar), 201–4, *203*
  magnetic, *see* magnetic sails
St. Louis, Mo., 56
Sandburg, Carl, 21
Santarius, John, 84
satellites:
  communications, 39–40, 43, 54–55, 58, 59, 69, 75–76
  constellations of, 39–40, 43, 54–55, 59
  power relay, 74
  robotic, 58, 69
  servicing of, 66–69
  solar powered, 70–74, 82, 83
Satevod, 40
Saturn, 163, 164, 170, 173, 174, 176, 182, 187, 243
  see also Titan
Saturn V rocket, xv, 42, 102
Scaled Composites, 56
science, xviii, 8–9, 280–81
SDIO (Strategic Defense Initiative Organization), 16, 31, 32, 45
Sea Launch system, 34–35
"Sea People," 7
Search for Extraterrestrial Intelligence (SETI), 259–61
self-replicating automatons, 235–40
Senate Appropriations Committee, 120
SERENDIP program, 259
sex, in zero gravity, 63
Siberia, 129
Sidey, Hugh, 11
single-stage-to-orbit (SSTO) vehicles, 30–34, 38, 41–49, 69, 72, 149
skyhook, 99, 108
Skylab space station, 10, 28, 67–68, 109, 142
Smith, Fred, 50
Smolin, Lee, 281–82
social development, seven stages of, 8
Sojourner, 115, 117, 120
solar flares, 81, 101
solar power arrays, on Moon, 82–84
Solar Power Satellite (SPS) systems, 70–74, 82, 83
solar (light) sails, 201–4, *203,* 229–30
solar system:
  age of, 140
  as humanity's natural environment, 151–52
solar system, outer, xvii
  data transmission rates from, 177–79, *179*
  early astronomical exploration of, 173
  energy resources in, 159–60, *160*
  exploration of, 173–75
  main obstacles to settling of, 180–81
  nuclear power and exploration of, 176–80
  robotic exploration of, 173–75
  settling of, 157–84
  transportation in, *148*
solar wind, 201, 205, 206

sound barrier, 43–44
Soviet Union, 7, 9–10, 11, 13, 14, 23, 26, 114, 129, 193, 228
Space Access, 54
space activist organizations, 120
space asset servicing, 66–69
space business parks, 64–65
space colonies, 72–75, 96, 190
Space Exploration Initiative, 104
spacefaring, fundamental reason for, 79
spacefaring civilization, *see* Type II (spacefaring) civilization
spaceflight and rocketry, fundamentals of, 35–38, *37*
Space Frontier Foundation, 120
SpaceHab, 59
space initiatives, entrepreneurial vs. governmental, 74–76
Space News, 48
space program, U.S., 9–15
  cost of, 10, 14
  decline of, xv, 9–10, 14
  goal needed for, 10–12
  Outer Space Treaty's effect on, 12–14
Space Resources (Lewis), 146
Space Shuttle, 9, 33, 35, 59, 69, 105, 142
  annual program cost of, 27
  design faults of, 29–30
  economics and politics of, 27–28, 30
  first flight of, 25
  Hubble repairs and, 66
  launch cost of, 21–22, 27
  MMU unit of, 143
  on-board computers of, 25
  zero gravity rat execution experiment conducted on, 116
Space Station, *see* International Space Station
space tourism, 62–64
space transportation systems, four generations of, 180–81
species maturity, 215–16
specific impulse (ISP), definition of, 35
Spectra, 100, 106
Spenser, Edmund, 157
Sponable, Jess, 31
SPS (Solar Power Satellite) systems, 70–74, 82, 83
Sputnik, xiv, 9
Sridhar, K. R., 154
SSTO (single-stage-to-orbit) vehicles, 30–34, 38, 41–49, 69, 72, 149
Stapledon, Olaf, 246
STARDUST mission, 141
Starmaker, The (Stapledon), 246
stars, 216–23
  classifications of, 231–32, 253, 263
  lighting of, 243–46
  Milky Way's population of, 113
  plasma as substance of, 84
  unknown types of, 209–15
  see also interstellar travel; Milky Way
starships, detection of, 262–69, *266, 267, 268*
Starsys, 40
Star Trek, 278
State Department, U.S., 12, 13
statites, 233
"Steady State" theory, 250
Stoker, Carol, 120
Strategic Defense Initiative Organization (SDIO), 16, 31, 32, 45
Sun, 189–90, 201, 212
  age of, 279
  brightness of, 150, 232, 243, 245
  classification of, 232
  energy/distance ratio for, 176
  type of fusion reaction in, 204
superconductivity, high-temperature, xv, 15
Surveyor missions, 10, 250
"synthesis gas," 154

TAI (transasteroid injection), 142
Tamburlaine (Marlowe), 58
TANSTAAFL, 38
Tau Ceti, 259
Taylor, Ted, 192
technology, 4, 6, 8
Teets, Peter, 40
Teledesic, 40, 43, 49, 54, 68
telescopes, 173, 247, 267
terawatt, definition of, 188
Terminator 2, 238
terraforming, 225–46
  definition of, 225
  lighting stars, 243–46
  of Mars, 108–10, 227–28
  technology for, 235–42
  of tidally locked worlds, 231–35
  of Venus and Venus-like planets, 228–31
Test Ban Treaty (1963), 193
Texas, University of, 15
"There's Plenty of Room at the Bottom" (Feynman), 56
thermal emissions, 211–12
thermonuclear fusion reactors, 80, 160, 161
Thiokol Star 48, 47, 48
Third World, 70, 74, 159

Thomas-Keprta, Kathie, 112
Thor-Delta rocket, 23
tidally-locked worlds, terraforming of, 231–35
Time, 11
Titan, 163–66, *165,* 173, 179, 181, 230
Titan rocket, 23, 24–25, 26, 27, 29–30, 33, 40, 49, 68, 142
Tokyo University, 63
Tombaugh, Clyde, 173
tonne, definition of, 22
tourism, 62–64
Toutatis, 140
transasteroid injection (TAI), 142
trans-Atlantic flights, 52, 55, 57
"triangle trade," 149–50
tritium, 86–87, 88
Triton, 173, 243
Truth, Sojourner, 115
TRW, 53
Tsiolkovsky, Konstantin, 3, 11
"tuna can" habitats, 106
Tunguska, 129
12th apparent magnitude objects, *211*
two-stage-to-orbit (TSTO) vehicles, 31, 42
Type I (global) civilization, 146, 189, 256, 271
  completing humanity's status as, xvi, 1–76
  definition of, xiv, 280
Type II (spacefaring) civilization, xvii, 76, 188–89, 202, 231, 235, 240–41, 256, 271
  definition of, xiv, 280
  humanity's transition to, xvi, 77–184
Type III (galactic) civilization, xvi, xvii
  definition of, xiv, 280
  destruction of, 256–58
  human diversity and, 276–78
  humanity's transition to, 185–283
  as imperative, 272–73, 275–76, 283
  as network, 269–72
  potential results of encounter with, 272
  probable existence of, 248
Type IV civilization, 280

Ulam, Stanislaw, 192
United Space Alliance, 33
Universal Space Lines, 54
universe, age of, 279
universes, within black holes, 281–82
Uranus, 133, 163, 170, 173, 174, 176, 187
U.S. News and World Report, 40

Van Allen, James, 166
"VentureStar" launch system, 33
Venus, 172, 242
  terraforming of, 228–31
vertical takeoff/vertical landing (VTVL) system, 41, 43, 44, 45
Vesta, 133
Viking missions, 10, 26, 111, 114, 115, 116, 174
Vladivostok, 129
von Braun, Wernher, 11, 23, 42, 138
Von Neumann, John, 235
Voyager missions, 10, 26, 158, 168, 173–74, 176, 179, 187
V-2 rocket, 23, 26

Wagner, Richard, 120
Wang, John, 42
war:
  humanity's potential end from, 118
  progress driven by, 6–7
  species maturity and, 215–16
Washington, University of, 155
Wasser, Alan, 12
water, 17, 113
  on Europa, 174–75
  from iceteroids, 183
  on Mars, 101, 106, 108, 115
  on Mercury, 91–92
  on Moon, 16, 91–95
  proportional weight of components of, 103
Webb, Walter Prescott, 123
Wehrmacht, 23, 26
Wells, H. G., 274
West Indies, 149–50
white dwarfs, 210, 212, 232–33
Whitmire, Daniel, 205
Wild 2, 141
Williams, John, 155
Wisconsin, University of, 84
Worden, Simon "Pete," 31–32, 92
Wright Patterson Air Force Base, 99

X-Prize Foundation and competition, 56–57
X-33 program, 32–33, 34, 41
X-34, 48

Young, Edward, 187
Young, Larry, 142
Yucatan, 131
Yung Lo, 18

Zare, Richard, 112
Zenit boosters, 34
zero gravity, 53, 59, 60, 63, 116, 142–43, 145
Zubrin, Maggie, 120

ABOUT THE AUTHOR

Dr. Robert Zubrin is an internationally renowned astronautical engineer and the acclaimed author of *The Case for Mars,* which Arthur C. Clarke called "the most comprehensive account of the past and future of Mars that I have ever encountered." NASA recently adapted Zubrin's humans-to-Mars mission plan. A former senior engineer at Lockheed Martin, Zubrin is president of the Mars Society and founder of Pioneer Astronautics, a successful space exploration and development firm. He lives with his family in Indian Hills, Colorado.